INSIGHTS INTO
MODERN MATHEMATICS

Twenty-Third Yearbook

THE NATIONAL COUNCIL OF TEACHERS
OF MATHEMATICS

WASHINGTON, D. C.   1957

Preface

This book has been designed to serve the needs of teachers of secondary mathematics. The editors and contributing authors, however, sincerely hope that it will also prove useful to scholars of varied interests. It has been written specifically to provide reference and background material for both the content and the spirit of modern mathematics. Its basic plan is posited on the premise that a person cannot teach something which he does not comprehend. Its development was carried forward on the recognition that many teachers of secondary mathematics have never had an opportunity to become acquainted with the more recent developments in mathematics. The content, thus, is directed to the task of providing an introduction to several areas of modern development in mathematics. Suggestions for further study are to be found in the many references cited at the end of the several chapters. It is hoped that teachers of secondary mathematics will seek opportunities for study of the YEARBOOK as a desirable professional activity, a challenge to invigorating mental exercise, and a stimulating prerequisite for introducing mathematics to the youth of the twentieth century.

During recent years, dating approximately from the middle of the nineteenth century, mathematics has experienced a remarkable metamorphosis. The foundations have been, and continue to be, subjected to critical examination. Some traditional techniques have been discarded as no longer efficient, others have undergone radical modification. Familiar definitions of many basic concepts have been dropped as passé and inexact, to be replaced by qualifications more modern in import. New structures have been postulated and subjected to critical examination, some to be discarded, some to be modified, some to be retained. Many new concepts and modern techniques are now considered as significant common property among mathematicians.

The choice of topics treated in the book has been the result of considerable discussion on the part of the editors; it is their belief that the volume provides a treatment of representative topics in modern mathematics that are of considerable value for teachers of secondary mathematics. The expositions have been written with care, and have been developed upon the assumption that the reader has had about two years of college mathematics. Moreover, each chapter has been criticized by selected high school teachers to safeguard the level upon which it has been written to assure maximum utility of the resulting materials.

Obviously it is not expected that the actual content of this volume will become the subject matter of the high school curriculum. However, it would be disappointing indeed if the work did not have some effect upon the choice of material in the secondary mathematics program and did not bring about some changes in emphasis; it should also provide motivation for a review of the types of organization and exposition now employed. Of course, the real significance of the book is for the teacher himself whose enthusiasm and confidence before a class and whose inspiring leadership of young students depend strongly upon the accuracy and breadth of his understanding of the subject that he teaches.

The members of the editorial committee desire to express their appreciation to the authors of the several chapters; these men, already heavily loaded with teaching and research, undertook without exception the additional burden of organization, research, and writing assigned to them. Moreover, we extend our thanks to the many critics and readers who have performed invaluable service in assisting the editorial staff to develop a volume which, it is hoped, will prove to be of great usefulness to the mathematics teachers of this country. In particular, we wish to express our thanks to Miss Caroline Lester, New York State College for Teachers, Albany, New York; M. H. Ahrendt, Executive Secretary of the National Council of Teachers of Mathematics; the Division of Publications of the National Education Association; and the Waverly Press for their able assistance in seeing the book through to publication. For all of us the joy of participating in such a worthy project has been its own reward.

The Editorial Committee
CARL B. ALLENDOERFER
SAUNDERS MACLANE
BRUCE E. MESERVE
CARROLL V. NEWSOM
F. LYNWOOD WREN, *Chairman*

Contents

Special Index of Symbols

$x \in A$	x is a member of the set A, 38
$x \notin A$	x is not a member of the set A, 38
$A = \{x_1, \cdots, x_n\}$	A is a set containing the members $x_1, \cdots, x_n$ and no others, 38
$A \subset B$	A is a subset of B, 38–39
$B \supset A$	B is a superset of A, 39
$A \cap B$, $A \cdot B$, or AB	Intersection of the sets A and B, 40, 315–16
$\bigcap_{A \in X} A$	Set of objects which belong to A whenever $A \in X$, 40
$A \cup B$, or $A + B$,	Union of sets A and B, 40, 315
$\bigcup_{A \in X} A$	Set consisting of all objects which belong to one or more sets A of the set of sets X, 40
$A - B$	Relative complement of B in A, 41
$\sim A$	Complement of set A, 42
$s(x)$	A condition on elements x of a given set, 42
$\{x \in A \mid s(x)\}$	Set of all x's in A which satisfy $s(x)$, 42
$\emptyset$ or 0	The empty, or null, set, 43, 316
2^A	The set whose members are all the subsets of A, 43
$A \times B$	Cartesian product of the sets A and B, 43–44
(a, b)	Ordered pair of elements a and b, 45
$x \, R \, y$	x is related to y, 54
$\aleph_0$	Aleph-null, the transfinite cardinal, 63
$a \circ b$	a operating on b, 72
$p \wedge q$	Conjunction of propositions p and q, 76–77
$p \vee q$	Disjunction of propositions p and q, 77
$p \rightarrow q$	Proposition p implies proposition q, 78
$p \leftrightarrow q$	Proposition p is equivalent to proposition q, 79
$\sim p$	Negation of proposition p (not p), 80
$\forall_x[p]$	Proposition p applies to all x, the universal quantifier, 86
$\exists_x[p]$	Proposition p applies to at least one x, the existential quantifier, 86
$-a$	Additive inverse of a, 103
a^{-1}	Multiplicative inverse of a, 108
$f: G \rightarrow H$	f maps G into H, 110
$\lvert a \rvert$	Absolute, or numerical, value of a, 110
$a \equiv b \pmod{n}$	a is congruent to b modulo n, or $a - b$ is divisible by n, 125
$T: X \rightarrow X$	A function which assigns to each element in set X a unique element $T(a)$ in X, 127
$S \circ T$	Composite transformation in which the transformation T is first applied, then S, 131

I

Introduction

CARROLL V. NEWSOM

NUMEROUS articles and books have been written, especially in modern times, that consider the elements which are involved in sound mathematical exposition. However, it may be questioned whether there is adequate awareness of the fact, even on the part of some teachers, that the major difficulties inherent in the teaching of modern mathematics arise fundamentally from two factors. First, mathematics, by definition, is an abstract discipline. Upon all except the most advanced levels the successful teacher of mathematics must give attention to applications of the subject, and students must become acquainted with those manipulative techniques that are a part of mathematical knowledge. Nevertheless, a course becomes something other than one in mathematics if attention is not given to those abstract concepts that characterize it. In the second place, mathematics is a dynamic, growing science. The very rapidity with which mathematics is changing has led to the fact that many courses on the subject, especially upon the more elementary levels, are a curious combination of the significant, the outmoded, the incorrect, and the trivial. Yet it would be quite inappropriate to hold the busy teacher responsible for this condition, for it is virtually impossible to keep abreast of developments.

ABSTRACT NATURE OF MATHEMATICS

To understand the abstract nature of mathematics it is desirable to consider briefly some ideas from natural philosophy. The subject matter of any natural science is a collection of sense-experiences, expressed as data, which appear originally as a chaotic variety. In attempting to interpret such a collection of data, and in trying to answer appropriate questions, science seeks some pattern or model to which the data appear to conform. Thus, for example, it might be ascertained in the study of a freely falling body in a vacuum that readings upon time t in seconds and distance s in feet, ignoring approximations, provide the following as-

1

sociated numbers (t, s): $(0, 0)$, $(1, 16)$, $(2, 64)$, $(3, 144)$, etc. When these number-pairs are graphed upon a rectangular coordinate system with axes designated as t and s it is evident that a certain second-degree parabola "fits the data." This information is of invaluable help in furthering the study of the behavior of the falling body, for a pattern has been determined to which the data appear to conform. This illustration is capable of generalization, for one of the purposes of natural science is the development of mechanisms, a mechanism being simply a man-made pattern or model that purports to relate a set of data in an understandable manner. A mechanism may be pictorial, as is the conventional atomic model portrayed to elementary students of physical science, or it may be diagrammatic like the device employed by the chemist to display the manner in which a large number of atoms cling together to form a complex molecule. Just as the architect's blueprint possesses a correspondence to the finished house, so the mechanism of the scientist is made to correspond to some part of nature.

The modern social and natural sciences, by contrast with much of classical science, have discovered that the symbolic formulations of mathematics possess great significance in their apparent ability to correlate data obtained from our social and natural universe and, as a consequence, in their power to answer fundamental questions of scientific importance. For example, one might return to the illustrative problem of the falling body to ascertain that the mathematical equation of the parabola discussed in connection with the problem is $s = \frac{1}{2}gt^2$, where g is approximately 32. The physicist prefers this mathematical formula to other devices that might be employed to relate data obtained from a study of falling bodies. However, such a mathematical formulation when applied as a correlating agent to the data of a science merely becomes a mechanistic device, and must be regarded as such by the scientist. In considering the use of such mathematical devices, or forms, it is noteworthy that by changing the meaning that is associated with the undefined symbols it commonly is true that a particular form may be used to interpret what appear superficially to be utterly dissimilar natural phenomena. Thus, for example, $s = \frac{1}{2}gt^2$ might also be used to find the volume of a box of square base t^2 and depth, $\frac{1}{2}g$, or it might be used to find the total force exerted by a gas on a plane surface of area t^2 if the pressure of the gas is $\frac{1}{2}g$ per square unit.

It does not require much mathematical ability to deduce that the particular proposition, $s = \frac{1}{2}gt^2$, is a rather direct consequence of the proposition that the acceleration of gravity g is taken as a constant. The deduction that is involved provides a fact of considerable significance to the physicist. Mathematics, in general, provides us tools of great power in dealing with natural phenomena not only because of the generality of its symbolic formulations but also because of the fact that individual propositions or formulas are studied in their logical relationship to other propositions or formulas. Fundamentally the mathematician is concerned with chains of propositions in which successive members of the chain, except the first, are deduced logically from those that precede. That is, briefly stated, a mathematical system is a collection of two classes of propositions, the first class being composed of propositions that may be described as axioms and the second class being composed of those which may be deduced logically from the axioms. The term *axiom* as used in this context means only that the proposition thus described is not proved within the system under consideration. In somewhat similar fashion it is possible to classify into two collections the symbols with which a mathematical system is concerned, namely, those that are not defined within the system and those that are defined in terms of the symbols that are undefined.

MATHEMATICAL SYSTEMS

It should be apparent immediately that a mathematical system as just described becomes merely an abstract form, and, as already implied, the same system may occur frequently as the underlying pattern in many diverse real and ideal situations. Thus there is validity in the assertion that the mathematician is less concerned with the solution of specific problems than he is with the development of general patterns that have widespread applicability in the study of particular situations.

The origins of the concept of mathematical system, as just described, are to be found in the studies of the early Greek philosopher-mathematicians, notably in Euclid's *Elements*, written about 300 B.C. This monumental work, in spite of some crudities, established a mathematical tradition that may be traced quite clearly up to the present time. From the very first it appears that the organization of Euclid's work was the subject of considerable discussion and controversy. The nature of the axiomatic basis of a mathematical system received clarification when in the eighteenth and nineteenth centuries, chiefly through the work of such men as Bolyai, Lobachevski, and Gauss, and later of Riemann, the independent nature of the famous Euclidean fifth postulate became apparent. It was ascertained that one may create alternative and logically consistent geometries when a suitable replacement is made for the fifth postulate. The truly formal nature of a mathematical system and its utilization in any study of reality was made clear in the nineteenth-century writings of Grassmann, Pasch, and others. Finally, the classical work of Hilbert on the foundations of geometry, published in 1899, along with similar works by several other mathematicians of the same period, provided the mathematical world with formal, rigorous developments of Euclidean geometry as models for other mathematical systems. This story of the contribution of geometers to our understanding of the nature of mathematics is told in some detail in Chapter IV and in Chapter IX.

Simultaneously with these latter accomplishments in geometry, algebra was undergoing a considerable rejuvenation through the extensive use of the postulational-deductive method introduced by geometers and through the utilization of the ideas of abstract group theory. Although many of the basic notions of the newer algebra that resulted go back to Kronecker, Dedekind, and Steinitz, and to several English students of the subject, major credit for its development should probably be given to the great woman mathematician, Emmy Noether. The work, *Modern Algebra*, by one of her students, B. L. Van der Waerden, provides the classical account of modern algebra as Emmy Noether conceived it. Some of the finest studies of recent years have continued to expand our knowledge of the subject so that it is acknowledged generally at the present time that modern algebra is one of the most advanced of mathematical disciplines.

Unfortunately, the remarkable change that has taken place in algebra appears to have had little effect upon the precollege and early college curriculum in mathematics. The rather common argument that geometry is essential in the high-school program because it illustrates the use of deductive procedures deserves reconsideration, for algebra, as presently conceived, may be a better vehicle for exhibiting the use of the postulational-deductive method. Certainly geometry is no longer the only mathematical subject in which deductive logic is used; it is used everywhere in mathematics. A brief introduction to some of the fundamentals of modern algebra is given in Chapter V, and many of its basic concepts are treated incidentally in other chapters, notably Chapters II, III, and VI.

MODERN DEVELOPMENTS IN MATHEMATICS

G. C. Evans has stated that mathematics, in the modern sense, "is typified by a free spirit of making hypotheses and definitions rather than a mere recognition of facts." Certainly all the classical branches of mathematics have profited from the new freedom, and significant new branches have been created. For instance, G. Cantor recognized in the latter part of the nineteenth century that an adequate theory of sets must be developed in order to clarify and make possible a solution of many problems in function theory and analysis. The basic nature of the concept of set for mathematics generally is now acknowledged, and an introduction to set theory is becoming increasingly common in elementary courses in mathematics. Chapter III provides an introduction to this important subject.

The early part of the twentieth century saw an extension of the theory of sets to the theory of abstract spaces. Also, as the result of extensive research, topology evolved rapidly into a basic mathematical discipline of far-reaching consequence. The development of topology has taken place generally along two lines, namely, the set-theoretic and the combinatorial. Point-set theoretic topology, as the name implies, is concerned with infinite sets of points, especially with the properties of such a set in the neighborhood of a particular point; through its use it has been possible, for example, to make a careful structural study of the continuous curve. Combinatorial topology, on the other hand, evolved naturally from earlier studies of Riemann and Poincaré; it is concerned with polyhedrals of a finite number of faces, especially with the manner in which these faces are joined together. Modern topology, with emphasis upon point-set theoretic topology, is the subject of Chapter X.

Modern analysis, which includes the calculus, also moved ahead rapidly in the latter part of the nineteenth century and in the first part of the twentieth century. Although there were many reasons for this progress, the development of set theory, including associated theories of measure, was extremely influential in providing a basis for rigorous developments. Also for the first time mathematics had available a general definition of "function."[1] Previous to the time of Fourier (1768–1830) a function was defined only by a single formula; for instance, s is a function of t by virtue of the formula, $s = \frac{1}{2}gt^2$. Fourier recognized, however, that such a restrictive definition was intolerable in the study of the trigonometric series that now bear his name. Dirichlet shortly thereafter approached the modern, broad definitions of function when he advocated, "y is a single-valued function of the variable x, in the continuous interval (a, b), when a definite value of y corresponds to each value of x such that $a \leq x \leq b$, no matter in what form this correspondence is specified." Chapter VII reviews some modern developments in

---

[1] No concept in mathematics is more used and less understood than the function concept. Even among mature mathematicians there is disagreement about definition, proper terminology and symbols. In this book each author has been permitted to present his own approach to the function concept in order that the reader may observe the variety to be found in modern treatments of this subject.

so-called mathematical analysis, and Chapter VIII treats certain elementary, but highly significant, functions from a modern standpoint.

Chapters XI and XII discuss some modern mathematical developments of major significance to the scientist and to the applied mathematician. Undoubtedly the new computing machine techniques will have a strong effect upon even the elementary curriculum; for instance, many traditional procedures for evaluating determinants and for solving equations of high degree are rapidly becoming obsolete. Similarly, an understanding of the modern foundations of probability theory requires more than the traditional empirical background; an understanding of set theory, for example, is quite essential.

It must be apparent from the previous discussion that mathematics has been characterized in recent years by great activity that should have considerable influence upon the curriculum of our schools and colleges. Some suggestions in this regard are treated in the last chapter. Moreover, the new concept of mathematics as the science of abstract form has given the subject a distinctive and indispensable role in all of knowledge; undoubtedly the subject must be accepted as fundamental in what is now called "a general education." This basic role was recognized in the influential Report of the Harvard Committee of 1945, entitled *General Education in a Free Society*, which stated: "Mathematics may be defined as the science of abstract form. . . . The discernment of structure is essential no less to the appreciation of a painting or a symphony than to understanding the behavior of a physical system; no less in economics than in astronomy. Mathematics studies order abstracted from the particular objects and phenomena which exhibit it, and in a generalized form."

II

The Concept of Number

IVAN NIVEN

THE number concept as it is now viewed by mathematicians is the result of an evolution in thought over many centuries. A complete statement of the modern position, including a careful, rigorous development of the real and complex number systems, would necessitate a volume in itself. In the face of this, we shall pursue the subject from two directions. First we shall outline the central number systems of mathematics, from the natural numbers to the complex numbers. This will be largely a description from an intuitive standpoint, but with proofs of some of the key points. Then, by drawing attention to some of the limitations of this presentation, we shall indicate the desirability of a more rigorous analysis of number systems, and give some illustrations of the nature of a more careful development.

NATURAL NUMBERS

The numbers 1, 2, 3, 4, 5, $\cdots$ are called the *natural numbers* because it is generally felt that they have in some philosophical sense a natural existence independent of man. The most complicated of the number systems, by way of contrast, are regarded as intellectual constructions of man. This infinite set of numbers has been represented in various ways in the course of history, another well-known representation being the system of Roman numerals I, II, III, IV, V, $\cdots$. The natural numbers are of course an abstract concept, independent of the nomenclature used to represent them. The most common representation, 1, 2, 3, $\cdots$, is sometimes called the Arabic system, but more often today is termed the Hindu-Arabic or Indo-Arabic system because of the invention of this notation in India. Knowledge of this way of writing numbers was transmitted from India to the Western world via the Arabs.

This notation for the natural numbers,

$$1, 2, 3, 4, 5, 6, 7, 8, 9, 10, 11, 12, \cdots ,$$

7

is also called the *decimal system* (from the Latin *decem*, meaning ten) because ten symbols or digits are employed to represent all numbers. The compactness of this excellent notation is due to its being a *place* or *positional* system. That is, in such a number as 543, the digit 5 represents 5 hundreds; the digit 4, 4 tens; the digit 3, 3 units. In contrast the Roman numeral XX for 20 has two symbols X of equal value, ten. Thus the Hindu-Arabic system can be used to represent numbers of any size with ten symbols, whereas a nonpositional notation like the Roman numerals would require new symbols for reasonable brevity in the representation of larger and larger numbers.

The decimal system is said to have *base* ten, this unit probably originating in an anthropological sense from our having ten fingers for use in counting. But ten is not the only possible base; we could use any natural number greater than one. For example, if we were to use eight as a base, we would employ only the digits 1, 2, 3, 4, 5, 6, 7, and 0. The sequence of natural numbers would then be written as

$$1, 2, 3, 4, 5, 6, 7, 10 \ 11, 12, 13, 14, 15, 16, 17, 20, 21, \cdots.$$

The notation 543 in this system would denote 3 units plus 4 eights plus 5 of the square of eight, so that 543 to base eight is the same as $5(64) + 4(8) + 3 = 320 + 32 + 3 = 355$ to the usual base ten.

With the advent of electronic digital computers much use is now made of the base two, which formerly found its role largely in theoretical mathematics. This is the base of the *binary* system, in which all numbers are represented by the use of the two digits 1 and 0. The sequence of natural numbers is represented in the binary system as

$$1, 10, 11, 100, 101, 110, 111, 1000, 1001, \cdots.$$

Thus the number 110011 in the binary system is the same as 51 in the decimal system since

$$1 \cdot 2^5 + 1 \cdot 2^4 + 0 \cdot 2^3 + 0 \cdot 2^2 + 1 \cdot 2^1 + 1 = 51.$$

Changing a number from one base to another is an operation involving some calculation in most cases. However, the change from base two to base eight or vice versa is trivial, because eight is a power of two. That is, any integer written to base two can be rewritten to base eight by the simple process of grouping the digits from the right end in blocks of three, and then replacing the blocks 000, 001, 010, 011, 100, 101, 110, 111 by 0, 1, 2, 3, 4, 5, 6, 7 respectively. For example the number 110011 in the binary system becomes 63 when written to base eight. In the other direction, the number 543 to base eight is the same as 101100011 in the binary system. Of course it is not always the case that a number has exactly three times as many digits in the binary representation as in the base eight representation. For example, the number 345 to base eight takes the form 11100101 in the binary system. This idea of the easy switch from either of these bases to the other is used by persons working with electronic digital computers because of the greater compactness of the representation to the larger base. While the machine performs its operations on numbers written to base two, the operator can himself use base eight in order to reduce significantly the total number of symbols.

The base twelve has been suggested from time to time as a more practical base than ten. Without discussing the merits of the case, let us say that the use of base twelve would result in a system (the duodecimal system) wherein symbols would be needed to represent the numbers ten and eleven. If we employ, say, the symbols t and e for this purpose, then the natural numbers would be written as

$$1, 2, 3, 4, 5, 6, 7, 8, 9, t, e, 10, 11, 12, 13, 14,$$

$$15, 16, 17, 18, 19, 1t, 1e, 20, 21, \cdots .$$

MATHEMATICAL INDUCTION

In mathematics a general proposition, as the word "general" implies, comprises a collection of special propositions, often with one corresponding to each natural number. For example, the proposition that the sum of the first n odd numbers equals n^2 is equivalent to the propositions P(1), P(2), P(3), P(4), $\cdots$, as follows:

$$P(1): \quad 1 = 1^2$$

$$P(2): \quad 1 + 3 = 2^2$$

$$P(3): \quad 1 + 3 + 5 = 3^2$$

$$P(4): \quad 1 + 3 + 5 + 7 = 4^2$$

$$P(5): \quad 1 + 3 + 5 + 7 + 9 = 5^2, etc.$$

These can be written compactly in the form

$$P(n): \quad 1 + 3 + 5 + 7 + \cdots + (2n - 3) + (2n - 1) = n^2,$$

since $2n - 1$ is the nth odd number, and $2n - 3$ is the preceding, or $(n - 1)$th odd number. (We note for reference in a moment that $2n + 1$ is the $(n + 1)$th odd number.) Now how can we go about proving such a proposition, if it is true? We cannot prove a general proposition by verifying a number of special cases, say $n = 1$, $n = 2$, up to $n = 1000$. For the proposition might be false for the next case, $n = 1001$. It has actually happened that mathematical results suggested by considerable numerical evidence have turned out to be incorrect upon deeper examination.

One method of proof, employed throughout mathematics, is called *mathematical induction*, and is based on the *principle of induction*. The principle of induction asserts the truth of a given collection of propositions P(1), P(2), P(3), $\cdots$, provided that the following two conditions are satisfied: (a) that P(1) is true; (b) that for each natural number n, P(n) implies P($n + 1$), *i.e.* that if P(n) holds, so does P($n + 1$). This principle is an assumption[1] about the natural numbers. It is sometimes suggested that the principle of induction is analogous to the physical situation of an infinite row of dominoes all of which fall over when (a) the first one is pushed over, because (b) they are so placed that when any one, say the nth, falls, it pushes over the next one, the $(n + 1)$th. But whatever the value of such an analogy, it does not suffice to prove the principle.

Let us apply the method of mathematical induction to the proposition cited above as an illustration, that the sum of the first n odd numbers equals n^2. Condition (a) is obviously satisfied, since P(1) is the assertion that $1 = 1^2$. Condition (b) is that the first of the following equations implies the second:

P(n): $1 + 3 + 5 + 7 + \cdots + (2n - 3) + (2n - 1) = n^2$;

P($n + 1$): $1 + 3 + 5 + 7 + \cdots$

$$+ (2n - 3) + (2n - 1) + (2n + 1) = (n + 1)^2.$$

Condition (b) is satisfied because if we add $2n + 1$ to both sides of the first of these two equations, we get

$$1 + 3 + 5 + 7 + \cdots + (2n - 3) + (2n - 1) + (2n + 1)$$
$$= n^2 + (2n + 1) = (n + 1)^2,$$

so that P(n) does indeed imply P($n + 1$).

Both conditions (a) and (b) must be established in proofs by mathematical induction. As an example to show the necessity of establishing

---

[1] Although our position is that the principle of induction is an assumption or axiom, and hence cannot be proved, the reader may note that a proof is given in Chapter III (page 53). However, the proof is based on a further set of more basic axioms. There is no contradiction between the two viewpoints, because there is always more than one way of formulating axioms for a mathematical system. One man's axiom may be another man's theorem.

condition (a), consider the proposition that the sum of the first n natural numbers equals $\frac{1}{2}(n^2 + n + 2)$,

$$P(n): \quad 1 + 2 + 3 + \cdots + n = \frac{1}{2}(n^2 + n + 2).$$

It can be established that $P(n)$ implies $P(n + 1)$, so that condition (b) is satisfied. However it happens that $P(1)$ is false, since $P(1)$ says that $1 = \frac{1}{2}(1 + 1 + 2)$. Indeed $P(n)$ is false for every value of n.

As an example to show that the condition (b) above is necessary in proofs by mathematical induction, consider the proposition $P(n)$ that the sum of the first n natural numbers equals $n^3 - 2n^2 + n + 1$,

$$P(n): \quad 1 + 2 + 3 + \cdots + n = n^3 - 2n^2 + n + 1.$$

Setting $n = 1$, we verify that $P(1)$ is true, but $P(n)$ is false in general, as shown by the case $n = 3$, for example.

The word "induction" is used here in quite a different sense from empirical induction or inductive reasoning in natural science. In this latter sense a general law in science is concluded from a series of particular observations, in much the same way we conclude that the apples on a tree are ripe from the fact that seven particular apples which we picked from various parts of the tree turned out to be ripe. This situation has its analogy in mathematics in the way that propositions or theorems are discovered, but not in the way that they are proved. For example, from the immediately verifiable cases

$$1 = 1^2, \quad 1 + 3 = 2^2, \quad 1 + 3 + 5 = 3^2,$$

$$1 + 3 + 5 + 7 = 4^2, \quad 1 + 3 + 5 + 7 + 9 = 5^2,$$

one might be led to make the conjecture that the sum of the first n odd numbers equals n^2. This is the manner in which some mathematical discoveries are made, but the discovery is complete only when the conjecture is followed by proof.

INTEGERS AND THE RATIONAL NUMBERS

The natural numbers are extended to the set of *integers*

$$\cdots, -4, -3, -2, -1, 0, 1, 2, 3, 4, \cdots$$

by the introduction of zero and the negatives of the natural numbers. Thus the positive integers are the natural numbers. The integers are sometimes called the *whole numbers*, and sometimes in advanced works the *rational integers*. As the latter usage suggests, there are "integers" in other more abstract systems of numbers, a point to which we shall return later in this chapter.

Let us consider the four elementary operations of arithmetic, namely addition, subtraction, multiplication and division. These are called *binary* operations because at least two elements are employed in any addition, subtraction, etc. A mathematical system is said to be *closed* under a specified binary operation if the result of the operation on every ordered pair of elements is an element in the system. Thus the natural numbers are closed under addition, because the sum of any two natural numbers is a natural number. Similarly, the natural numbers are closed under multiplication. However, the natural numbers are not closed under subtraction because, for example, the number resulting from the subtraction of 5 from 3 is not a natural number. The integers, on the other hand, are closed under subtraction as well as under addition and multiplication.

But neither the integers nor the natural numbers are closed under division, which suggests the need for a further extension. We obtain closure under division, except division by zero, by introducing "quotients" of ordered pairs of integers. The *rational numbers* consist of the collection of all ordered pairs of integers a/b, with b not zero, satisfying the following definitions:

$$Equality: \frac{a}{b} = \frac{c}{d} \quad \text{if and only if} \quad ad = bc;$$

$$Addition: \frac{a}{b} + \frac{c}{d} = \frac{ad + bc}{bd};$$

$$Multiplication: \frac{a}{b} \cdot \frac{c}{d} = \frac{ac}{bd}.$$

(The notation here, a/b or $\frac{a}{b}$ for the ordered pair of integers a and b, is the traditional one. On page 30 there is a discussion of a more modern notation, namely (a, b), which, although not used in numerical work, is commonly employed in discussions of the foundations of the number system.) In case $b = 1$ the rational number a/b has the form $a/1$ which is identified with, or equated to, the integer a. By the definition of equality, $0/b = 0/d = 0/1$, and these relations show the possible notation for the integer 0 within the system of rational numbers.

This system is closed under division, except by zero, because with b, c, and d different from zero the problem of the division $\frac{a}{b} \div \frac{c}{d}$ is equivalent to the problem of finding a rational number x/y such that

$$\frac{a}{b} = \frac{c}{d} \cdot \frac{x}{y}.$$

By the definitions of multiplication and equality, this equation is satis-fied by $x = ad$ and $y = bc$. It must be verified of course that y is not zero; this follows from our assumptions that $b \neq 0$ and $c \neq 0$.

The rational numbers can be represented geometrically as points on a straight line. Select two points to correspond to the numbers 0 and 1; the distance between these two points is the unit of length. The other rational numbers correspond to the points obtained by using the so-called commensurable ratios of this unit length.

These are called the rational points, and we speak of "the point -2" for example in referring to the point corresponding to the number -2 in this geometric representation. If a point a is to the left of b, as for example -2 is to the left of $\frac{3}{2}$, then the corresponding numbers have the property that a is less than b, or b is greater than a, with the nota-tion $a < b$ or $b > a$ for this inequality relationship.

Between any pair of distinct rational points a and b there is another rational point, in fact infinitely many rational points. For the midpoint of the line segment joining a and b is $(a + b)/2$, also a rational point, and this process of taking midpoints can be repeated between a and $(a + b)/2$ etc. to give infinitely many rational points between a and b. Thus we say that the rational points are "dense," but this does not mean that all points on the line are rational points.

IRRATIONAL NUMBERS

The ancient Greek mathematicians were confronted with a dilemma when they attempted to calculate, in terms of rational numbers, the ratio of the length of the diagonal of a square to the length of one side. For if the side of a square is taken as the unit length, the Theorem of Pythagoras implies that the diagonal is of length $\sqrt{2}$. And the Greeks proved that the number $\sqrt{2}$ is not rational, and they therefore called it irrational. The following argument will establish the irrationality of $\sqrt{2}$.

To prepare the way for the argument, we make a few observations concerning the odd positive integers 1, 3, 5, $\cdots$, and the even positive integers 2, 4, 6, $\cdots$. An integer m is said to be divisible by an integer n provided there is an integer q so that $m = nq$. An even integer, being an integer divisible by 2, is one of the form $2q$, and an odd integer is of the form $2q + 1$. The square of any even integer is again an even integer, since $(2q)^2 = 4q^2 = 2(2q^2)$. Similarly the square of any odd integer is odd, since

$$(2q + 1)^2 = 4q^2 + 4q + 1 = 2(2q^2 + 2q) + 1.$$

Now let us suppose, contrary to what we want to establish, that $\sqrt{2}$ is rational, say $\sqrt{2} = a/b$ where a and b are positive integers. We may specify that a and b are not both even integers, for if they were both even we could rewrite the rational number with a replaced by the integer $a/2$ and b by the integer $b/2$. For example $2828/2000$ could be rewritten $1414/1000$, which in turn could be replaced by $707/500$. This specification, that a and b are not both even, is crucial in the proof. If we square both sides of the equation $\sqrt{2} = a/b$ and simplify, we get $2b^2 = a^2$. Hence the integer a^2, being 2 times b^2, is even, so that a itself is even by the argument of the preceding paragraph. Thus a is of the form $2q$, where q is also an integer. Replacing a by $2q$ in the relation $2b^2 = a^2$ we get $2b^2 = 4q^2$ or $b^2 = 2q^2$. This reveals that b^2 is even, and hence b itself is even. Thus we have the conclusion that a and b are even integers, contrary to our earlier specification. The contradiction so obtained establishes that $\sqrt{2}$ is irrational. This procedure is an example of an indirect proof, or proof by contradiction.

The number π, the ratio of the circumference of a circle to its diameter, is another example of an irrational number. This result, more difficult to prove, was established in 1761.

THE FUNDAMENTAL THEOREM OF ARITHMETIC

We digress from the general development of number systems to discuss some ramifications of the above proof of the irrationality of $\sqrt{2}$. The proof can be extended readily to establish the irrationality of $\sqrt{3}$, with this difference, that the matter no longer turns on the question of evenness and oddness. The issue centers on the question of divisibility by 3. That is, it would be assumed that $\sqrt{3} = a/b$ where this time it would be specified that not both a and b are divisible by 3. The argument goes through with this adjustment, because it can be established that the square of any integer is divisible by 3 if and only if the integer is divisible by 3. The latter assertion can be seen from two observations: First that any integer must take one of three forms, $3q$, $3q + 1$, or $3q + 2$; second that

$$(3q)^2 = 3(3q^2), \qquad (3q + 1)^2 = 3(3q^2 + 2q) + 1,$$

$$(3q + 2)^2 = 3(3q^2 + 4q + 1) + 1.$$

But suppose we wish to generalize, not simply from $\sqrt{2}$ to $\sqrt{3}$, but to $\sqrt{n}$ where n is a positive integer which is not a perfect square, that is, not of the form m^2 where m is an integer. This generalization can be obtained, but the solution of the problem is deeper because it is not easy to generalize the above assertions about divisibility by 2 and by 3. To indicate what is involved here, we must define prime numbers. Any positive integer p, larger than 1, is called a *prime* if it is divisible by no positive integers other than 1 and p. The sequence of primes begins 2, 3, 5, 7, 11, 13, 17, $\cdots$. If an integer n, larger than 1, is not a prime, it is called *composite*. The sequence of composite numbers begins 4, 6, 8, 9, 10, 12, 14, $\cdots$. Any integer, larger than 1, can be factored into primes; for example $504 = 2 \cdot 2 \cdot 2 \cdot 3 \cdot 3 \cdot 7 = 2^3 \cdot 3^2 \cdot 7$. In general any positive integer n will have a factorization into primes of the form

$$n = p_1^{\alpha_1} p_2^{\alpha_2} p_3^{\alpha_3} \cdots p_r^{\alpha_r},$$

where the symbols p_1, p_2, p_3, $\cdots$, p_r denote distinct prime numbers, and the exponents α_1, α_2, α_3, $\cdots$ α_r are positive integers. In the example $n = 504$, this notation would be interpreted as $r = 3$, $p_1 = 2$, $p_2 = 3$, $p_3 = 7$, $\alpha_1 = 3$, $\alpha_2 = 2$, $\alpha_3 = 1$. Now is it possible to factor a number in more than one way? Of course we can write $504 = 2^3 \cdot 3^2 \cdot 7 = 2 \cdot 3 \cdot 2 \cdot 7 \cdot 2 \cdot 3$, but the latter product is merely a trivial rearrangement of the same factors. The answer to the question is No: *Any positive integer n can be factored into primes in only one way*, apart from the order in which the factors are written. More briefly, *there is unique factorization of the integers into primes*. This proposition, which we do not prove here, is called the *fundamental theorem of arithmetic*.

With this fundamental theorem we can readily establish the general result that $\sqrt{n}$ is irrational if n is not a perfect square. For if we suppose that $\sqrt{n} = a/b$, we can square and simplify this equation to get $nb^2 = a^2$. The unique factorization of a^2 implies that if a has the form

$$a = p_1^{\alpha_1} p_2^{\alpha_2} \cdots p_r^{\alpha_r},$$

then a^2 has the unique form

$$a^2 = p_1^{2\alpha_1} p_2^{2\alpha_2} \cdots p_r^{2\alpha_r}.$$

Similarly every exponent in the factorization of b^2 into primes is even. Hence if we conceive the equation $nb^2 = a^2$ written in completely factored form, we can conclude that every exponent in the factorization of n into primes is also even, whence n is a perfect square.

The significance of the fundamental theorem of arithmetic is increased by our knowledge of more abstruse systems of "integers" which lack unique factorization. We shall not set forth any of these systems here. However, in case it is felt that the fundamental theorem of arithmetic is in some sense intuitively obvious, the following example is given to show that factorization need not be unique. We limit our attention to the set E of even positive integers, 2, 4, 6, 8, $\cdots$, a set which is closed under multiplication. We consider the factorization of elements of the set E into smaller elements of E; *e.g.*, $60 = 6 \cdot 10$. We ignore such equations as $6 = 2 \cdot 3$ and $10 = 2 \cdot 5$, because 3 and 5 are not elements of E. Let us call a number of E a *pseudo-prime* if it cannot be factored into two or more smaller numbers of E. Thus the sequence of pseudo-primes in E begins

$$2, 6, 10, 14, 18, 22, 26, 30, 34, 38, \cdots .$$

Any number in E can be factored into pseudo-primes, but this factorization is not unique in every case. For example $180 = 6 \cdot 30 = 10 \cdot 18$, so that there are two ways of factoring the number 180 into pseudo-primes of E.

DECIMAL EXPANSIONS

In addition to our earlier usage, the word "decimal" has another sense, as when we say that $\frac{1}{2}$ is equal to the decimal 0.5. Similarly the rational numbers 7 and $1\frac{2}{5}$ are equal to the decimals 7.0 and 2.4. These three instances, 0.5, 7.0, and 2.4, are examples of *terminating* decimals, as contrasted with the *nonterminating* decimal for $\frac{1}{3}$, namely $0.333 \cdots$. Any terminating decimal is a rational number since it can be written as the quotient of two integers, the denominator being a power of ten; for instance $7.3564 = 73564/10000$. Conversely, however it is not true that any rational number is a terminating decimal; the rational number $\frac{1}{3}$ suffices to prove this.

The procedure for converting any rational number a/b to its decimal form is well known: Divide the denominator into the numerator. Either the process terminates, as in the cases

$$\tfrac{6}{5} = 1.2, \qquad -\tfrac{1}{8} = -0.125,$$

or it does not, as in the cases

$$-\tfrac{13}{6} = -2.1666 \cdots , \qquad \tfrac{1}{7} = 0.142857142857142857142857 \cdots ,$$

$$40587/99900 = 0.406276276276 \cdots .$$

These examples illustrate the general proposition that *the decimal expansion of any rational number is periodic*; *i.e.*, that apart from some initial digits, a particular digit or block of digits will repeat itself infinitely often. This definition will be interpreted to say that any terminating decimal is periodic; for example 0.5 may be written in periodic form with an endless succession of zeros, $0.5000 \cdots$. To prove the general proposition we investigate the decimal expansion of a/b obtained by dividing b into a. In the division process the remainder at every stage is one of the numbers 0, 1, 2, $\cdots$, $b - 1$. If the remainder 0 occurs, then the decimal expansion is terminated at this point, and we have a periodic decimal. If the remainder 0 never occurs, then the decimal expansion is nonterminating, and the remainders are among the values 1, 2, $\cdots$, $b - 1$. Since we have only a finite number of possible values for the remainders, some remainder r must turn up for a second time, and the division process will repeat from that point to give a periodic decimal.

The converse of the proposition just proved is also true: *Any periodic decimal equals some rational number.* To prove this we will use the idea that multiplication of the periodic decimal by suitable powers of 10 produces identical expansions to the right of the decimal point. To begin with an example, let us consider

$$x = 0.123456345634563456 \cdots .$$

Multiplying by 10^2 and 10^6 we get

$$10^2 x = 12 + 0.345634563456 \cdots , \qquad\qquad \text{and}$$

$$10^6 x = 123456 + 0.345634563456 \cdots ,$$

and we subtract to get:

$$10^6 x - 10^2 x = 123456 - 12, \text{ whence } x = \frac{123444}{999900}.$$

In general consider a periodic decimal

$$x = 0.b_1 b_2 \cdots b_r c_1 c_2 \cdots c_n c_1 c_2 \cdots c_n \cdots .$$

The symbols b_1, b_2, $\cdots$, c_1, c_2, $\cdots$ denote digits, and the block of digits $c_1 c_2 \cdots c_n$ is repeated indefinitely. We are taking x positive for convenience since the proof for the negative values follows the same lines. We are also taking x to be a number between 0 and 1, since the deletion of the integral part does not affect the question of rationality. We multiply by 10^r to get

$$10^r x = b_1 b_2 \cdots b_r + 0.c_1 c_2 \cdots c_n c_1 c_2 \cdots c_n \cdots ,$$

where the notation $b_1 b_2 \cdots b_r$ denotes an integer with r digits. We multiply this equation by 10^n to get

$$10^{n+r} x = b_1 b_2 \cdots b_r c_1 c_2 \cdots c_n + 0.c_1 c_2 \cdots c_n c_1 c_2 \cdots c_n \cdots .$$

Then we subtract to remove the common infinite decimal,

$$10^{n+r} x - 10^r x = b_1 b_2 \cdots b_r c_1 c_2 \cdots c_n - b_1 b_2 \cdots b_r .$$

This is a linear equation in x, and if we divide both sides by $10^{n+r} - 10^r$, we will have expressed x as a rational number.

The application of this process to a periodic decimal with an infinite succession of nines is revealing. For example consider $x = 0.4999 \cdots$. We can write the equations

$$10x = 4 + .999 \cdots, \qquad 100x = 49 + .999 \cdots,$$

$$100x - 10x = 49 - 4, \qquad 90x = 45, \qquad x = \tfrac{1}{2}.$$

Thus the number $\tfrac{1}{2}$ has two (and only two, it turns out) decimal expansions, 0.5 and $0.4999 \cdots$. Any terminating decimal has such an alternative form as a nonterminating decimal with a succession of nines; for example $7.248 = 7.247999 \cdots$. Again, if we multiply the equation

$$\tfrac{1}{3} = 0.333 \cdots \text{ by } 3, \qquad \text{we get} \quad 1 = 0.999 \cdots.$$

REAL NUMBERS

The real numbers are defined as the class of all decimals

$$\pm a_1 a_2 \cdots a_k . b_1 b_2 b_3 \cdots.$$

Two real numbers are equal if and only if (a) their decimal expansions are identical, digit by digit, or (b) one is a terminating decimal, say $\pm a_1 a_2 \cdots a_k.b_1 b_2 \cdots b_r$, and the other is the nonterminating decimal obtained from this by replacing b_r by $b_r - 1$ followed by an infinite succession of nines. As we proved in the preceding section, the rational numbers can be characterized as the periodic decimals, and so the nonperiodic decimals are the irrational numbers.

We established on page 14 that $\sqrt{2}$ is an irrational number. Its decimal expansion can be constructed to as many places as we wish by a series of successive approximations:

$$1^2 = 1 < 2 < 4 = 2^2$$

$$(1.4)^2 = 1.96 < 2 < 2.25 = (1.5)^2$$

$$(1.41)^2 = 1.9881 < 2 < 2.0164 = (1.42)^2$$

$$(1.414)^2 = 1.999396 < 2 < 2.002225 = (1.415)^2$$

$$(1.4142)^2 = 1.99996164 < 2 < 2.00024449 = (1.4143)^2, \text{ etc.}$$

The decimal expansion, $\sqrt{2} = 1.4142 \cdots$, is not periodic, and no explicit formula is known for the successive digits.

That the real numbers and all points on a straight line stand in one-to-one correspondence is suggested by the following heuristic considerations. (To get a more rigorous formulation, we would have to turn to that branch of mathematics called *topology* for an inquiry into the innocent-sounding phrase "points on a line.") Let us use the decimal expansion 1.4142 $\cdots$ for illustration. Consider the infinite sequence of intervals I_1, I_2, I_3, $\cdots$, corresponding to the successive approximations above:

I_1, from 1 to 2;

I_2, from 1.4 to 1.5;

I_3, from 1.41 to 1.42;

I_4, from 1.414 to 1.415;

I_5, from 1.4142 to 1.4143; etc.

(By *interval* we mean the two points and all points between.) These are called *nested intervals*, because each is contained in the preceding one, and the length of the general interval I_n tends to zero as n increases indefinitely. We shall assume that a sequence of nested intervals defines a point, the one point which is common to all the intervals.

Conversely, suppose we are given any point P on the line, and we wish to determine the decimal to which it corresponds. For convenience we shall assume that P is a point to the right side of zero. If P is at one of the integral points 1, 2, 3, 4, $\cdots$, then P is a rational point. Otherwise, P can be located between two integers

$$a_1 a_2 \cdots a_k \quad \text{and} \quad a_1 a_2 \cdots a_k + 1,$$

where the a's are the digits. Define I_1 as the interval of length one joining these integral points, and divide I_1 into 10 sub-intervals each of length 10^{-1}. If P is the end point of any one of these sub-intervals, then P is a rational point; for example, if P is the left end point of the third interval, then P is the rational point $a_1 a_2 \cdots a_k . 2$. Otherwise we can locate P in precisely one of these 10 sub-intervals, and we define I_2 as that sub-interval. We repeat the process by splitting I_2 into 10 sub-intervals of length 10^{-2}, and so we construct a series of intervals I_1, I_2, I_3, $\cdots$ of lengths 1, 10^{-1}, 10^{-2}, $\cdots$. If at any step in this procedure the point P is an end point of an interval, we have at this step a terminating decimal corresponding to P. If P is not an end point of any interval, then the left end points of the intervals I_1, I_2, I_3, $\cdots$ give us a sequence

$$a_1 a_2 \cdots a_k, \qquad a_1 a_2 \cdots a_k . b_1, \qquad a_1 a_2 \cdots a_k . b_1 b_2, \cdots,$$

and the infinite decimal expansion corresponding to P is created by extending the digits $b_1b_2b_3 \cdots$ indefinitely.

On page 18 we mentioned the ambiguous representation of a terminating decimal, for instance 0.5 or 0.4999 $\cdots$. If the point P in the above process corresponds to such a number, the procedure outlined would lead to the terminating representation 0.5, not 0.4999 $\cdots$.

ALGEBRAIC AND TRANSCENDENTAL NUMBERS

Besides the separation of the real numbers into the two categories, rational and irrational, there is another significant classification: algebraic and transcendental numbers. We say that a real number α is *algebraic* if there is some equation

$$a_0x^n + a_1x^{n-1} + \cdots + a_{n-1}x + a_n = 0,$$

with integers a_0, a_1, $\cdots$, a_n as coefficients, having α as a root. For example, $\sqrt{2}$ is algebraic because it is a root of $x^2 - 2 = 0$. There are other equations having $\sqrt{2}$ as a root, some of which satisfy the conditions of the definition, such as $x^4 - 4 = 0$ and $3x^3 + 7x^2 - 6x - 14 = 0$, and some of which do not, such as $x - \sqrt{2} = 0$ and $\sqrt{3}x^2 - 2\sqrt{3} = 0$; but the number is algebraic provided there is at least one equation of the specified kind. Every rational number a/b is algebraic, because it satisfies the equation $bx - a = 0$.

Any real number which is not algebraic is said to be *transcendental*. It is not obvious from these definitions that there are any transcendental numbers. Their existence was established in 1851, and in 1882 it was proved that π, the ratio of the circumference of any circle to its diameter, is transcendental. This latter result, the proof of which is beyond the scope of this exposition, cleared up an ancient problem, *squaring the circle*. The problem is this: Given a geometric square, is it possible to construct by the straight edge and compass procedures of Euclidean geometry a circle of the same area? The answer is No, because on the one hand the problem is equivalent to constructing a line of length π from a given unit length, and, on the other hand, it can be demonstrated that all lengths which can be constructed by the permitted procedures are algebraic numbers; but π is not algebraic.

Other examples of transcendental numbers are the common logarithms (i.e. logarithms to base ten) of most numbers. Now log 100 = 2, which is an algebraic number, so we certainly cannot say that all numbers have logarithms which are transcendental. In detail the theorem is this: *If c is any positive rational number not of the form 10^n for some integer n, then log c is a transcendental number.* This is part of a general result established in 1934; another special case of the general result is that $2^{\sqrt{2}}$ is a transcendental number.

The values of the trigonometric functions can serve as examples of both algebraic and transcendental numbers, depending on the unit of measure used for the angles. On the one hand, using the system of degrees, minutes and seconds, we say that such values as $\sin 7°$, $\cos 14°3'$, and $\tan 111°4'5''$ are all algebraic numbers. The general statement is this: The trigonometric function of any angle whose ratio to $90°$ is rational is an algebraic number. (This statement is not completely correct as it stands, because such cases as $\tan 90°$ that are not real numbers should be eliminated from consideration.) On the other hand, if radian measure of angles is used, such values as $\sin 1$, $\cos 2$, $\tan 3$, are transcendental. In general, given a non-zero angle whose radian measure is a rational number, any trigonometric function of this angle is a transcendental number.

RATIONAL APPROXIMATIONS

The trigonometric values and common logarithms which we have been discussing are tabulated in most elementary textbooks in mathematics. But such tables give only approximations of these values to a few decimal places, and these rational approximations suffice for computation. Indeed all numbers used in computation are rational. All measurements in the physical world are approximations, made with varying degrees of precision and accurate to different numbers of significant digits. The number π, which is not rational, may occur in a formula, but in specific numerical calculations with the formula, π is replaced by some rational approximation; 22/7 is one well-known approximate value for π. Furthermore, as was proved on page 13, the rational numbers are dense on the straight line, and so no physical means can be used to determine whether a given length or ratio is rational. In view of this, and since all computations are performed with rational numbers, it might seem that irrational numbers are not needed in the application of mathematics to the physical world.

So far as the actual computations are concerned, it is the case that only the rational numbers are needed. But in the theoretical mathematical models which are used to describe the real world, there is a need for a "complete" system of numbers to give us all limiting values, a so-called *continuum* of values. Calculus, as we know it, could not exist without the irrational numbers to provide limits. We shall illustrate the above remarks with one example. (For a detailed discussion of the concept of limits see Chapter VII, page 200.) Just as the sequence of increasing rational numbers

$$0.3, 0.33, 0.333, 0.3333, \cdots$$

tends to the limit $\frac{1}{3}$, which is rational, in a similar way the sequence of increasing rational numbers

$$(1 + \tfrac{1}{2})^2, \quad (1 + \tfrac{1}{3})^3, \quad (1 + \tfrac{1}{4})^4, \quad (1 + \tfrac{1}{5})^5, \cdots$$

tends to a limit which is irrational. This limit, denoted by the symbol e, is one of the most significant numbers in mathematics; and it is central in calculus, and more generally in that part of mathematics called analysis.

COMPLEX NUMBERS

The formula for the solution of the equation $ax^2 + bx + c = 0$ given in elementary algebra books,

$$x = \frac{-b \pm \sqrt{b^2 - 4ac}}{2a},$$

does not give real numbers in the event that the so-called discriminant, $b^2 - 4ac$, is negative. The symbol i, called the imaginary unit, is introduced and given the basic property $i^2 = -1$. In this way the real number system can be extended so that such an equation as $x^2 - 6x + 25 = 0$ has two roots, $x = 3 + 4i$ and $x = 3 - 4i$. Numbers of the form $a + bi$, where a and b are real numbers, are called *complex numbers*, and we say that a is the real part and b the imaginary part. *Equality, addition,* and *multiplication* of complex numbers are defined thus:

$$a + bi = c + di \quad \textit{if and only if} \quad a = c \textit{ and } b = d;$$

$$(a + bi) + (c + di) = (a + c) + (b + d)i;$$

$$(a + bi)(c + di) = (ac - bd) + (ad + bc)i.$$

The real number system is a subclass of the complex number system: those numbers $a + bi$ having $b = 0$.

Subtraction and division are defined in terms of addition and multiplication. First, $(a + bi) - (c + di)$ is the complex number $x + yi$ satisfying the equation

$$(x + yi) + (c + di) = a + bi.$$

Applying the definitions of addition and equality, we get the equations $x + c = a$ and $y + d = b$, so that $x = a - c$ and $y = b - d$, whence

$$(a + bi) - (c + di) = (a - c) + (b - d)i.$$

Next, the quotient $(a + bi) \div (c + di)$ is the complex number $x + yi$ satisfying the equation

$$(x + yi)(c + di) = a + bi.$$

Applying the definitions of multiplication and equality here, we get

$$xc - yd = a, \qquad xd + yc = b.$$

These equations have solutions if and only if $c^2 + d^2$ is not equal to zero, written $c^2 + d^2 \neq 0$. This condition is equivalent to $c + di \neq 0$, and if this is satisfied, we conclude that

$$x + yi = \frac{ac + bd}{c^2 + d^2} + \frac{bc - ad}{c^2 + d^2} i.$$

We introduced the complex numbers in order to solve quadratic equations having no real roots, such as $x^2 - 6x + 25 = 0$. Some equations, such as $x^3 - 8 = 0$, are not without real roots, but also have complex roots which are not real. The equation $x^3 - 8 = 0$ has the real root $x = 2$ and the complex roots $x = -1 + \sqrt{3}\, i$ and $x = -1 - \sqrt{3}\, i$. These three roots are exhibited in the complete factorization of $x^3 - 8$ into linear factors,

$$x^3 - 8 = [x - 2][x - (-1 + \sqrt{3}\, i)][x - (-1 - \sqrt{3}\, i)].$$

This result for the special polynomial $x^3 - 8$ of degree 3 generalizes to polynomials of any degree with complex numbers for coefficients.

The generalization is known as the *fundamental theorem of algebra*, as follows: *Consider the equation of degree n,*

$$x^n + a_1 x^{n-1} + a_2 x^{n-2} + \cdots + a_{n-1}x + a_n = 0,$$

with coefficients a_1, a_2, $\cdots$ which are complex numbers. Then there is a unique factorization into n linear factors

$$(x - r_1)(x - r_2) \cdots (x - r_n) = 0,$$

the r's being complex numbers. Thus the equation has n roots, $x = r_1$, $x = r_2$, $\cdots$, $x = r_n$, and no others. The theorem indicates that the separation into linear factors is possible, but it does not tell us how to find the values r_1, r_2, $\cdots$, r_n. These roots need not be distinct, as, for instance, in the case of the equation $x^2 - 6x + 9 = 0$, with factorization $(x - 3)(x - 3) = 0$, and roots $x = 3$ and $x = 3$. The theorem covers all cases of equations in one unknown with complex numbers for coefficients. For while it might appear that we are dealing with a special case in taking the term of highest degree, x^n, to have coefficient unity, any other value c for this coefficient could be divided out to reduce the equation to the case we have discussed. (This process of dividing the equation throughout by c would fail in case $c = 0$, but in this case the degree of the equation would be less than the presumed value n.)

At the start of this section the well-known quadratic formula for solving equations of degree two is exhibited. There are similar formulas, not reproduced here, for equations of degree three (cubic equations). All general formulas for solving cubic equations in one unknown have an interesting feature in common: Any such formula expresses the roots in terms of non-real complex numbers, even in case all three roots of the equation are real. In this case of equations with three real roots, attempts were made for many years to find a formula, or set of formulas, that would express the roots in terms of real numbers only. Finally it was established that no such formula is possible.

It has been shown that the real numbers are separated into two categories, algebraic and transcendental numbers. This classification extends readily to complex numbers, and we can now give the general definitions of algebraic and transcendental numbers. An *algebraic number* is one which satisfies some equation

$$a_0 x^n + a_1 x^{n-1} + \cdots + a_{n-1} x + a_n = 0$$

with coefficients which are integers; any complex number which is not algebraic is called a *transcendental number*. An algebraic number is called an *algebraic integer* if it satisfies some such equation with $a_0 = 1$. For example $\sqrt{2}$ is an algebraic integer because it satisfies the equation $x^2 - 2 = 0$. The algebraic number $1/\sqrt{2}$ is not an algebraic integer, but this is not obvious from the above definition. It is not even apparent from the definition that such numbers as $\frac{1}{2}$, $\frac{1}{3}$, $\frac{2}{3}$ and $\frac{1}{5}$ are not algebraic integers. We state without proof that the only rational numbers which are algebraic integers are the integers 0, ± 1, ± 2, $\cdots$. The affirmative half of this assertion is clear from the definition: Any integer n satisfies $x - n = 0$ and so n is an algebraic integer. Thus the integers 0, ± 1, ± 2, $\cdots$ are often called the *rational integers* in advanced works on mathematics. The unique factorization theorem for rational integers does not extend to more general systems of algebraic integers in all cases, and this is the reason in part for its label, the *fundamental* theorem of arithmetic.

THE GEOMETRICAL REPRESENTATION OF COMPLEX NUMBERS

Just as the real numbers can be put into correspondence with the points on a line, so the complex numbers can be put into one-to-one

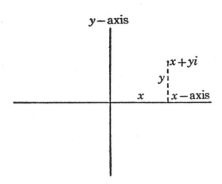

correspondence with the points in a plane. Let the number $x + yi$ correspond to the point with abscissa x and ordinate y as in coordinate geometry. The abscissa x is the perpendicular distance from the point to the y-axis, and the ordinate y is the perpendicular distance from the point to the x-axis, subject to the usual convention of signs: positive abscissas for points in the right half of the plane, negative for points in the left half; positive ordinates for points in the upper half of the plane, negative for points in the lower half.

The complex number $x + yi$ thus corresponds to the point commonly designated as (x, y) in analytic geometry; the number 0 corresponds to the origin $(0, 0)$ of the coordinate system. Take any two complex numbers and their sum: The three corresponding points together with the origin form a parallelogram. This suggests the parallelogram law for the addition, or "composition," of forces or velocities in mechanics. That is,

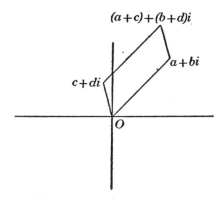

we regard $a + bi$ as denoting a "vector," a mathematical object having both magnitude and direction, as velocity does: the distance from 0 to the point $a + bi$ indicates both the magnitude and the direction. Then the line from 0 to $(a + c) + (b + d)i$ gives the magnitude and direction of the sum of the vectors $a + bi$ and $c + di$. These ideas are treated in more detail in Chapter VI.

The multiplication of complex numbers also has a geometric interpretation, and this is more revealing if formulated in *polar* coordinates. The polar coordinates of a point are denoted by (r, θ), where r is the distance from the point to the origin, and θ is the angle between (a) the line from the point to the origin and (b) the positive end of the x-axis, measured from (b) to (a) counterclockwise. Apart from the origin where

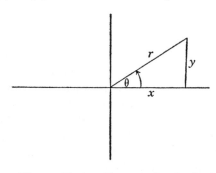

$r = 0$, we take r positive so that r, like x and y, is single-valued, meaning that one point has exactly one value for r. On the other hand θ is multiple-valued because the addition or subtraction of 360° does not change the location of the point, for example the point with rectangular coordinates $x = 0$ and $y = 1$ has $r = 1$ and $\theta = 90° + k \cdot 360°$ where k is any integer.

The equations $x = r \cos \theta$ and $y = r \sin \theta$ relate the two coordinate systems, and these equations are often used to define the trigometric functions $\cos \theta$ and $\sin \theta$. Hence the complex number $x + yi$ can be written in polar, or trigonometric, form

$$x + yi = r (\cos \theta + i \sin \theta),$$

and r is said to be the *modulus*, or *absolute value*, of the complex number, and θ its *amplitude*. Given two complex numbers with moduli r_1, r_2 and amplitudes θ_1, θ_2, the product formula

$$[r_1(\cos \theta_1 + i \sin \theta_1)] \cdot [r_2(\cos \theta_2 + i \sin \theta_2)]$$

$$= r_1 r_2 [\cos (\theta_1 + \theta_2) + i \sin (\theta_1 + \theta_2)]$$

can be established by elementary trigonometry. This equation may be formulated verbally: The product of two complex numbers has modulus which is the product of the moduli, and amplitude which is the sum of the amplitudes.

If, in this product formula, we consider the special case where $r_1 = r_2$

and $\theta_1 = \theta_2$, and so use the simple labels r and θ for these, we have

$$[r(\cos \theta + i \sin \theta)]^2 = r^2(\cos 2\theta + i \sin 2\theta).$$

This is a special case of a formula which holds for any real number n as exponent,

$$[r(\cos \theta + i \sin \theta)]^n = r^n(\cos n\theta + i \sin n\theta),$$

a result known as *de Moivre's theorem*.

To show the strength of this result, let us consider the question of finding the roots or radicals of a complex number. In the case of real numbers, the question of evaluating say $\sqrt[3]{5}$ can be handled by the approximation scheme on pages 18–20. It lies between 1 and 2, since $1^3 = 1$ and $2^3 = 8$; next, it lies between 1.7 and 1.8 since $(1.7)^3 = 4.913$ and $(1.8)^3 = 5.832$; next, it lies between 1.70 and 1.71 because $(1.70)^3 = 4.913,000$ and $(1.71)^3 = 5.000211$; etc. When we turn to complex numbers, there is no such simple process of approximation, and so we turn to de Moivre's theorem with $n = \frac{1}{3}$:

$$\sqrt[3]{r(\cos \theta + i \sin \theta)} = [r(\cos \theta + i \sin \theta)]^{1/3}$$

$$= r^{1/3}\left(\cos \frac{\theta}{3} + i \sin \frac{\theta}{3}\right).$$

For example to calculate $\sqrt[3]{3 + 4i}$, we write $3 + 4i$ in polar form with $r = 5$ and $\theta = 53°8'$,

$$3 + 4i = 5(\cos 53°8' + i \sin 53°8'),$$

this being an approximate equation with the angle to the nearest minute. Hence a cube root is

$$1.70998 \, (\cos 17°43' + i \sin 17°43').$$

At first glance we appear to be getting just one answer to the problem of solving $z^3 = 3 + 4i$. The other answers stem from the fact that the amplitude θ is multiple-valued, so that $\theta = 53°8'$ can be replaced by $\theta = 53°8' + k \cdot 360°$ where k is any integer. Taking $k = 1$ and 2 we get the approximate equations

$$3 + 4i = 5 \, (\cos 413°8' + i \sin 413°8') \qquad \text{and}$$

$$3 + 4i = 5 \, (\cos 773°8' + i \sin 773°8'),$$

and so we obtain two other cube roots,

$$1.70998 \, (\cos 137°43' + i \sin 137°43') \qquad \text{and}$$

$$1.70998 \, (\cos 257°43' + i \sin 257°43').$$

Giving further values to k does not give any new results, merely repetitions of the three answers.

CRITIQUE OF OUR APPROACH

Modern mathematics has revealed many limitations in such a survey of number systems as we have just given. Much of our exposition has been descriptive, assertions being frequently made without proof. This in itself is not a serious criticism, for it is easy to get bogged down in so much detail that the larger view recedes and perhaps disappears. But we discover, if we try in a careful way to fill in all the details of proof, that the going might be easier if we had used slightly different approaches to our number systems.

Indeed, a different approach to complex numbers can be seen to be desirable without even getting into proofs of the deeper results. We introduced the symbol i because we could not solve equations such as $x^2 = -1$ and $x^2 - 6x + 25 = 0$ in real numbers. We postulated the existence of a mathematical object, i, in order to get solutions to equations which we could not otherwise solve. Do we always invent or postulate new symbols or new systems when the old ones will not solve our problems? If we cannot find a number x to satisfy the equation $1^x = 3$, or the equation $0^x = 3$, shall we postulate some solution? Or if we cannot solve simultaneously the two equations $x + y = 6$ and $x + y = 7$, do we create an extended number system to obtain a solution? The answer to each of these questions is No. It is possible to formulate the system of complex numbers without using the so-called imaginary unit i, and this we now do.

The clue lies in our representation of $a + bi$ as a point with coordinates (a, b). Disregarding the idea of a point, what we have here is simply an ordered pair of real numbers. So let us regard the complex number system as the collection of all ordered pairs of real numbers (a, b). The definitions of *equality*, *addition* and *multiplication* are merely a rewriting of the definitions as they were given on page 22:

$$(a, b) = (c, d) \quad \textit{if and only if} \quad a = c \textit{ and } b = d;$$

$$(a, b) + (c, d) = (a + c, b + d);$$

$$(a, b) \cdot (c, d) = (ac - bd, ad + bc).$$

The number pair $(0, 0)$ plays the role of zero, and $(1, 0)$ plays the role of unity, because by the above definitions

$$(a, b) + (0,0) = (a, b), \qquad (a, b) \cdot (1, 0) = (a, b).$$

Subtraction and division,

$$(a, b) - (c, d) \quad \text{and} \quad (a, b) \div (c, d),$$

are defined by solutions (x, y), if they exist, of the equations

$$(x, y) + (c, d) = (a, b), \qquad (x, y) \cdot (c, d) = (a, b).$$

Again we have a mere restatement of the equations

$$(x + yi) + (c + di) = a + bi, \qquad (x + yi)(c + di) = a + bi$$

of pages 22 and 23, and the results are as expected:

$$(a, b) - (c, d) = (a - c, b - d), \frac{(a, b)}{(c, d)} = \left(\frac{ac + bd}{c^2 + d^2}, \frac{bc - ad}{c^2 + d^2} \right),$$

the latter being valid provided $c^2 + d^2 \neq 0$, which is equivalent to $(c, d) \neq (0, 0)$.

We seem to be getting the usual results concerning complex numbers by reformulating the old $a + bi$ as a number pair (a, b). However, in this new formulation the real numbers do not appear specifically as a subclass of the complex numbers. For the subclass with $b = 0$ is the set of pairs $(a, 0)$ where a is any real number. In place of such real numbers as 1, 2, 3, $\sqrt{2}$, 7, we get the pairs $(1, 0)$, $(2, 0)$, $(3, 0)$, $(\sqrt{2}, 0)$, $(7, 0)$. We say that the subclass of pairs of the form $(a, 0)$ is *isomorphic* to the class of real numbers. This simply means that the subclass of pairs $(a, 0)$ is the class of real numbers in a new guise: After each real number write a comma, then a zero, and enclose the whole thing in parentheses.

ISOMORPHIC SYSTEMS

To give a formal definition of the word *isomorphic*, we say that two number systems S and S' are isomorphic if (i) there is a one-to-one correspondence between the elements of S and the elements of S' such that (ii) if elements α and β of S correspond to elements α' and β' of S', then $\alpha + \beta$ must correspond to $\alpha' + \beta'$, and $\alpha\beta$ must correspond to $\alpha'\beta'$.

To apply this definition to the situation in the last paragraph of the previous section we may let S be the set of real numbers, and let S' be the set of pairs $(a, 0)$ subject to our earlier definitions of the addition and multiplication of the general number pairs (a, b). The condition (i) is satisfied when we set up the obvious correspondence $a \leftrightarrow (a, 0)$. With this correspondence, the condition (ii) is satisfied because any two real numbers a and c in S correspond to $(a, 0)$ and $(c, 0)$ in S', and by the definitions we have

$$(a, 0) + (c, 0) = (a + c, 0), \qquad (a, 0)(c, 0) = (ac, 0).$$

To show that it is not enough to satisfy condition (i) in the above definition of an *isomorphism* between two systems, let us create a one-to-one correspondence between the set S of real numbers and the set S' now defined as consisting of all pairs $(0, b)$. Thus S' is a subclass of the complex numbers. Thus condition (i) is satisfied when we establish the correspondence $b \leftrightarrow (0, b)$. And the first part of condition (ii) is satisfied, because we have $d \leftrightarrow (0, d)$

$$(0, b) + (0, d) = (0, b + d),$$

by definition. But the last part of condition (ii) for an isomorphism between these two systems fails, because by definition

$$(0, b) \cdot (0, d) = (-bd, 0),$$

and $(-bd, 0)$ is not in general an element of S'.

The concept of isomorphism is basic in mathematics: Isomorphic systems have essentially the same mathematical structure, although outwardly they may appear to be different. An almost trivial example is the isomorphism between the Roman numerals I, II, III, IV, $\cdots$ and the Hindu-Arabic numerals 1, 2, 3, 4, $\cdots$. It is not always so easy to recognize isomorphic systems, because they may arise quite differently. For example, in some cases the question is whether a certain system defined and established algebraically is isomorphic to another system of geometric origin.

NUMBER PAIRS

The outline of number systems, as given on pages 7–28, developed from natural numbers to integers, to rational numbers, to real numbers, and finally to complex numbers. We have seen that the last of these four steps can be regarded as the formulation of a system of ordered pairs of real numbers. This device of creating number pairs to pass from one system to a larger one also applies in the first two steps in our development of number systems, from natural numbers to integers, and from integers to rational numbers.

That the rational numbers can be regarded as ordered pairs of integers is apparent from the ordinary notation used, 3/5, 7/4, *etc.* In order to keep the number pair concept in the foreground, let us say write (a, b) for a/b, and say that the set of rational numbers is the collection of all pairs of integers (a, b) with $b \neq 0$. This restriction on b is simply the ruling out of division by zero. *Equality*, *addition* and *multiplication* of these number pairs are defined as on page 12, namely:

$$(a, b) = (c, d) \quad \textit{if and only if} \quad ad = bc;$$
$$(a, b) + (c, d) = (ad + bc, bd);$$
$$(a, b) \cdot (c, d) = (ac, bd).$$

The system is closed under addition and multiplication because $b \neq 0$ and $d \neq 0$ implies $bd \neq 0$.

These definitions contrast sharply with those of page 28 in the development of complex numbers as ordered pairs of real numbers. One marked difference may be noted in the definition of equality: in the case of complex numbers, different number pairs stood for different numbers in the system, whereas in the present case any pair (a, b) in the system is equal to an infinite set of such pairs. For example, the pair $(3, 7)$ is equal to each of $(-3, -7)$, $(6, 14)$, $(30, 70)$, $\cdots$; the list is endless. Thus one rational number is denoted by an infinity of pairs of integers. When we speak of the element $(3, 7)$, or the rational number $(3, 7)$, we shall mean the infinite class of pairs all of which are equal to $(3, 7)$ by definition. We are confronted at once with a logical question: Is the definition of addition (or multiplication) consistent with the definition of equality in the sense that an equal answer is obtained for the sum if (a, b) and (c, d) are replaced by other representatives of their respective classes? Do we get equal answers to the additions $(3, 7) + (2, 11)$ and $(6, 14) + (20, 110)$? This specific question is quickly answered:

$$(3, 7) + (2, 11) = (47, 77); \qquad (6, 14) + (20, 110) = (940, 1540).$$

Now $(47,77)$ and $(940, 1540)$, while not identical term by term, are equal by definition, since $47 \cdot 1540 = 77 \cdot 940$.

Checking a single example is not sufficient proof for the general case, however. To prove the consistency of the definition of addition, we let (a, b) and (a', b') be two representatives of the same class, so that $(a, b) = (a', b')$ and $ab' = a'b$. Similarly let $(c, d) = (c', d')$ so that $cd' = c'd$. Then we must prove that

$$(a, b) + (c, d) = (a', b') + (c', d').$$

To do this, we apply the definition of addition to get

$$(a, b) + (c, d) = (ad + bc, bd);$$
$$(a', b') + (c', d') = (a'd' + b'c', b'd').$$

By the definition of equality, these are equal provided

$$(ad + bc)(b'd') = (a'd' + b'c')(bd) \qquad\qquad \text{or}$$

$$adb'd' + bcb'd' = a'd'bd + b'c'bd.$$

To establish this we note that $ab' = a'b$ can be multiplied by dd' to give $adb'd' = a'd'bd$, and similarly $cd' = c'd$ implies that $bcb'd' = b'c'bd$.

A similar proof can be worked out to establish the consistency of the definition of multiplication, but we omit the details.

Just as the complex numbers contained a subclass of numbers isomorphic to the real numbers, so the rational numbers contain a subclass of elements $(a, 1)$ isomorphic to the integers. If we set up the correspondence $(a, 1) \leftrightarrow a$, where a is any integer, we can establish that this is an isomorphism by the simple observations

$$(a, 1) + (c, 1) = (a + c, 1), \qquad (a, 1)\cdot(c, 1) = (ac, 1).$$

ANOTHER LOOK AT THE INTEGERS

The number-pair concept is a fairly obvious idea in the cases of complex numbers and rational numbers. For after all, the standard symbols for numbers in these systems, $a + bi$ and a/b, are simply ways of writing the ordered pair of numbers a and b. It is not so apparent that the integers can be obtained as an extension of the natural numbers by the number-pair device. This extension is obtained by writing (a, b) to designate what we ordinarily write as $a - b$. We prefer the notation (a, b) to $a - b$ because our starting point is the system of natural numbers in which $a - b$ does not always have meaning: for instance $4 - 7$.

The definitions of equality, addition and multiplication are easily written from this identification of (a, b) with what is usually written $a - b$. We must avoid the use of the minus sign at this early stage of the theory, and thus the definitions are:

$$(a, b) = (c, d) \quad \textit{if and only if} \quad a + d = b + c;$$

$$(a, b) + (c, d) = (a + c, b + d);$$

$$(a, b) \cdot (c, d) = (ac + bd, ad + bc).$$

The whole collection of elements is separated by the definition of equality into classes of equal pairs:

$$(1, 1) = (2, 2) = (3, 3) = (4, 4) = \cdots;$$

$$(1, 2) = (2, 3) = (3, 4) = (4, 5) = \cdots;$$

$$(1, 3) = (2, 4) = (3, 5) = (4, 6) = \cdots; \textit{etc.}$$

As with the rational numbers, it may be verified that the definitions of addition and multiplication are consistent with the definition of equality, so that a class of equal pairs may be represented by any one of its members. Observing that every class of equal pairs has one representative containing the symbol 1, we see that the whole system can be reduced to the infinite set of unequal pairs

$\cdots$, $(1, 5)$, $(1, 4)$, $(1, 3)$, $(1, 2)$, $(1, 1)$, $(2, 1)$, $(3, 1)$, $(4, 1)$, $(5, 1)$, $\cdots$.

Any pair (a, b) of natural numbers is equal to one of these.

Now let the natural numbers 1, 2, 3, 4, 5, $\cdots$ be put into one-to-one correspondence with the pairs $(2, 1)$, $(3, 1)$, $(4, 1)$, $(5, 1)$, $(6, 1)$, $\cdots$; in general n corresponds to $(n + 1, 1)$. To verify that this correspondence is an isomorphism, we add and multiply the pairs $(n + 1, 1)$ and $(m + 1, 1)$, using the definitions:

$(n + 1, 1) + (m + 1, 1) = (n + m + 2, 2) = (n + m + 1, 1)$;

$(n + 1, 1) \cdot (m + 1, 1) = (nm + n + m + 2, n + m + 2)$

$$= (nm + 1, 1).$$

These results prove that we have an isomorphism, because the correspondence n to $(n + 1, 1)$ implies:

$$n \leftrightarrow (n + 1, 1), \qquad m \leftrightarrow (m + 1, 1),$$

$$n + m \leftrightarrow (n + m + 1, 1), \qquad nm \leftrightarrow (nm + 1, 1).$$

The element $(1, 1)$ is the "zero element" of the system, because

$$(a, b) + (1, 1) = (a + 1, b + 1) = (a, b).$$

In fact if we simply change notation and replace

$\cdots$, $(1, 5)$, $(1, 4)$, $(1, 3)$, $(1, 2)$, $(1, 1)$, $(2, 1)$, $(3, 1)$, $(4, 1)$, $(5, 1)$, $\cdots$

the symbols

$$\cdots , \; -4, \; -3, \; -2, \; -1, \; 0, \; 1, \; 2, \; 3, \; 4, \; \cdots ,$$

we have the integers as they are usually written.

The question may be asked, "Why go to all this bother when it is so much easier to introduce the simpler notation at the outset?" Part of the answer is that, although the number-pair notation is too clumsy for actual work with the integers, it gives us an excellent foundation for proving the fundamental propositions about integers without recourse to vague analogies to profit and loss, temperatures above and below zero, rectangles split into subrectangles, and the like.

Consider, for example, the proposition that $c \cdot 0 = 0$ for any integer c. To prove this we translate the symbols into the corresponding number pairs: We write $(1, 1)$ in place of 0, and, say, (a, b) in place of c. Then we get

$$(a, b) \cdot (1, 1) = (a + b, a + b) = (1, 1)$$

by straightforward use of the definitions. As a second example, consider the proposition that $(-1)(-1) = 1$. To prove this we replace -1 by the number pair $(1, 2)$, and so we have

$$(1, 2) \cdot (1, 2) = (5, 4) = (2, 1),$$

again by application of the definitions. This completes the proof because $(2, 1)$ corresponds to the integer 1. These proofs are simply a couple of examples which illustrate the general situation: that viewing the integers as pairs of natural numbers enables us to obtain the basic properties of the integers from the properties of natural numbers with no further assumptions.

REMARKS ON THE FOUNDATIONS

In the last few sections we have been trying to indicate the desirability of constructing the number systems so as to maximize the number of properties which we are able to prove, and to minimize the number of properties which must be assumed. Mathematicians working on the foundations of numbers have carried this program to the point where all the known properties of the systems we have discussed can be proved on the basis of a set of five assumptions about the natural numbers. (We shall not list these five assumptions, the so-called Peano postulates, but we do mention that the principle of finite induction (pages 9–11) is one of them. It should also be stated that the basic assumptions can be selected in various ways, and there are alternative schemes for building the whole structure up to complex numbers.) Such a program is no small undertaking; for example, the basic properties of the various systems up to complex numbers constitute 301 theorems in one book on the subject, Edmund Landau's, *Foundations of Analysis* (New York, Chelsea, 1951).

The most difficult link in the chain of development from natural numbers to complex numbers is the step from rational numbers to real numbers. This step, unlike the others, cannot be handled by the number-pair device; a deeper limit process is involved. It may be noted that in the previous section on real numbers we did define equality of real numbers, but made no attempt to define addition and multiplication. As might be expected, the definition of the product, for example, of two infinite decimals is a fairly involved matter. Then, granting that we have some definition, can such basic properties as $\sqrt{2} \cdot \sqrt{3} = \sqrt{6}$ be proved? They can be proved, but the proofs lie beyond the scope of this discussion. One further remark: However useful geometric analogies may be in pedagogical explanations, the real number system should be constructed independently of any geometric considerations. For example, any "proof" by means of Euclidean geometry that $\sqrt{2} \cdot \sqrt{3} = \sqrt{6}$ would create logical difficulties if we then went on to use this property of real numbers in discussions of non-Euclidean geometry. We want the properties of real numbers to be independent of the axioms of Euclidean, or any other, geometry.

We have confined our attention in this chapter to the most basic of the number systems of mathematics. The possibility of further extensions is implicit in the number-pair approach, indeed is one of the advantages of that approach. For there is no reason to stop at pairs: We could also look into triples, quadruples, and in general n-tuples of numbers. Moreover, by varying the definitions of equality, addition and multiplication we might obtain other algebraic structures of interest and of use. The continuing study of such possibilities has revealed a wide variety of mathematical systems for further investigation.

BIBLIOGRAPHY

1. ALLENDOERFER, C. B., and OAKLEY, C. O. *Principles of Mathematics.* New York: McGraw-Hill Book Company, 1955, Chapter 2.
2. BIRKHOFF, GARRETT, and MACLANE, SAUNDERS. *A Survey of Modern Algebra.* Revised edition. New York: The Macmillan Company, 1953. Chapters 1 to 5.
3. COURANT, RICHARD, and ROBBINS, HERBERT. *What Is Mathematics?* New York: Oxford University Press, 1941. Chapters 1 and 2.
4. DUBISCH, ROY. *The Nature of Number.* New York: The Ronald Press Company, 1952.
5. HOBSON, E. W. *Squaring the Circle.* New York: The Macmillan Company, 1913. (Reprinted, Chelsea Publishing Co., New York, 1953.)
6. KLINE, MORRIS. *Mathematics in Western Culture.* New York: Oxford University Press, 1953. Chapters 3 and 7.
7. MESERVE, BRUCE E. *Fundamental Concepts of Algebra.* Cambridge, Mass.: Addison-Wesley Publishing Company, 1953.
8. ORE, OYSTEIN. *Number Theory and Its History.* New York: McGraw-Hill Book Company, 1948.
9. WAISMANN, FRIEDRICH. *Introduction to Mathematical Thinking.* New York: Frederick Ungar Publishing Co., 1951.
10. WEISS, M. J. *Higher Algebra for the Undergraduate.* New York: John Wiley & Sons, 1949, Chapters 1 and 2.

III

Operating with Sets

E. J. McSHANE

BERTRAND RUSSELL has said that "the obvious is difficult to prove and often wrong." The idea of a set is ancient and obvious—and Russell's remark applies to it. There is certainly nothing unusual in forming the mental concept that results from thinking of several things simultaneously, as a coherent whole. As far back as recorded language goes, it has contained collective nouns, such as a "swarm" of bees and a "flock" of sheep. And a set (or class or collection—we use them as synonyms) is just this mental concept formed by thinking of several things as forming a coherent single. Perhaps it is because the idea of set seemed so simple and obvious that it did not emerge as a clear-cut part of mathematics until less than a century ago. But by now the habit of thinking in terms of sets has become routine for most mathematicians, and has also become part of the every-day thinking of many other scientists too. It is worth while to look at some examples of this.

EXAMPLES OF SETS

In classical Euclidean geometry it is habitual to keep points, lines and planes more or less in separate mental pigeon-holes. Given a line, in plane or solid geometry, one does not tend to think of it first and foremost as a set of points; it is a single entity in its own right. But if we shift our point of view and choose to think of the line as a set of points, it becomes natural to think of a "rigid motion" of the plane as a method of matching up each point P of the plane with a point P' (often called "moving P to P'") in such a way that the distance between any two points P and Q is the same as the distance between the points P' and Q' which match up with them.

But now it becomes clear that there may be other ways of matching points too, in which the points on a line may "move to" (that is, match

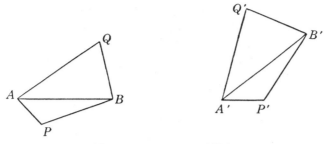

Rigid motion; $AP = A'P'$, etc.

up with) points that are not on a straight line. This leads us into newer geometries; a look at Chapter X of this book will show at once the importance of the concept of set in the study of topology.

Another important application of set-thinking is in statistics and probability. The very foundation of life-insurance mathematics is the subdivision of a population into age groups, which have been found each to have a consistent life expectation. More recently, the study of atomic phenomena has forced physicists into a probabilistic point of view. A single atom is not treated as an isolated object, but as a member of a set of atoms of similar type, and the typical statement is not "the distance from innermost electron to nucleus is k," but instead "there is a probability of 97 per cent that the experiment of measuring the distance from the innermost electron to the nucleus will give a result between h and k." This has meaning only for the atom as a member of a large set of atoms, not for a single atom isolated from all others.

These examples, and others which could be added to them, show that the idea of set is not too trivial to be of use. The other error to guard against is that the idea of set is so simple that it can be brushed off with a remark that it should be clear to everybody. At this moment we content ourselves with an indication; later we give more detailed reasons. When we say that some property of sets is obvious, we mean something like this: I have had experience with a number of sets, and each one was easily seen to have that property, so I shall believe that all sets have that property. But the sets that we meet in everyday experience are always sets with only a few elements. The "set of fingers on my left hand" is easily comprehended and studied. The "set of all human beings alive today" is not really visualized by us; we mentally form an image composed of a relatively few samples of humans, and by analogy apply to it the ideas that worked with finite sets, such as counting. But this analogy becomes far-fetched when we try to discuss really large sets.

For instance, if we say that we are thinking of the set consisting of all material objects and all ideas, we are deluding ourselves. We are thinking of a few material objects and a few ideas, and hoping to be able to proceed by the same kind of thinking that was successful with small sets. Soon we shall prove that we cannot; the analogy does not work, and we have invited hopeless confusion.

Since the uncritical self-confident approach can put us in difficulties, we are faced with the problem of making up a safe way of discussing sets that will be inclusive enough to let us do the operations needed in the situations where sets arise, and not inclusive enough to take in the dangerous territory just mentioned. But we cannot do this intelligently until we see what kinds of operations are needed in studying sets, and which are the important relationships between sets. So now we begin to look at those relationships and those operations, or "maneuvers."

OPERATION WITH SETS

The basic ideas involved are merely "set," "object," and "is a member of." We think of the word "object" as more inclusive, so that all sets are thought of as objects, but not necessarily conversely. As a rule we shall use capital letters to mean sets. The verb "is a member of" is symbolically indicated by the symbol ϵ, and its denial by $\notin$; thus "$x \epsilon A$" is read "x is a member of A," and "$y \notin B$" is read "y is not a member of B." Another notational convention, by no means necessary but often convenient, concerns the symbol for a set having only a few members, which can be listed in detail. If, for instance, A has only the two members x and y, it is convenient to write $A = \{x, y\}$; if A has the members $x_1, \cdots, x_n$ and no others, we write $A = \{x_1, \cdots, x_n\}$.

Here we digress a moment to caution against an error sometimes made. Even if A has only a single member, say x, it is not at all true that A is x. By the notation just explained, we can write $A = \{x\}$, but this is quite different from $A = x$. For instance, the Federation of Women's Clubs is a set whose members are clubs, not women. If the Sagebrush Women's Society for Motet Singing has just two members, it can be a member of the Federation. If one of the two women leaves, and only Mrs. Smith is left in the Society, the Society can still belong to the Federation; Mrs. Smith cannot. If Mrs. Smith, discouraged, permits the Society to die, although the Society is the same as the set {Mrs. Smith} it may still be true that Mrs. Smith retains perfect health and vigor.

Suppose now that A and B are sets. It is possible that every object which belongs to A also belongs to B; that is, for all x, if $x \epsilon A$ then $x \epsilon B$. In this case we say that A is contained in B, or that A is a *subset* of B; and we also say that B contains A, or that B is a *superset* of A. In symbols this is expressed by $A \subset B$ or $B \supset A$. Pictorially, we can represent this relation by means of a "Venn diagram." A *Venn diagram* is a symbolic representation of a set, obtained by letting the points of the plane represent the objects under consideration, and letting each set be represented by the points inside of some closed curve. Thus, given a set A, we can draw a circle, and assert that the points inside the circle stand for members of A, and the points outside for nonmembers of A. In this way the relation $A \subset B$ is represented pictorially by the diagram:

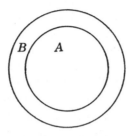

It should be emphasized that the relations ϵ and $\subset$ are quite different. One difference is obvious; the sentence "$x \ \epsilon \ A$" can be correct even if x is not a set, while "$A \subset B$" cannot be correct unless A and B are both sets. But even when both A and B are sets the meanings are still different. Let A stand for the set of all citizens of Virginia and B for the set of all citizens of the United States. Then $A \subset B$, for every citizen of Virginia is a citizen of the United States. But the sentence "$A \ \epsilon \ B$" is false; the set of all citizens of Virginia is not a citizen of the United States.

The sentence "A and B have the same members" can be broken down into "every member of A is a member of B, and every member of B is a member of A." Being thus broken down, it can be written symbolically in the form "$A \subset B$ and $B \subset A$." We want nothing to enter into the specification of a set beyond the naming of its members; that is, a set should be completely identified when we know just which objects are its members. This we can state as an axiom, which we shall assume henceforth:

If A and B are sets, and $A \subset B$ and $B \subset A$, then $A = B$. For example, let A be the set of all real numbers x such that $x^2 - 5x + 6 = 0$, and let B be the set $\{2, 3\}$. Each member of B is in A, since $2^2 - 5 \cdot 2 + 6 = 0$ and $3^2 - 5 \cdot 3 + 6 = 0$; and each member of A is in B, since the equation has no real solutions except 2 and 3. Hence $A = B$.

Suppose again that A and B are sets. From the two of them we can make a new set consisting of just those objects which simultaneously belong to A and to B. This new set we call the *intersection* of A and B, and we indicate it by the symbol $A \cap B$. For instance, if A consists of all rational numbers and B of all numbers between 0 and 1, $A \cap B$ will consist of all numbers which are rational and between 0 and 1, that is of all proper fractions. This can easily be extended to more than two sets. Thus if A, B, C and D are all sets, the intersection $A \cap B \cap C \cap D$ will consist of just those points which are members of all four sets. In fact, the idea of intersection can be extended without much trouble to quite arbitrary collections of sets; however, the notation needs a little changing, since we cannot go on writing the symbol $\cap$ between consecutive members of very large finite collections, let alone infinite collections. Suppose that X is a non-empty set whose members are themselves sets. (Sometimes, to help keep this in mind, we shall call X a set-of-sets.) By the *intersection of the sets of X* we shall mean the set consisting of all those objects which belong to every set A in the set-of-sets X. For this we use the symbol $\bigcap_{A \epsilon X} A$. If X is the set of all clubs in a given school, the members of X are clubs, which are themselves sets of persons. The intersection $\bigcap_{A \epsilon X} A$ is the set of persons (the perfect joiners) who are members of every club in the school. Notice that in the intersection symbol the letter A is a "dummy"; the sets $\bigcap_{A \epsilon X} A$ and $\bigcap_{B \epsilon X} B$ are the same. The first means "the set of objects which belong to A whenever $A \epsilon X$," and the second means "the set of objects which belong to B whenever $B \epsilon X$," and both of these can be worded "the set of all objects which belong to all sets in the set-of-sets X." In the last form it is clear that the choice of letter A or B is unimportant, since they have completely disappeared.

Second, if A and B are sets there is a set whose members are just those objects which belong to at least one of the sets A and B. This new set is called the *union* of A and B, and is indicated by the symbol $A \cup B$. Thus if A is the set of all chairs in one room of a two-room apartment and B is the set of all chairs in the other room, $A \cup B$ is the set of all chairs in the apartment. This idea can be extended to more than two sets, just as in the case of intersection. The union of A, B and C is the set $A \cup B \cup C$ consisting of all objects which belong to one or more of the sets A, B, C. Furthermore, if X is a set-of-sets, the *union of the members of X* is the set $\bigcup_{A \epsilon X} A$ consisting of all objects which belong to one or more of the sets A in the set-of-sets X. For example, the National League is a set N with eight members, each of which is a baseball team and therefore is a set of men. The American League is a set A of eight members, each a baseball team. The "Majors" is a set M with two members, the American League and the National League. Then $\underset{T \epsilon A}{\cup} T$ is the set of all objects (men) which are members of one or more (in this case, of course, just one) of the teams in the American League. The set $\underset{L \epsilon M}{\cup} L$ is the set of all objects (teams) which are members of one or more of the sets in M, that is, it is the same as $A \cup N$, and consists of 16 baseball teams. The set $\underset{T \epsilon A \cup N}{\cup} T$ is the set of all major-league baseball players.

The operation of union is an "enlarging" operation, by which we mean that if A is any set and B any set, then $A \cup B$ contains A. Likewise, intersection is a "contracting" relation, for if A and B are any sets then $A \cap B$ is contained in A.

These ideas can be expressed pictorially by Venn diagrams. In the diagrams below, A, B and C are represented by the points inside the circles so labelled, while $A \cap B$, $A \cup B$ and $(A \cap B) \cup C$ are represented by the shaded parts.

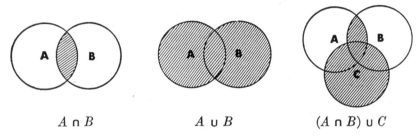

$A \cap B$ $\qquad$ $A \cup B$ $\qquad$ $(A \cap B) \cup C$

The reader might find it instructive to prove the following. Let K consist of all the points in the plane which are in or on a given square Q. Let X consist of all circular disks (circumference together with interior) which contain Q. Then $\underset{A \epsilon X}{\cap} A = Q$. Also, if K is a circular disk (circumference and interior) and Y consists of all triangles inscribed in the circumference of K, then $\underset{A \epsilon Y}{\cup} A = K$.

The idea of union clearly has at least a vague resemblance to addition. There is an operation with an even vaguer resemblance to subtraction. If A and B are sets, there is a set consisting of just those members of A which do not belong to B. This set is called the *relative complement of B in A*, and indicated by the symbol $A - B$. (Most people read this "A minus B," which is harmless provided we do not confuse its peculiarities with those of numerical subtraction.) Often we are considering only subsets of some one given set. Thus in plane geometry we study only sets which are subsets of the plane. In this case the one given set, containing all the other sets in the discussion, is called the *universal set*. Of course it is the universal set only for that particular study; the plane which is a universal set in studying plane geometry can no longer serve as universal set when we go to solid geometry. Suppose then that in some particular study there is a universal set U. Then for every set A (by agreement, contained in U) the relative complement of A in U is $U - A$. This is often called simply the *complement* of A, and indicated by $\sim A$, for the sake of brevity.

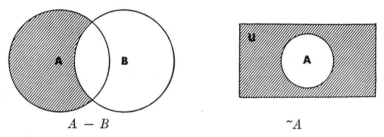

$$A - B \qquad\qquad\qquad \sim A$$

Next suppose that A is a set, and that $s(x)$ is an incomplete sentence with one unspecified noun x in it, and that whenever we replace x by any member of A we obtain a complete sentence, true or false. Then the sentence $s(x)$ is called a *condition* in A. For instance, if A consists of the real numbers, and $s(x)$ is "$x > 5$," then $s(x)$ as it stands is neither true nor false; it is merely incomplete. But as soon as we replace x by a real number we obtain a sentence. If we replace x by 9, we obtain the true sentence "$9 > 5$"; if we replace x by 2, we obtain the false sentence "$2 > 5$."

If A is a set and $s(x)$ is a condition in A, then $s(x)$ can be used to make a distinction among the elements of A, for it distinguishes the elements for which $s(x)$ is true (these are said to *satisfy* the condition $s(x)$) from the others, for which $s(x)$ is a false sentence (these are said *not to satisfy* $s(x)$). If we keep the elements of A which satisfy $s(x)$ and reject those which do not, we are left with a subset of A. This subset is denoted by the symbol $\{x \in A \mid s(x)\}$, for which we may use the rather lengthy reading "the set of all x's in A which satisfy $s(x)$."

As an example, let A be a plane, and let r be a positive number and C a point in the plane A. For $s(x)$ we choose the sentence "the distance from x to C is equal to r." Then the set $\{x \in A \mid s(x)\}$ or, in more detail, $\{x \in A \mid (\text{distance from } x \text{ to } C) = r\}$, is the circle with center C and radius r. Again, let A be the set of all complex numbers, and let $s(x)$ be the sentence "$x^2 + x - 12 = 0$." Then the set $\{x \in A \mid s(x)\}$, or in full $\{x \in A \mid x^2 + x - 12 = 0\}$, is the set of all solutions of the equation, namely -4 and $+3$. The process of solving an equation $f(x) = 0$ consists of determining the members of the set $\{x \in A \mid f(x) = 0\}$. If $f(x) = 0$ is an identity on A, the set $\{x \in A \mid f(x) = 0\}$ is all of A, and conversely.

One special case deserves attention. If the sentence $s(x)$ happens to be false whenever x is replaced by any member of A—for example, if $s(x)$ is the sentence "x is not a member of A"—then the set $\{x \in A \mid s(x)\}$ has no members. This unusual but important set is called the *empty set*, and often denoted by $\emptyset$. There are some mathematicians who object to calling this a set at all. It is logically possible to avoid calling $\emptyset$ a set, just as in algebra it would be logically possible to avoid ever calling 0 a number. But most mathematicians are quite content to call $\emptyset$ a set, and they thereby gain noticeably in convenience of wording. For instance, we can define the statement A and B are *disjoint sets* to mean that for every object x, if $x \in A$ then $x \notin B$. But the same meaning is more compactly expressed by the definition that A and B are *disjoint* if $A \cap B = \emptyset$. One peculiarity of $\emptyset$ is that it is a subset of every set A. For if $\emptyset$ were not a subset of A, there would be an object x which is a member of $\emptyset$ but not of A, and this is impossible; there is no object x which is a member of $\emptyset$. If A is a *non-empty* set, meaning $A \neq \emptyset$, then there exists some object x such that $x \in A$.

As soon as a set is given, we can think of the collection formed of its subsets. For example, if A has just three members, x, y, and z, the subsets of A are $\emptyset$, $\{x\}$, $\{y\}$, $\{z\}$, $\{x, y\}$, $\{x, z\}$, $\{y, z\}$, and $\{x, y, z\}$, eight in all. If A had four members it would have 16 subsets; if A had n members it would have 2^n subsets. This is the motivation for the name frequently given to the set whose members are all the subsets of A, namely 2^A.

As an example, let A be the Euclidean plane. The definition of circle in geometry is a set definition, with the help of the idea of distance; the sentence "K is a circle" has the meaning "there exists $C \in A$ and there exists a positive number r such that $K = \{x \in A \mid \text{distance } x \text{ to } C = r\}$." The set of all circles in the plane is the same as the set of all those objects x which are subsets of the plane and which satisfy the condition "x is a circle." Hence the set of all circles in the plane A is the set $\{x \in 2^A \mid x \text{ is a circle}\}$.

The last operation with sets that we wish to describe is the formation of the *cartesian product*. If A and B are sets, we can think of another set whose members are all the ordered pairs whose first members belong to A and whose second members belong to B. For example, if a team A of three chess players a, b and c is to play a team B of four players x, y, z and w, each member of each team to play one game with each member of the other, the possible games are (a, x), (a, y), (a, z), (a, w), (b, x), (b, y), (b, z), (b, w), (c, x), (c, y), (c, z), (c, w). (It will be noticed that the number of games is 3 times 4.) This set, consisting of all the ordered pairs with first member belonging to A and second member belonging to B, is called the *cartesian product* of A and B, and is denoted by $A \times B$. As an old and important special case, suppose that A and B are both the same as the set R of real numbers. Then $R \times R$ is the set of all ordered pairs of real numbers. It was René Descartes who noticed that these pairs can be used to represent all the points of the plane, and thereby initiated the study of analytic geometry, and it is in his honor that $A \times B$ is called the *cartesian product*.

The idea of cartesian product can obviously be extended to more than two sets. For instance, the cartesian product $A \times B \times C$ can be defined as the set of all ordered triples (x, y, z) in which $x \in A$, $y \in B$ and $z \in C$. However, since $A \times B$ is itself a set we could define $A \times B \times C$ to mean $[A \times B] \times C$; that is, the result of first forming the cartesian product $A \times B$, and then forming its cartesian product with C.

Now we have listed the operations with sets that mathematicians need in using sets. For example, from the set R of real numbers R we first form $R \times R \times R$, which consists of all triples of real numbers and therefore represents three-dimensional space. The idea of distance can be introduced by defining the distance from (x, y, z) to (x', y', z') to be $\sqrt{(x - x')^2 + (y - y')^2 + (z - z')^2}$. The set of all subsets of the space $R \times R \times R$ is $2^{R \times R \times R}$. From this we select the configurations (surfaces, curves, etc.) of particular interest by means of appropriately chosen conditions. Thus our operations allow us to describe all the kinds of figures that are of interest in solid geometry.

In the preceding descriptions of operations we used many ideas connected with sets, but we also smuggled in two other ideas that seem not to be expressible in the language of sets. One of these is the idea of *sentence*, used in discussing conditions and the formation of sets of the type $\{x \in A \mid s(x)\}$. The other is the idea of *ordered pair*, used in forming the cartesian products of sets. The former is in fact a question of logic rather than of set theory. We need to be able to describe just what basic elements can go into a properly constructed sentence, and just how these elements can be put together. However, we shall not go any further into this question than to say that it really is a question, and that it can be logically answered. The other idea, *ordered pair*, can be handled within the framework of set theory without bringing in outside ideas

such as *to the left of* or *preceding*. One such way is as follows. To describe
the ordered pair (a, b) we need to say which two elements are members
of the pair, and which of them is to be regarded as the *first* element. This
can be done by naming the set composed of the two elements, and then
naming the set composed of the first element alone. Both of these are
named in the set $\{\{a, b\}, \{a\}\}$. This has all the properties of *ordered pair*
that are ever needed. So we shall content ourselves with saying that the
idea of *ordered pair* can be defined entirely within the range of ideas
that belong to set theory, and it will have all the properties that we expect it to have, and henceforth we shall not bother about the device by
which it is defined in set-language.

FUNDAMENTAL ASSUMPTIONS

Now we know the relations and operations that are to form our tool
chest, and we want to adopt some set of axioms that will let us perform
these operations freely enough to let us handle the sets occurring in
ordinary mathematical situations. There are two courses open to us.
The one that would naturally be chosen by the specialist is to attempt to
pare down the axioms to the barest minimum. This is cautious, because the less we assume the less likely we are to have assumed too
much; and it is esthetically satisfying, because when it is carried through
we finish up with a vast structure of mathematics all deduced logically
from an astonishingly scanty supply of assumptions. However, to do
this one must pay a price in length and difficulty of proofs, and it is a
price that in this book we cannot wisely pay. So we shall take instead
the other course, and make stronger assumptions, thus making it easier
to prove conclusions. In fact, what we are about to list as axioms amount
to the statement that if we start with sets, and do any of the operations
previously discussed, the result is again a set. Obviously this will not
hamper our operations; we cannot prove that it will never lead to contradiction, but at least what we are saying is taken from a system in
which no one has yet shown any internal inconsistency.

One axiom we have already mentioned:

(1) *If A and B are sets, and $A \subset B$ and $B \subset A$, then $A = B$.*
The next five allow all the operations we wish:

(2) *If A is a set, and $s(x)$ is a condition in A, there is a set $\{x \in A \mid s(x)\}$
whose members are all the objects in A which satisfy the condition $s(x)$.*

(3) *If A is a set, there is a set 2^A whose members are all the subsets of A.*

(4) *If A and B are sets, there is a set $A \times B$ whose members are all the
ordered pairs (a, b) with $a \in A$ and $b \in B$.*

(5) *If X is a set, and each member A of X is itself a set, then there is*

a set $\underset{A \epsilon X}{\bigcup}$ *A whose members are all those objects which belong to one or more of the sets which belong to X.*

(6) *If X is a non-empty set, and each member A of X is itself a set, then there is a set* $\underset{A \epsilon X}{\bigcap}$ *A whose members are all those objects x which belong to every set which is a member of X.*

(If the reader is at all interested in keeping down assumptions to a minimum, he might be interested to find that this sixth axiom need not be assumed at all; it can be proved to be a consequence of the preceding ones.)

None of these six axioms guarantee that there are any sets at all; they merely permit us to make up new sets if we have some sets to start with. We need a set to start with. In order to have a specific and useful set, we assume

(7) *There is a set R whose members are the real numbers.*

Earlier in this chapter we remarked that if we were over-optimistic about how many objects we could mentally sweep together into a set, we would create confusion. We now have the symbolism needed to make this precise. The too-optimistic idea is that if we have some description satisfied by some objects, there is a set whose members are just the objects satisfying the description. This description would then be a condition, but not just a condition in a set A; it would be a condition with unlimited applicability. Thus if $s(x)$ is the sentence "x is a set," the corresponding set S would have as members all the objects which are sets and no others. Now use the condition in S expressed by "$x \notin x$"; that is, "x is not a member of itself." Applying the process in axiom (2) we obtain a set $N = \{x \epsilon S \mid x \notin x\}$. The members of N consist of all the sets x for which "$x \notin x$" is true, and none for which it is false. Now we test to see if the sentence "$N \epsilon N$" is true or false. If it is true, then by replacing the second "N" by its definition we find $N \epsilon \{x \epsilon S \mid x \notin x\}$. Hence N must be one of the sets which satisfy the condition $x \notin x$; that is, $N \notin N$, so that the sentence "$N \epsilon N$" is false, even though we assumed it true. On the other hand, suppose that "$N \epsilon N$" is false. Then N is a member of S, being a set; and by assumption it is false that $N \epsilon N$, so N satisfies the condition $x \notin x$. Therefore N must be a member of $\{x \epsilon S \mid x \notin x\}$, which is N; that is, $N \epsilon N$, contradicting the assumption that "$N \epsilon N$" is false. So if we assume that the sentence "$N \epsilon N$" is true we can prove it false, and if we assume it false we can prove it true, which is an intolerable situation. (This situation is known as the Russell paradox.) By listing the axioms above we have saved ourselves from the paradox, for starting with given sets we can never build up S as a set, and if S is not a set at all the paradox disappears.

FUNDAMENTAL FORMULAS

If we are to manipulate collections of sets with ease, we need some fundamental formulas, like the fundamental formulas in algebra that we use almost mechanically in calculations. Let us suppose that U is a set, and that all the other sets mentioned are subsets of U; that is, U is our *universal set*. As before, $\tilde{A}$ shall mean the set of all points in U which are not members of A. Then $\tilde{\;}(\tilde{A}) = A$; for if x belongs to $\tilde{\;}(\tilde{A})$ it is a member of U which is not a member of $(\tilde{A})$, so it is a member of A, while if x is a member of A it is a member of U which is not a member of $(\tilde{A})$, so it is a member of $\tilde{\;}(\tilde{A})$. By Axiom 1, $\tilde{\;}(\tilde{A}) = A$. Next, by a Venn diagram we indicate the proof of the formula $A \cap B = \tilde{\;}[\tilde{A} \cup \tilde{B}]$. With A and B indicated as the interiors of the circles, $\tilde{A}$ is the part shaded by strokes rising to the right, $\tilde{B}$ the part shaded by strokes falling to the right, $\tilde{A} \cup \tilde{B}$ the whole shaded part, and therefore $\tilde{\;}[\tilde{A} \cup \tilde{B}]$ is the unshaded part. This is the same as $A \cap B$.

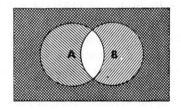

$\tilde{A}$ ▨ $\tilde{B}$ ▧

The formulas $A - B = A \cap [\tilde{B}]$, $A \cup [B \cup C] = [A \cup B] \cup C$, $A \cup B = B \cup A$, $A \cap \tilde{A} = \emptyset$, $A \cap U = A$ and $A \cup \emptyset = A$ are easy, and can be left as exercises. Not quite as easy is the formula $A \cup (B \cap C) = (A \cup B) \cap (A \cup C)$. Its proof is indicated by the Venn diagrams below. The shaded area in the first diagram is $A \cup (B \cap C)$, and is the same region as the cross-hatched part of the second diagram, which is $(A \cup B) \cap (A \cup C)$.

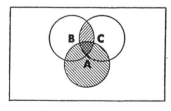   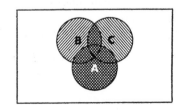

$B \cap C$ ▨, A ▧, $A \cup B$ ▨, $A \cup C$ ▧,

$A \cup (B \cap C)$ shaded $(A \cup B) \cap (A \cup C)$ ▦

For ease of reference we collect these nine formulas and fifteen others into one single list, and state the result as a theorem.

THEOREM. *Let U be a set, and let A, B and C be subsets of U. Then*

(1)* $A \cup B = B \cup A$ (1') $A \cap B = B \cap A$

(2)* $\sim(\sim A) = A$

(3)* $A \cup (B \cap C) = (A \cup B) \cup C$ (3') $A \cap (B \cap C) = (A \cap B) \cap C$

(4)* $\sim A \cap B = \sim[\sim A \cup \sim B]$ (4') $A \cup B = \sim[\sim A \cap \sim B\,]$

(5) $\sim[A \cap B] = \sim A \cup \sim B$ (5') $\sim[A \cup B] = \sim A \cap \sim B$

(6)* $A \cup (B \cap C) = (A \cup B) \cap$ (6') $A \cap (B \cup C) = (A \cap B) \cup$
$\quad\ (A \cup C)$ $\quad\ (A \cap C)$

(7)* $A \cap \sim A = \emptyset$ (7') $A \cup \sim A = U$

(8)* $A \cap U = A$ (8')* $A \cup \emptyset = A$

(9) $A \cup U = U$ (9') $A \cap \emptyset = \emptyset$

(10) $A \cup A = A$ (10') $A \cap A = A$

(11)* $A - B = A \cap (\sim B)$

(12) $A - \emptyset = A$ (13) $A - A = \emptyset$

(14) $(A \cap B) \cup (A - B) = A$ (15) $A \cap B \cap (A - B) = \emptyset$

Nine of these are marked with asterisks. These are the formulas already proved, but this is by no means the real reason for singling them out. These are in a sense fundamental formulas. They were deduced by using the meanings of the symbols $\sim$, $\cup$, etc. All the others follow from these nine by direct formal deduction; it is not necessary ever to refer back to the meanings of the symbols used, or even to think of the letters as meaning sets.

This has an important and interesting consequence. Let us say that the *dual* of a formula is the new formula resulting when we everywhere replace $\cup$ by $\cap$ and $\cap$ by $\cup$, U by $\emptyset$ and $\emptyset$ by U. In the theorem, the formulas (1'), (3'), $\cdots$, (10') are the duals of (1), (3), $\cdots$, (10). Now if any formula can be proved by use of the starred formulas, its dual can be beduced by the same steps from the duals of the starred formulas, which are themselves proved statements. Hence the whole theory of subsets of U is its own dual, in the sense that whenever we have proved a theorem, we can dualize each step of the proof and thus prove the dual of the theorem,

As an example let us deduce (4') from the starred statements.

$$\sim[\sim A \cap \sim B] = \sim[\sim\{\sim(\sim A) \cup \sim(\sim B)\}] \qquad \text{by (4)*}$$

$$= \sim[\sim\{A \cup B\}] \qquad \text{by (2)*}$$

$$= A \cup B \qquad \text{by (2)*}$$

establishing (4'). From (4) and (4'), with (2)*, we can easily prove (5) and (5′). With a little more effort we deduce (6′) from these and the starred statements, as follows.

$$A \cap (B \cup C) = {}^{\sim}[{}^{\sim}\{A \cap (B \cup C)\}] \qquad \text{by (2)*}$$

$$= {}^{\sim}[{}^{\sim}A \cup {}^{\sim}(B \cup C)] \qquad \text{by (5)}$$

$$= {}^{\sim}[{}^{\sim}A \cup ({}^{\sim}B \cap {}^{\sim}C)] \qquad \text{by (5′)}$$

$$= {}^{\sim}[({}^{\sim}A \cup {}^{\sim}B) \cap ({}^{\sim}A \cup {}^{\sim}C)] \qquad \text{by (6)*}$$

$$= {}^{\sim}({}^{\sim}A \cup {}^{\sim}B) \cup {}^{\sim}({}^{\sim}A \cup {}^{\sim}C) \qquad \text{by (5)}$$

$$= (A \cap B) \cup (A \cap C) \qquad \text{by (4)}$$

If we compare these rules with the basic operating rules for ordinary algebra (complex numbers), we obtain some very neat analogies by comparing the sets with complex numbers and the operations $\cap$ and $\cup$ with multiplication and addition respectively. Then (1) and (1′) have the same appearance as the commutative laws for addition and multiplication of numbers, and (3) and (3′) look just like the associative laws for ordinary addition and multiplication. Also, (6′) has the same appearance as the distributive law $a \cdot (b + c) = (a \cdot b) + (a \cdot c)$ for numbers. However, (6) tells us that we have entered new territory. For this would correspond with the formula $a + (b \cdot c) = (a + b) \cdot (a + c)$ for complex numbers, and this is a false formula. Formulas (6) and (6′) are called the *distributive* laws, but in comparison with real or complex numbers we are embarrassed by an oversupply.

In the list of formulas in the theorem there is no mention of the relations $\subset$ and $\supset$. This is appropriate, because the theorem makes statements about operations with sets, and $\subset$ and $\supset$ are relations between sets. However, the formulas in the theorem contain all the information needed to work with these relations, once we notice their fundamental connection with the operations, namely

$$A \subset B \quad \textit{if and only if} \quad A \cup B = B.$$

This can easily be transformed into $A \cap B = A$, which is equally serviceable. For instance, if $A \subset B$ and $B \subset C$, then $A \cup B = B$ and $B \cup C = C$, by definition. Hence $A \cup C = A \cup (B \cup C)$ by substitution, and this is equal to $(A \cup B) \cup C$ by (3) of the theorem; substituting B for its equal $A \cup B$ yields $A \cup C = B \cup C = C$. Then by definition $A \subset C$, and we have proved that if $A \subset B$ and $B \subset C$ then $A \subset C$ without ever going back of the theorem to the meanings of the symbols.

ALGEBRA OF SETS

Sometimes we have a set U and wish to study a selected collection of its subsets, not necessarily all its subsets. If a non-empty collection Z of subsets of U has the two properties: (A_1) *whenever A is a member of the collection Z, so is $\sim A$, and (A_2) whenever A and B are both members of the collection Z, so is $A \cup B$*, the collection Z is called an *algebra of sets*. Then we can at once prove that if A and B belong to Z, so does $A \cap B$; for then by (A_1) both $\sim A$ and $\sim B$ are members of Z, and by (A_2) $\sim A \cup \sim B$ is a member of Z, and by (A_1) again $\sim[\sim A \cup \sim B]$ is a member of Z, whereas by $(4)^*$ of the theorem this last set is $A \cap B$. Similarly, if A and B are members of Z, so is $A - B$. In particular, if A is any member of Z, so is $\sim A$, and then so are $A \cup \sim A$ and $A \cap \sim A$, which are U and $\emptyset$ respectively.

One interesting application of algebras of sets will have to be passed by with a mere mention. This is the application to logic. (Hence this paragraph will be more intelligible if it is read after Chapter IV.) To each proposition we can assign a set, namely the set of cases in which it is true. Then to composite propositions we assign sets constructed from those assigned to the parts of the composite statement; for instance, if a is a proposition and A the set of cases in which proposition a is true, and likewise b is a proposition and B the corresponding set, then to the proposition "a or b" we assign the set $A \cup B$, and to the proposition "a and b" we assign the set $A \cap B$. This program can be carried through consistently for all the standard manners of combining propositions, and thus the study of composite statements is transformed into the study of an algebra of sets. Then all the 24 formulas of the theorem, as well as all of the vastly many others that can be formed out of them by combining them, are transformed into formulas about composite statements. The theory of sets thus provides us with a technique for handling composite statements which is of an algebraic nature, and can be called an algebra of logic. This is a close relative of the ideas published a century ago by George Boole, whose object was to provide an algebra by which the study of logic could be formalized and handled by methods like those of ordinary elementary algebra. If the reader would like more details on this, he can find them in *Universal Mathematics*, Part II, Chapter 1, Sec. 4 or in *What Is Mathematics?* (see Bibliography).

There is also a connection between the theorem and the algebraic concepts in Chapter V. If we start with a set Z of undefined objects, together with an undefined operation $\sim$ that assigns to each A of Z another member $\sim A$ of Z and an undefined operation $\cup$ which assigns to each pair A, B of members of Z another member $A \cup B$ of Z, then we can define a third operation $\cap$ by means of $(4)^*$ of the theorem and a fourth operation — by (11)*. We now assume (1)*, (2)*, (3)*, (6)* as axioms, and we assume that there are two special members U, $\emptyset$ of Z with which (7)*, (8)* and (8′)* are true for all A in Z. Then the set Z with the operations named is a *Boolean algebra*. All the other assertions in the theorem can now be proved, because the starred statements are true by hypothesis and the others follow from the starred statements without reference to the meanings of the symbol involved. Boolean algebras have been interesting objects of study for mathematicians for some years; in particular, there are various sets of postulates much shorter than our set of starred statements which are still adequate to let us deduce the others. More recently, Boolean algebras have attained prominence in research in electricity. They enter in the study of switching circuits, and also in the study of information carried by messages. This is a complete subject of study in its own right, and we cannot in a short space present enough of it to be of any use, so we go no further in this direction.

INDUCTIVE SETS

One important and not obvious consequence of the axioms about sets is the principle of mathematical induction. In our last axiom we assumed that the real numbers form a set. However, we do not need to assume a complete knowledge of them. It is enough to know that they are ordered and that there is a special number 1 and an operation of adding 1, so that for each real number x there is a real number $x + 1$, and $x + 1 > x$. If we try to describe those real numbers which are *natural numbers*, or positive integers, we know that we want to include 1, $1 + 1$ (named 2), $2 + 1$ (named 3), and so on. But what is "and so on?" It is not possible to form a complete table of the natural numbers in this way, and nothing less would be sufficient. So we turn instead to a different device. We try to describe the *set* of natural numbers as a whole. Immediately we think of two essential properties of this set:

(1) *The number 1 is a member of it.*
(2) *If x is a member of it, so is $x + 1$.*

Whenever a set has these two properties we shall call it an *inductive set*. Now the set that we have in mind as the natural numbers is one example of an inductive set, but it is not the only one. For example, the set of all positive real numbers has both properties, so it is an inductive set. Likewise the set consisting of the number 1 and all numbers $x \geq 2$ is an inductive set. So the set of natural numbers is not fully described when we say that it is an inductive set. However, it will be noticed that the other examples of inductive sets that we gave contained all the natural

numbers and other things too. It is very plausible that this is true of
every inductive set of real numbers. But then this would show the set
of natural numbers as the smallest of all inductive sets. Consequently,
it would be natural to adopt the definition that the set of natural numbers is the smallest inductive set. In fact, we are going to do this, but
first there is a difficulty to clear up. The definition seems to define
something, but it can happen that a definition that seems to define
something actually does not define anything. One example was the "set
of all sets," which sounds as though it means something but does not.
Another easier example is "the smallest positive number." This sounds
sensible, but there is no such thing. If there were a smallest positive
number x, then $x/2$ would still be positive (being the product of the
positive numbers x and $1/2$) and it would be smaller than x, contradicting the definition of x. Hence "the smallest positive number" is a meaningless phrase. Until we show that the phrase "the smallest inductive
set" is meaningful, we cannot use it as a definition of anything. However, we can show that there is such a set. For, first, the set 2^R of all
subsets of R is a set, by Axiom 3. Let $s(A)$ be the sentence "A is an inductive set." Then $I = \{A \in 2^R \mid s(A)\}$ is a set by Axiom 2. It is the set
of all inductive sets. By Axiom 6, there is a set $J = \bigcap_{A \in I} A$; this is the
intersection of all inductive sets. We shall next prove that J is itself
an inductive set; that is, J has the two properties (1) and (2) just described. First, 1 is a member of every $A \in I$, because each such A is
inductive and therefore must have 1 as a member. Since 1 is in each set A
of the set-of-sets I, it is in their intersection J. That is, 1 is a member of
J, and so J has property (1). Next, let x be any member of J; we must
show that $x + 1$ is a member of J. Since $x \in J$, and J is the intersection
of all sets A of the set-of-sets I, x must be in each of these sets A. Consider any such A. Since x is in A, and A is inductive, $x + 1$ must also
be in A. Hence $x + 1$ is in A for each set A of the set-of-sets I, and is
in their intersection J. That is, J has property (2). This finishes the
proof that J is an inductive set. To show that it is the smallest of all
inductive sets, let A' be any inductive set. Then A' is a member of I,
and so it contains the intersection of all members of I, which is J. Now
J is an inductive set and is contained in every inductive set, so it is the
smallest of all inductive sets. This is then what we call the *set of natural
numbers*. Summarizing, we have shown that the essential peculiarities
of the set of natural numbers are (1) 1 is in the set, (2) if x is in the set so
is $x + 1$, and (3) it is the smallest set for which (1) and (2) are true. We
have shown that a set with these properties can in fact be constructed
by the principles contained in the axioms for sets, and we have given

the name J (or, better, *the set of natural numbers*) to the set thus constructed.

MATHEMATICAL INDUCTION

Now the principle of mathematical induction is easy to establish, not as a somewhat vague feeling that it ought to be so but as a provable theorem.

PRINCIPLE OF MATHEMATICAL INDUCTION. *Let $S(n)$ be a statement involving a variable n which becomes a sentence (true or false) whenever a natural number is put in place of n. If $S(1)$ is true, and whenever $S(n)$ is true $S(n + 1)$ is also true, then $S(n)$ is true for every natural number n.*

Let T stand for the set of natural numbers n for which $S(n)$ is true. By hypothesis, 1 is a member of T, for $S(1)$ is true. Also, if x is in T then $S(x)$ is true, so by hypothesis $S(x + 1)$ is true, and so $x + 1$ is a member of T. That is, T is an inductive set. But J is the smallest inductive set, so $J \subset T$. For each natural number, $n \in J$ by definition of natural number, so $n \in T$, which means that $S(n)$ is true. This completes the proof.

At this point we make a digression into a subject which is somewhat abstract and perhaps not easy to follow, but is of great interest to those who are interested in the foundations of mathematics. This paragraph can be omitted without interfering with later parts of the chapter. The assumption that the real numbers form a set is convenient, but quite strong. It can be replaced by other weaker assumptions which do not presuppose any idea of number at all. One of these has as a consequence that no set is a member of itself; $x \in x$ is always false. This means that whenever x is a set, we can construct another set different from x. For if x is a set so is $\{x\}$ (by one of those other hypotheses), and $x \cup \{x\}$ is also a set. The members of this last consist of all objects which are members of at least one of the sets x and $\{x\}$, and in particular x is included among them because x is a member of $\{x\}$. But x is not a member of x, so x and $x \cup \{x\}$ are definitely different sets. The set $x \cup \{x\}$ is called the *successor of x*, and denoted by $s(x)$. Now let us choose from all possible objects one Favored Object, F. (Mathematicians usually choose $\emptyset$ as this Favored Object, because it is so easily described and identified, but this is not essential.) We can say that a set A is an *inductive set* if (1) it contains $\{F\}$ and (2) whenever it contains x, it also contains $s(x)$. However, we now have to add in an axiom (which may be thought of as the very feeble remains of our previous Axiom 7) to the effect that there is at least one inductive set. This done, we can follow the proof in the preceding paragraphs and show that there is a smallest inductive set J, which is contained in all inductive sets. To this the principle of mathematical induction applies, by the proof already given. But now our set J has all those essential properties of natural numbers from which the possibility of defining sums and products, etc., can be deduced. As indicated at the end of the preceding chapter, by a somewhat long and laborious set of steps it is now possible to construct the real number system. But now we have the basic raw material of real and complex algebra, of calculus and all the mathematics built upon it. In this sense the axioms of sets are the axioms of all of analysis. Once they are assumed, no further axioms need be postulated. In these basically simple ideas we have everything needed for the huge structure of classical mathematics.

CONCEPT OF RELATION

The ideas of set theory can be used to clarify the subject of relations. Any particular relation in ordinary experience tends to be colored by connotations, sometimes of an emotional nature. For the needs of mathematics, we wish to define the concept of relation so as to have the essential properties of the somewhat vague concept in ordinary discourse, but to have no connotations at all. Consider the idea "is a descendant of" as applied to the whole human race, but with all ideas of family pride removed. What we want is a scheme by which, whenever X and Y are specified human beings, we can find an answer "yes" or "no" to the question "Is X a descendant of Y?" This could at least in concept be managed by first making a vast supply of cards, listing in every possible way the name of a person on the left half of a card and the name of a person on the right half. Now put in a cabinet all those cards for which the name on the left half of the card is that of a descendant of the person named on the right half; the other cards are to be thrown away. The cards in the cabinet now allow the construction of every possible family tree; given any two names X and Y, if the card with X on the left and Y on the right is in the cabinet, then X is a descendant of Y, otherwise not. The collection of cards allows us to answer the questions about descendancy, but without any connotation or shade of subjective feeling. Thus the essence of the idea of this relation has been reduced to a collection of cards, with an ordered pair of names on each card.

With this as guide, we set down our definition of relation: *A relation is a set of ordered pairs.*

Suppose now that R is a relation. If (x, y) is an ordered pair, (x, y) may belong to the relation R and it may not. If it does, we already have a notation for it, namely $(x, y) \in R$. But it is customary and convenient to prefer another notation for the same thing, namely $x\,R\,y$. For example, consider the relation consisting of all pairs of natural numbers

(m, n) such that for some natural number a we have $m + a = n$. This relation is named $<$. In particular $(2, 5)$ is a member of the relation $<$, because $2 + 3 = 5$. But rather than write this in the perfectly permissible form $(2, 5) \epsilon <$, we usually prefer the notation of the style $x \, R \, y$, which in this case takes the form $2 < 5$.

This idea of relation may seem somewhat abstract, and in fact the definition did arise in pure mathematics. Nevertheless it is clearly consistent with what is done in the highly practical activity of machine computation. For example, the sine function, usually written by an equation such as $v = \sin A$, is in IBM computing quite literally a set of ordered pairs. There is a set of cards with the punched equivalent of the value of A in one spot on the card and the punched equivalent of $\sin A$ in another spot. This set of ordered pairs (visible as cardboard) is the relation.

Given any relation R, we can immediately produce another relation by reversing each ordered pair in R. This produces a new relation R^{-1}, the *inverse* of R. Thus a pair (a, b) is in R^{-1} if and only if (b, a) is in R; in other words $a \, R^{-1} \, b$ if and only if $b \, R \, a$. If R is "is a descendant of," then R^{-1} is "is an ancestor of"; if R is $>$, R^{-1} is $<$.

Relations can have an assortment of properties and can be studied accordingly. For example, a relation R is *reflexive* on a set A if whenever a is a member of A, it is true that $a \, R \, a$. Thus $\subset$ and $\supset$ are reflexive relations on every set of sets, and $\geq$ is a reflexive relation on real numbers, but $>$ is not reflexive. A relation R is *transitive* if whenever $a \, R \, b$ and $b \, R \, c$ are both true, so is $a \, R \, c$. Thus whenever $A \supset B$ and $B \supset C$ it is true that $A \supset C$, so $\supset$ is transitive. Likewise $\subset$ is transitive, and among real numbers $\geq$ and $>$ are both transitive. A relation R is *symmetric* if whenever $a \, R \, b$, it is also true that $b \, R \, a$. For example, among geometric figures the relation "is congruent to" is symmetric; but $\supset$ and $\subset$ are not. A transitive relation with the property that for two different objects a and b the relations $a \, R \, b$ and $b \, R \, a$ are never both true is called a *partial ordering*. Thus sets are partially ordered by the relation $\subset$, for if A and B are different sets the two relations $A \subset B$ and $B \subset A$ cannot both be true, by Axiom 1. Partially ordered sets have received considerable attention from mathematicians in the last few years.

CONCEPT OF FUNCTION

The example of a relation with which we started, "is a descendant of," has the obvious property that there are many pairs having the same first element but having different second elements. For if X is any person, then $X\ R\ Y$ is true when Y is the mother of X, the father of X, any of the grandparents of X, and so on. However, we are also used to relations in which no two pairs have the same first element and different second elements. For example, a price list has this property. The first element is the name of an object, the second element is its price. In the list there are never two lines having the same name of object but different prices. There may be different objects with the same price, but that is another matter; we are thinking of relations in which, when the *first* element is given, there is no ambiguity about the *second*, because only one pair in the relation has the given first element. A relation having this property is called a *function*. Thus if a relation R is known to be a function, and it is known that (a, x) and (a, y) are both in R or, in other words, that $a\ R\ x$ and $a\ R\ y$, then it has to be true that x and y are the same thing.

When a relation happens to be a function, it is customary to use a slightly different notation. Suppose that F is a relation which is a function. If (x, y) is a pair which belongs to F, we already have the two possible notations $(x, y)\ \epsilon\ F$ and $x\ F\ y$. However, we now introduce still another notation, $F(x) = y$. For example, let Sq be the relation consisting of all pairs (x, y) in which x is a real number and y is $x \cdot x$. The pair $(3, 9)$ is in the relation Sq, so we can write $3\ Sq\ 9$. This relation is in fact a function, because if $a\ Sq\ x$ and $a\ Sq\ y$ then x is $a \cdot a$ and y is $a \cdot a$, so y is the same as x. So we can use the other symbolism and write $Sq(3) = 9$, which would usually be read "the square of 3 is 9." (Of course for this well-known function there is still another notation $3^2 = 9$, and this is more convenient in computation, but Sq is essentially the same as a very ancient notation for the function.)

Thus the central idea of function is that with a given first element there are never two different second elements. (It should be mentioned that this agreement, although widespread, is not universal. Some mathematicians use the name "single-valued function" for what we have called a function, and the name "function" as a synonym for what we have called a relation.) Thus, for example, in a strict sense the statement "A student's scholastic standing is a function of his intelligence" is incorrect. It would mean that given a student's intelligence his scholastic standing has only one single possible value, which is not the case since other complicating circumstances can be present. A correct statement would be that the student's scholastic standing has a high correlation with his intelligence; but this is a topic that properly belongs in the chapter on probability.

A function may be specified in any one of several different ways.

One common way is by means of a table, in which the entries are often the results of some experimental procedure. Thus the population of the United States is a function of time; it is presented as a table, the entries being the result of census-taking. Another way is by means of a graph. This may embody the results of experiment, or may be more direct; for instance, the curve drawn by a thermograph records the temperature at a fixed place as a function of the time. It is a function, because to each time there is only one single value of the temperature at that place. Notice that the function of the time does not have to "vary with the time." It is quite possible that the thermograph is sitting in a well-thermostated room, with no perceptible change of temperature. Then the temperature is a function of the time, but does not vary with the time. It is an example of a peculiarly simple but important kind of function, called a *constant*. More frequently in mathematics the function is defined with the help of a formula of some kind. This does not mean that the formula is the function; it is merely an aid in constructing the pairs of numbers which constitute the function. For example, let us form a relation by making an ordered pair for each circle; the first element of the ordered pair is the radius of the circle and the second element is the area of the circle. The resulting relation, or collection of pairs, is in fact a function, for when the first number in the pair (the radius) is given there is only one possible second number (area). This function (which we shall call F) expresses the area of a circle as a function of its radius. As we all know, if r is the radius the area is πr^2. But this does not mean that the formula $A = \pi r^2$, as an algebraic relation between A and r, is the function. The pair $r = -2$, $A = 4\pi$ satisfies the formula, but it does not belong to the function F. There is no pair $(-2, 4\pi)$ in the function F (that is, the statement $F(-2) = 4\pi$ is incorrect), because there is no circle of radius -2. This distinction between the function and the formula which is a help in computing the function is not merely a matter of hair-splitting. It has happened often that a function has been found by some experimental method, and then a formula devised that enabled one to express the statement of the function compactly. But all too often it has also happened that some one else has taken the formula for the function and used the formula to calculate pairs which were meaningless, just as the pair $(-2, 4\pi)$ or $4\pi = F(-2)$ was meaningless as a statement about areas of circles. This using of a formula "beyond its domain of validity" can lead to absurdities, not only in mathematics but in any science.

The *domain* of a function F is the set of all objects x for which $F(x)$ has meaning; in other words, an object x is in the domain of F if and only if there exists some y such that (x, y) is in F (that is, $F(x) = y$).

The *range* of the function F is the set of all objects y which are values of the function; that is, y is in the range of F if and only if there is an x for which the equation $F(x) = y$ is true.

CORRESPONDENCE BETWEEN SETS

This is not the place to make a detailed study of numerical functions; that will be done in Chapter VIII. Here it is enough to explain what we mean by the idea of function in its widest generality, and to make use of the idea insofar as it relates to sets. One such use has to do with "matching" sets, which is connected with the idea of counting. We have already seen that every relation furnishes an inverse relation, obtained by simply reversing each ordered pair in the relation. In particular, if F is a function it is a relation, so by reversing each pair we obtain the inverse relation. But this inverse relation is not necessarily a function. In a price list, the price is a function of the article, because each article determines just one price; but the inverse relation is not necessarily a function, because one price may correspond to several articles. At a given place, the temperature is a function of the time, because at each time there is just one temperature at the place; but the inverse relation is not necessarily a function, because a single temperature may correspond to several different times. The square function, in which to each real number x corresponds $Sq(x) = x \cdot x$, is a function; but the inverse relation (which by general notational principles would be Sq^{-1}, but is familiarly known as "square root") assigns to each positive number two different square roots, so it is not a function. On the other hand, it sometimes does happen that the inverse of a function is again a function. Thus the inverse of the cube function (for real numbers) is the cube root function. When this happens, F^{-1} is a new function, whose domain is the range of F and whose range is the domain of F. Let X be the domain of F and Y its range. To each x in X there is exactly one y in Y for which $F(x) = y$, and to each y in Y there is exactly one x in X such that $x = F^{-1}(y)$, which is the same as $y = F(x)$. This makes it natural to say that the sets X and Y are "in one-to-one correspondence," or "have been matched one-to-one by the function F." Repeating—two sets X and Y are "matched one-to-one by F" if and only if F is a function whose domain is X and whose range is Y, and its inverse F^{-1} is also a function; then necessarily the domain of F^{-1} is Y, and its range is X.

When two sets can be matched one-to-one, it is natural to think of them as being equally numerous. Thus the set consisting of Smith, Jones and Robertson is as numerous as the set consisting of January, February and March. But long ago the human family learned a better trick than this. It learned to label sets with numbers in such a way as to indicate their numerousness. In our wording, the two sets named are as numerous as the set $\{1, 2, 3\}$, and this last set is indicated by merely saying "three." To be more precise, let "Cardinal 3" stand for the set $\{1, 2, 3\}$, and more generally for each natural number n let *Cardinal n* stand for the set $\{x \in J \mid x \leq n\}$ consisting of all the natural numbers which do not exceed n. Then the statement "the set A has exactly n members" is defined to mean "the set A can be matched one-to-one with the set Cardinal n." (We usually omit the word "exactly" but it is always to be understood.) The operation of counting, which is the simplest and earliest of scientific experiments, is thus described purely in terms of the concepts of set theory.

This idea is important enough to need more explanation. Of course it is true that numbers entered language and thought by way of the process of counting, which is fundamentally experimental. But in the interests of clarity, both for pure mathematics and for applied, it is important to keep pure mathematics self-contained, developing it from its own axioms by purely logical processes. Thus in an earlier digression we showed that from the axioms for sets we could build up entities that had all the properties that we wished the natural numbers to have, and from this we could go on higher and higher into mathematics. But the natural numbers, whether introduced from these axioms or assumed as part of the real numbers, are essentially undefined objects with a few assumed properties. These assumed properties do not include any mention of counting. Now the fundamental device of applied mathematics is to make up a sort of dictionary by which certain material objects correspond to certain pure-mathematical structures, and certain physical procedures correspond to certain mathematical operations on those structures. Counting is just such a procedure. If I touch the fingers of my left hand, each just once, and say the names of the natural numbers in succession, "one, two, three, four, five," then as I say "five" I have finished touching the fingers. Thus I have set the fingers in one-to-one correspondence with the set Cardinal 5, which is $\{1, 2, 3, 4, 5\}$. Now in my dictionary of correspondences between material objects and mathematical concepts I put the set of fingers in correspondence with the number 5, and say "there are five fingers on my left hand." To the physical operation of "putting sets together," which means forming the set-union of disjoint sets, corresponds the arithmetic operation of adding the corresponding natural numbers. This correspondence is the right kind of correspondence, because it agrees with experience and permits prediction of experiments. If there are twelve eggs in one box and six in another, we are all so confident that the number of eggs in the whole set is 12 + 6 that we usually accept it without even verifying it. We could in fact prove the theorem:

If A is a set with n members, and B is a set with m members, and A and B are disjoint, then A ∪ B has n + m members. However, we shall not prove this, but shall merely sketch the proof of an even simpler and more basic theorem:

If a set A has n members, and m is different from n, then A does not have m members. (Recall that "A has n members" means "A has exactly n members.")

Suppose that this is false, and that there is some set which has n members and also has m members, with $m \neq n$. Then A can be matched one-to-one with Cardinal n by some function F, and can be matched one-to-one with Cardinal m by some function G. But then Cardinal n is matched one-to-one with Cardinal m. It is this which we must show is impossible. We prove by induction: If n is a natural number, Cardinal n cannot be matched one-to-one with Cardinal m for any natural number $m < n$. First, this is true for $n = 1$, since there are no natural numbers less than 1. Now suppose it true for $n = k$; we shall show it true for $n = k + 1$. If Cardinal $(k + 1)$ could be matched one-to-one with Cardinal m, where $m < k + 1$, then $k + 1$ would correspond to some number a in Cardinal m and some number b in Cardinal n would correspond to m in Cardinal m. We "switch these partners" so that $k + 1$ corresponds to m and b to a. This is still a one-to-one matching of Cardinal $(k + 1)$ with Cardinal m. Now take away the member $k + 1$ from the set Cardinal $(k + 1)$ and the member m from Cardinal m. If we assume (though we should and could prove it) that what is left is Cardinal k in one case and Cardinal $(m - 1)$ in the other, then we have a one-to-one correspondence between Cardinal k and Cardinal $(m - 1)$, where $m - 1 < k$. This has been assumed impossible. So the correspondence between Cardinal $(k + 1)$ and Cardinal m with $m < k + 1$ is also impossible, and if our assertion is true for $n = k$ it is true for $n = k + 1$. By the principle of mathematical induction it is true for every natural number n, and the theorem is proved.

A set is called *finite* if it can be put in one-to-one correspondence with Cardinal n for some natural number n. (This definition agrees with the idea that finite sets are those with some natural number of elements in them). As a result of the theorem just proved, it is easy to go on to show that if A is a finite set, and B is a subset of A which is not all of A, then the number of elements in A is not the same as the number of elements in B. It would seem silly to prove such an obvious statement, if it were not that similar statements that seem almost as obvious are in fact quite wrong. Consider for example the set J of all natural numbers and the set E of all even numbers. The function F which assigns to each x in J the functional value $F(x) = 2x$ has a range which consists of all the even numbers, and its inverse is the function $F^{-1}(y) = y/2$ $(y \; \epsilon \; E)$, so F furnishes us with a one-to-one matching of the set J and the set E, which is contained in J but is not all of J. This simple example has caused great mental trouble to those people who reason thus. "There are just as many numbers in the set E of even numbers as there are in the set J of all natural numbers. Therefore there aren't any odd numbers!" This is a clear case of using the conclusion of a theorem without checking that the hypothesis is satisfied. The set J is not finite, so the theorem that it cannot be matched one-to-one with a proper subset does not apply, since the theorem was proved only for finite sets. This also shows that the extension of the theorem to infinite sets, which some people seem to feel is as obvious and unworthy of serious proof as the finite case, is in fact entirely wrong. In very much the same way we can in fact show that if A is a set which is not finite (so that we call it an *infinite set*), then there is a set B contained in A but not all of A which can be matched one-to-one with all of A. So our theorem about matching with subsets turns out to be a peculiarity of the finite sets, not shared by any infinite set.

But now we can generalize the idea of counting beyond the finite sets, with the help of the idea of matching. The set J of natural numbers is an infinite set, and there are also other infinite sets which can be matched one-to-one with J. None of these can be finite sets. On the other hand, we shall see soon that there are other infinite sets which cannot be matched one-to-one with J. Hence the sets which can be matched with J are in a sense (which can be made perfectly precise) the "smallest" infinite sets. A set which can be matched one-to-one with J is said to be *countably* (or *denumerably*) infinite. If A is countably infinite, and F is a function which maps the set J of natural numbers on A, then there is some member of A which corresponds to 1, and is thus $F(1)$. This we re-name a_1. Likewise there is a member of A which corresponds to 2, and is thus $F(2)$. We rename this a_2. In this way we obtain a sequence of elements of A, namely a_1, a_2, a_3, $\cdots$ and because F maps J onto all of A, this sequence has every member of A in it. So a countably infinite set can be arranged in a sequence. Conversely, a set that can be arranged in a sequence is countable, since the arrangement automatically matches the members of the set with the natural numbers.

This makes it fairly easy to show that if A is countably infinite and B

is a subset of A, then B is either finite or countably infinite. For let A be arranged in a sequence, a_1, a_2, a_3, $\cdots$. From this sequence strike out all members which do not belong to B. The first surviving element of the sequence we rename b_1, the next b_2, and so on. If this process ends, B is matched with some Cardinal n and is finite. If it does not end, B is matched with the set of all the natural numbers and is countably infinite. Incidentally, we have proved that any set which can be matched one-to-one with a subset of the natural numbers is either finite or countably infinite.

A very useful theorem on combining countably infinite sets is the following. *Let E_1, E_2, E_3, $\cdots$ be a finite or countably infinite set, each member of which is a finite or countably infinite set. Then the union of the sets E_1, E_2, $\cdots$ is finite or countably infinite.* To prove this we recall that there are infinitely many prime numbers, and that each natural number can be factored into primes in only one way, apart from re-arrangements in the order of factors. We match the k-th member of E_n with the k-th one of those natural numbers which are the products of n primes. Thus, for example, the products of three primes are 8, 12, 18, 20, 27, 28, 30, 42, $\cdots$. So the fifth member of E_3 is matched with 27 and the seventh member of E_3 with 30. In this way each member of the union is matched with a natural number, and no natural number is used more than once. More than one natural number may be matched with the same object, since some member of say E_1 can occur again as a member of E_3 and of E_n for other n. In this case we reject all but the first occurrence. Then the members of the union are matched one-to-one with some of the natural numbers, and the union is finite or countably infinite.

The negative integers form a countably infinite set, since they match the natural numbers by change of sign. Since the set consisting of 0 alone is finite, the set of all integers is countably infinite (it cannot be finite). Each natural number d furnishes us with a countably infinite set E_d of rational numbers, namely those fractions $0/d$, $1/d$, $-1/d$, $2/d$, $-2/d$, $\cdots$ which have d as denominator. There are countably infinitely many sets E_d, so the union is countable. Thus we find that there are only countably infinitely many rational numbers. This is at first somewhat astonishing, since one feels that there are far more rational numbers than there are integers. But as we have shown, there are no more; we can arrange the rationals in a sequence r_1, r_2, r_3, $\cdots$ which omits none of them. Of course this sequence has a total disregard for the natural order of the rationals; the important thing is that the rationals are matched up with the integers with no left-overs.

Any one who is agile at jumping at conclusions may be expected at this stage to jump to the conclusion that all infinite sets are countably infinite. However, this can be shown false, and not by any strange device; for the real numbers are neither finite nor countably infinite, as we now prove. In fact, we show that the real numbers between 0 and 1 are not countably infinite. To do this we assume that they are countably infinite and show that we are then forced into a contradiction. Suppose then that the reals between 0 and 1 were countably infinite. We could arrange them in a sequence $r_1 , r_2 , r_3 , \cdots$, omitting none of them. Each of them can be written as a decimal:

$$r_1 = .a_{11}a_{12}a_{13}a_{14} \cdots$$

$$r_2 = .a_{21}a_{22}a_{23}a_{24} \cdots$$

$$\cdots\cdots\cdots\cdots\cdots\cdots$$

$$r_n = .a_{n1}a_{n2}a_{n3}a_{n4} \cdots .$$

Now we define $b_1 , b_2 , \cdots$ as follows. If a_{11} is 0, 1, 2, 3, or 4 we choose $b_1 = 8$; if a_{11} is 5, 6, 7, 8 or 9 we choose $b_1 = 2$. Similarly, if a_{22} is 4 or less we choose $b_2 = 8$; if a_{22} is 5 or more we choose $b_2 = 2$. This we do for each n; if a_{nn} is 4 or less then $b_n = 8$, if $a_{nn} = 5$ or more then $b_n = 2$. Now consider the number expressed decimally as $B = .b_1b_2b_3 \cdots$. This is a real number between 0 and 1, so it has to be one of the numbers $r_1 , r_2 , r_3 , \cdots$, since these are all the reals between 0 and 1. But the equation $B = r_n$ is false for every n, because the digit in the nth place of r_n is a_{nn} and the digit in the nth place of B is b_n , and by the way we defined b_n these have to be different.

It follows at once that every interval $a \leqq x \leqq b(b > a)$ has uncountably infinitely many numbers in it; for this interval can be matched one-to-one with the interval from 0 to 1 by means of the function F, where $F(t) = a + t(b - a), 0 \leqq t \leqq 1$. Incidentally, this shows that all intervals of real numbers are equally numerous. It is more difficult, but still possible, to show that every interval of real numbers and every rectangle in the plane can be matched one-to-one. But the set of all subsets of the real number system is more numerous, and cannot be matched one-to-one with the real numbers. This should serve as an indication that the concept of matching can be used to classify sets, finite or infinite, according to "numerousness," and that just as for the finite sets, so also for the infinite sets, we can introduce symbols indicating how "numerous" a set is. These symbols are called *transfinite cardinals*. As a sample, the *transfinite cardinal* that labels the countably infinite sets is $\aleph_0$, read "aleph-null." The statement that the union of two countably infinite sets is countably infinite is written as the equation $\aleph_0 + \aleph_0 = \aleph_0$; the statement that the union of countably many countably infinite sets is again countably infinite can be written $\aleph_0 \cdot \aleph_0 = \aleph_0$. This suggests, as is in fact the case, that there is a whole arithmetic of transfinite cardinals. Even the principle of mathematical induction has an extension to these transfinite cardinals. It probably would be unwise to try to show any more details of this theory. But what we have already seen is enough to indicate that in the theory of sets, "infinite" is not a word meaning something vague and beyond the grasp of the human mind; there is an assortment of sizes of infinite sets, of which we have specifically exhibited two and mentioned the existence of others, and these various sizes of infinite sets can be characterized by a collection of symbols called transfinite numbers (or cardinals) which can be studied as having an arithmetic in their own right.

BIBLIOGRAPHY

1. Although the literature of the theory of sets is large, it is mostly written by specialists for specialists. One serious attempt to write up the theory in a form accessible to a college student is to be found in Part II of *Universal Mathematics*. This is a collaborative effort of several mathematicians directly or indirectly connected with the Committee on the Undergraduate Mathematical Program of the Mathematical Association of America, but particularly of Professor W. L. Duren, Jr. This book is obtainable from the University of Kansas Book Store, Lawrence Kansas.

2. A brief but readable account of the elements of set theory, with some indication of applications to logic and to probability, is given on pages 108–116 of *What Is Mathematics?*, by Richard Courant and Herbert Robbins, Oxford University Press, 1941.

3. The connection between Boolean algebras and switching circuits is discussed in a paper by Franz Hohn, entitled "Some Mathematical Aspects of Switching," *American Mathematical Monthly*, vol. 62, 1955, p. 75–90. In this article there are also references to other papers on the subject.

IV

Deductive Methods in Mathematics

CARL B. ALLENDOERFER

THE study of demonstrative geometry has been an essential part of the training of educated men from the earliest beginnings of our Western civilization. Its discovery not only was a contribution to the progress of mathematics, but also was a major factor in shaping our methods of thought in fields such as philosophy, religion, political science, and literature. It is not our purpose here to outline the widespread nature of the influence of this subject upon our culture, for this is one of the major topics treated in the recent book by Morris Kline, *Mathematics in Western Culture*. We shall rather discuss the essential nature of the methods of Euclid, their influence on mathematics, their place in modern mathematical thought, and their importance in the education of high-school students.

Let us first examine the course in demonstrative geometry as it is commonly taught to students in the tenth grade. This course may be considered as having three main objectives: (1) to teach some of the important facts about geometry as such, such as the properties of triangles, circles, parallel lines and the like; (2) to teach the deductive method as it is applied to mathematical reasoning, and thus to give the students a first taste of the nature of mathematical proof; (3) to teach logical reasoning *per se* and to show the students how it can be applied in nonmathematical situations. This is a great deal to expect of any one course, and so it is not surprising that it has been subject to criticism, watering-down, and lack of popularity. A good deal of the difficulty stems from the fact that the course combines geometry with logical reasoning in such a way that many students have the idea that logical reasoning is restricted to geometric situations. When they have trouble with geometry, they also build up resentment against logic in general.

As mathematics has developed today there seems to be no compelling reason for using Euclidean geometry as the principle example of the deductive method of logical reasoning. As a matter of fact, there are reasons to believe that Euclidean geometry is even an unfortunate example to use with beginning students. Euclidean geometry is quite a complicated mathematical system, and as presented in most textbooks is not even completely logical. Euclid begins with a rather large number of axioms and postulates, some of which are not particularly elementary in character; but modern scholarship has shown that even these are not sufficient for the completely rigorous development of the subject. Although it is doubtful that many students are bothered by Euclid's logical slips, they are certainly subject to confusion because of the complexity of the subject, and they are likely not "to see the forest for the trees." The reason that Euclidean geometry has been used traditionally as the prime example of logical reasoning is that until recent times there was no other example to which teachers could turn. There are now available, however, a number of other mathematical examples of deductive systems which lend themselves much more effectively to this purpose.

One of the simplest examples of a deductive theory which the teacher can utilize is that of a "group." This is an algebraic notion which has close connections with ordinary arithmetic and algebra as well as with more complicated mathematical disciplines. For the details of this subject the reader is referred to Chapter V of this book. Another very simple example is a "finite projective geometry." The ideas of such a geometry can easily be introduced into a tenth-grade geometry course, and should be much more understandable to students than the usual Euclidean ones. We shall discuss the details of such a system later in this chapter. More complicated deductive systems occur in all branches of mathematics, and two of these, Fields and Boolean algebra (the Algebra of Sets), are treated in this book in Chapters V and III respectively.

Since the deductive method is an essential part of modern mathematical thinking, the teacher should use every opportunity to illustrate it in every aspect of her work. Illustration, however, is probably not enough to teach the students the essential structure of a deductive system. At some stage in the high-school mathematics curriculum there should be a serious discussion of deductive systems *per se*, and later applications of this to mathematics and to nonmathematical situations should be used to reinforce the understanding of the students about deductive methods. Perhaps the tenth grade is the place for this, but no firm statement of this kind should be made until more experimental teaching has been carried out. It is toward the development of such a portion of the curriculum that the main part of this chapter is devoted.

ESSENTIALS OF A DEDUCTIVE SYSTEM

Undefined Words. The first requirement for an understanding of any branch of knowledge is that the student should know the meanings of the words which he is employing. In early life a child learns the meaning of words such as "dog" when his mother points to an animal and calls it a "dog." When he grows older he is taught to look up the meanings of unfamiliar words in a dictionary, where he will presumably find them explained by the use of words with which he is already acquainted. He thus comes to believe that every word has a definition, and that properly he should start a new subject by learning the definitions of the special words which have technical meanings in that subject.

Because of this experience, it is a hard blow to a young student when his mathematics teacher tells him that certain words are "undefined." Not only is this contrary to his whole previous experience, but also he can find so-called definitions of these words in his dictionary. The fact is, however, that the starting point for any deductive system is the choice of a set of words which are taken to be completely undefined. Indeed this is one of the essential places in which a deductive system differs from everything that a student has learned before. It is therefore very important that he understand the reason behind this novel procedure.

A little experience with a dictionary will show a student that the dictionary is useless if he does not know the meanings of any words at the outset. For without knowing some words he cannot understand even the simplest definition. If he examines the definition of any word he will find other words used; then he must look up these other words and find their definitions; then he must look up further definitions, etc. Eventually he will find himself going in circles such as the following:

point: the intersection of two *lines*;

line: the shortest distance between two *points*.

The discovery of such circular chains of definitions is a good exercise for the students. Another good exercise is to give the students a foreign language dictionary (intended solely for natives of that country) and to ask them to find the meaning of some particular word in that language using only this particular dictionary.

By these devices we can convince our students that the process of forming definitions is a circular one which is of no use to us unless we can cut into the circle and make a beginning somewhere. In a deductive system this is accomplished by choosing a small number of words which will be undefined—to which we will give no meanings at all. We may use common words such as "point" and "line" as undefined words, but when we do so we must strike from our minds all the connotations which we ordinarily associate with these words. Perhaps it would be better to start with nonsense words such as "pont" and "gerd." The choice of the undefined words, of course, will depend upon the particular deductive system which we are constructing. Later on we shall see some examples of how they are chosen in particular cases.

Defined Words. Now that we have a basic vocabulary of undefined words we can proceed to define other words in terms of them. In doing so we must assume that we have some ordinary, non-technical language to help us. Using the undefined words and non-technical words from our common language (English for us) we proceed to define additional technical words in our deductive system. Let us suppose that in geometry "point," "line segment" and "join" are undefined. Then we can define a *triangle* as the figure consisting of three points (not joined by a single line segment) and the three line segments which join these points in pairs. Notice that we have used a number of common English words such as "consisting," "three," "pairs," etc., since these do not have a technical meaning in geometry as such.

Propositions. When we have built up an adequate vocabulary of technical words—both undefined and defined—we are ready to form sentences which use these words and also non-technical English words. These sentences must follow the usual rules of good grammar and must also be so clearly stated that it is meaningful to call them "true" or "false." For example, the following statement is too vague to be permissible: "A triangle consists of three lines and three points." Of course, we do not have to know whether one of our sentences is true or false in order to use it; one of our problems will be to settle the truth or falsehood as our work proceeds. Most sentences which ordinary people write down fail to have this latter property, for often they are poorly stated so that their truth or falsehood is subject to qualification. Mathematicians are careful, however, to use language with greater precision than is usual in common speech. Even so it is possible to get into difficulties. Consider the following sentence, which we call "*S*" for short:

S: "Every sentence which contains more than nine words is false." If we assume that *S* is true, then since *S* contains ten words it is also false. Thus we see that *S* is not at all a proper sentence for us to use in our deductive system. Logicians have given quite careful rules concerning the formation of sentences which will be allowed in their discussions, but these are too complicated to be given here.[1] If the student

---

[1] See Kemeny, J. G.; Snell, J. L.: and Thompson, G. L. *Introduction to Finite Mathematics*, Englewood Cliffs, N. J.; Prentice-Hall, 1957. pp. 1–53.

is careful about his use of language and avoids trick sentences like S, he is not likely to run into difficulties of this sort. We now give the technical name *propositions* to the sentences which are allowed into our discourse.

We have assumed that every proposition can be meaningly called true or false, but our discussion of these words has been somewhat misleading. These words "true" and "false" have connotations which are based upon our particular philosophy or experience, and we must discard these connotations at this point. It is better for us now to forget the usual meanings of these words and to treat them as undefined "tags" which we hang onto propositions. The connection between these tags and the usual meanings of the words will become apparent when we come to discuss the connection between deductive systems and their applications to nature at a later stage in this chapter.

Our basic task, now, is to hang a tag of true or false onto as many of the propositions of our system as we can, or at least onto those propositions in which we have a particular interest. As in the case of words and their definitions, we can make no progress in this direction unless we have a starting place. We must take some, and preferably a relatively few, propositions and arbitrarily call them true or false. It is customary to phrase these propositions so that all of the corresponding tags are "true." When we have selected such a set of initial "true" propositions, we have set down a set of *axioms*. For emphasis we state that:

An *axiom* is an initial proposition which is assumed to be true.

In Euclid there is a division of the initial propositions into axioms and postulates, but the difference between these two types of propositions is highly artificial, and we shall ignore this point and call all propositions of this type by the single name of "axiom."

In many introductory treatments it is said that an axiom is a self-evident truth, and this statement will need clarification later on in this chapter. We shall not wish to adopt this point of view. Any set of axioms can be used as the starting point for a deductive system; they need not be self-evident at all, and there is no harm even if we have questions of the validity of some or all of them. For the purposes of the system which we are building we shall assume them to be true. If we have chosen a bad set of axioms, our deductive system may run into difficulties such as inconsistency, or it may be of no use to us in any applications. (A deductive system is called *inconsistent* if there is some proposition in the system which can be proved to be both "true" and "false.") Still we have a right to choose any axioms that we please, as long as we are willing to suffer the consequences of this choice.

Once we have our axioms established, we proceed to combine these into more complicated propositions, and wish to have rules for hanging tags onto these. These rules amount to what is usually called the "propositional calculus," and they are the basic tools of logic. To some extent these rules are arbitrary, but there is one set which is used by nearly all mathematicians, and we shall discuss this in full below. By using these rules to determine the proper tags for more and more complicated propositions we develop the superstructure of our deductive system. Those propositions which turn out to be "true" are called *theorems*.

Summary. A deductive system consists of a collection of undefined words, other words defined in terms of these and common English words, initial propositions called axioms which we assume to be true, and a collection of propositions called theorems which we determine to be true on the basis of our axioms and the assumed laws of logic. We now turn to a more detailed discussion of the various components of a deductive system.

AXIOMS

It was stated above that any set of axioms may be used as the beginning of a deductive system. We would be foolish, however, to give ourselves free rein in the choice of these axioms, for most of the deductive systems so established would be of no value to us. The choice of a profitable set of axioms is a delicate matter requiring great intuition and creative power. It is difficult to reduce strokes of genius to a pat formula, but at least we can illustrate how good systems of axioms may be discovered.

Most sciences as well as other branches of knowledge begin with a descriptive, observational phase. The interested investigator looks about him and records what he finds. At the outset his observations are likely to be isolated and uncoordinated, but if he is diligent and perceptive he will begin to find some pattern emerging in his subject. When this pattern has taken tentative shape, he is ready to try setting up his deductive system and to choose his undefined words and axioms. Let us take two examples of this kind of process.

There is a legend that plane geometry grew out of the practical needs of farmers in ancient times who wished to establish boundaries for their fields and to solve other problems of land measurement. They may well have observed some elementary facts about fields such as the statements that "two boundary lines meet in a corner; there is a boundary line between each two corners," and the like. In setting up their deductive system they might then have constructed an array such as the following:

Practical Situation	*Undefined Words*
Corner	*Point*
Boundary	*Line*
Field	*Plane*

Although no meaning is to be given to "point," "line," and "plane," in the applications of this deductive system to the practical problem, these words will later be translated into "corner," "boundary," and "field" respectively. Having now obtained the undefined words (or at least some of them) our farmers would proceed to set up their axioms. These will be abstract statements which are derived from observed relationships in the practical situation. Thus they may have developed the array:

Practical Situation	*Axioms*
Two boundaries meet at a corner	*Two lines intersect at a point*
A boundary lies between a pair of corners	*A line is determined by two points*

In such a way the axioms of plane geometry may well have been invented. They are not self-evident statements about our physical world, but are abstractions from observations of practical men who were interested in this subject. We should note that observers may well come up with different axiom systems when their practical situations are different. In ancient times men did not go very far from their homes, and their experience might have led them to assume as an axiom that:

The sum of the interior angles of a triangle is 180 *degrees.*

However, other more adventurous men discovered that the earth is indeed a sphere, and this observation might have led them to the axiom:

The sum of the interior angles of a triangle is greater than 180 *degrees.*

There is no point in arguing which of these axioms is correct; each is a description of what the observers found in their investigations, and each can be used as an axiom for a useful geometry.

As a second example let us consider an abstract algebraic system which is motivated by our experience with the addition of pairs of natural numbers (*i.e.* positive whole numbers). We observe that the following statements are true:

(1) We can add any pair of natural numbers and obtain a natural number.

(2) The sum of two numbers, such as $2 + 7$ is equal to the sum of these numbers taken in reverse order, such as $7 + 2$.

(3) Successive addition of three numbers such as $(3 + 6) + 4$ gives the same result as if the addition had been performed in the order $3 + (6 + 4)$.

(4) If the addition of a number, say 4, to a number a gives the same result as adding 4 to a number b, then a must be the same as b.

In algebra we phrase these observations more compactly as follows: *Let a, b, c represent arbitrary natural numbers. Then:*

(1) $a + b$ is a natural number c

(2) $a + b = b + a$

(3) $(a + b) + c = a + (b + c)$

(4) *If $c + a = c + b$, then $a = b$.*

We get our true abstraction from these observations by taking a, b, and c to be completely undefined symbols, by supposing that "o" is an operation which can be used to combine two of our undefined elements, and then by taking as our axioms, the propositions:

(1′) *a o b is one of our undefined elements, c.*

(Read: a operating on b.)

(2′) a o $b = b$ o a

(3′) $(a$ o $b)$ o $c = a$ o $(b$ o $c)$

(4′) *If c o $a = c$ o b, then $a = b$.*

Observe that these no longer refer to the addition of numbers, but to the properties of the undefined operation "o" upon the undefined symbols a, b, c. Of course, we may interpret a, b, and c to be natural numbers and "o" to be addition, but we may make other interpretations as well. For instance we may interpret a, b, and c as natural numbers, and "o" as multiplication. Then we arrive at a new set of statements about multiplication, which also happen to agree with our experience. This illustrates one point of great importance and value:

The same deductive system may be abstracted from a number of quite different practical situations. The theorems obtained in this system will then have applications to this great variety of situations.

This process of looking at nature and then abstracting our observations and axioms has been carried out most systematically in the physical sciences. The deductive systems so obtained come under the general heading of "theoretical physics," and they have been extraordinarily effective in helping us to understand this aspect of our world. In other subjects such as economics, psychology, sociology, and warfare the use of the deductive method is still a relative novelty. Such usage, however, appeared in many places during and after World War II, and there is a growing group of scholars who are attempting to apply the axiomatic method to a variety of fields of knowledge. In their terminology they are constructing "mathematical models" of portions of their subjects. A *mathematical model* is nothing more than a deductive system in which the undefined words and axioms represent certain aspects of the observed realities. In the construction of a model it is always necessary to ignore certain aspects of nature and to concentrate upon those aspects which seem to be most important. For instance, in celestial mechanics the earth is considered to be a sphere even though we know that it has mountains and valleys and is flattened at the poles. In other branches of science, however, these aspects of the earth have to be taken into account. Model building, then, is quite like painting; for an artist does not make a photographic reproduction but selects those details of the scene before him which are most important for his purposes.

To give an illustration of a mathematical model in the social sciences let us consider the problem of selecting a political official, say the mayor of a small town. Let us suppose that there are five candidates and that each of the 500 voters of the town can put the five candidates into his own preferential order. Who should be elected mayor? There are a number of different procedures that might be adopted such as electing the candidate with the greatest number of first preferences, nominating the two candidates with the greatest and next greatest number of first preferences respectively and then having a run-off election, or having each voter number his choices 1, 2, 3, 4, 5, add up the scores for each candidate and declare elected the candidate with the lowest total score. How shall we choose among alternatives of this sort? We can do so only by formulating our philosophy of what is socially desirable in such a situation. At the outset we will set down a number of social objectives which are proposed by various interested persons. At first these will be in rather vague terms such as the desirability of "majority rule," or the "protection of minorities." We may wish to discuss questions such as: Is majority rule desirable if a very large minority are strongly opposed to the candidate favored by the majority? Or would it be better to select the second choice of the majority if he is more acceptable to the minority? As discussion continues these ideas become sharper, and we start to build our model. Since the names of the voters and the names of the candidates and their qualifications are unimportant, we represent these by abstract symbols—these are our *undefined* words. We introduce an undefined inequality relationship which will represent the order of preferences expressed by the voters. Then in terms of these symbols we write down the various principles of the various philosophies which have been expressed. These are our *axioms*. If we are extremely lucky we can deduce from these axioms a single method of election which we will then adopt. Two other possibilities, however, exist: (1) There may be several methods of election which are consistent with our axioms. Then we may use any of these, or we may introduce additional axioms until only one method remains. (2) Our axioms may turn out to be inconsistent, so that there is no method of election which is consistent with all of them. Then we have to make a social judgment: On the basis of which axioms do we wish to proceed? Mathematics here will not tell us what choice to make, but it will clarify the situation by telling us what choices are possible and by expressing these in unmistakable terms. We will therefore have a better basis for our decision than if we had not gone through the process of constructing our model. It is clearly impossible to give the details of such a model here, but for a systematic study of problems of this sort the reader is referred to the book by Kenneth Arrow, *Social Choice and Individual Values*.

In discussing the social model just above we noted that axiom systems may turn out to be inconsistent, and a few words of explanation are needed about this concept. A system is *inconsistent* if it is possible to prove that some particular proposition is both true and false. Such systems of axioms are of no value to us and must be discarded at once. The real question is: How do we know whether a given system of axioms is inconsistent? If we happen upon an inconsistency, the question is settled. We may, however, spend many years developing the theorems in our system without finding an inconsistency; and yet we have no assurance that on the next day we will not find one of these. In other words, how can we be sure that any deductive system is consistent? In mathematics we proceed largely by faith with respect to this question. The system of real numbers has been used so long and so extensively by so many people that we believe that it is consistent. We, therefore, frequently accept this as a working hypothesis and proceed as if it were firmly established. Suppose now that we have before us some particular deductive system, say that based on the axioms (1') to (4') of page 72. To show that it is consistent, we observe that the real numbers satisfy these axioms if "o" is identified as "+." If we assume that the arithmetic of these numbers is consistent, it follows that there can be no inconsistency among our axioms. Although this does not prove the consistency of the given system, it at least gives us a very strong reason for believing in its consistency, and we proceed on this assumption. It is unwise to proceed too far in the development of a deductive system without first checking on the consistency of the axioms in some such fashion; for an inconsistency discovered at some later time may invalidate all the work which we have done.

It would be misleading to suggest that all axiom systems used by mathematicians are developed as models of some practical situation. Although model building is the source of many of our mathematical disciplines, some branches of mathematics have no such source. A common method of obtaining other axiom systems is to take some standard set of axioms (constructed as a model) and then to modify or omit one of the axioms. For instance, in algebra, we may assume all the usual rules except that we do not require that $a \times b = b \times a$. Such an algebra is said to be *non-commutative*. In geometry, the most familiar example of this sort concerns the Fifth Postulate of Euclid which is equivalent to the statement that there exists a unique line parallel to a given line and passing through a given point not on the given line. This axiom was modified in two distinct ways by assuming (1) that no such parallel line exists, and (2) that more than one such parallel line exists. These *non-Euclidean* geometries later turned out to be models of concrete physical situations, but their existence as abstract systems did not depend upon these later discoveries. Axiom systems of this kind have become commonplace in modern mathematics and have served two purposes: (1) by examining the consequence of modifications of existing theories, light has been shed upon the structure of these theories, (2) new theories have been developed which later turned out to be of great importance in applications which were quite unforeseen when the system was first examined. It would be too much to expect that every axiom system of this kind would turn out to be valuable, and many of them have been shelved after a cursory examination. But out of this chaff, some very important ones have emerged and the development of these justifies the time spent on the others.

We should now be in a position to understand the relationship between mathematics and the everyday world. Mathematics itself has nothing to do with reality and can prove nothing about the world. It has been called a "game in which we do not know what we are talking about, nor whether our conclusions are true." Although this description of mathematics is strictly correct, few would wish to study mathematics if this were the whole story. Our abstract mathematical theories are abstractions from the real world; they are models of the relationships which we have observed experimentally. The conclusions which we draw from these models are conjectures about new relationships in the real world.

In order to put substance into these conjectures we must examine them to see if they correspond to reality. The value of our mathematics is that without it we would seldom have been able to guess the relationships which our mathematics has suggested to us. Mathematics has been particularly valuable in physical science, because the mathematical model of this subject has turned out to be remarkably true to life. We therefore have great confidence in the physical truth of our mathematical conclusions. In the newer applications of mathematics to the social sciences the mathematical models are not so highly developed, and we have less confidence in the conclusions derived from them. As the models improve, however, the usefulness of mathematics in these subjects will increase by leaps and bounds.

Because of the closeness between the mathematical models of physics and the observed relationships, it has been said by some that "God is a mathematician." I would rather put it another way and say that "God has endowed man with such great inventive capacities that he can construct mathematical models of such variety that at least one of them will apply to any part of the world of nature."

LOGIC

It is now time to discuss the rules of the propositional calculus and to see how these can be used to decide what tags (true, or false) we are to hang upon propositions derived from our assumed system of axioms.

There are four basic operations which we will use to combine two propositions into a third proposition. Assuming that we know the tags assigned to the given propositions we must adopt rules for hanging a tag upon the combined propositions. These operations are known as *conjunction, disjunction, implication* and *equivalence*.

Conjunction. In a *conjunction* we combine the two given propositions by placing an "and" between them. Thus the conjunction of:

Prices are high

People are starving

is:

Prices are high, and people are starving.

By analogy with ordinary reasoning, we say that a conjunction is true when both of its components are true and that it is false when either or both of its components are false.

For future use it is desirable to introduce some symbolism which will express these ideas in a compact fashion. Let us denote propositions by the letters p, q, r, s. Then we shall write the conjunction of p and q as the proposition $p \wedge q$, read "p and q." In order to define how the tags T (true) and F (false) are to be hung onto $p \wedge q$, we construct the following "truth-table."

p	q	$p \wedge q$
T	T	T
T	F	F
F	T	F
F	F	F

CONJUNCTION

This table is easy to remember; it should be memorized at this stage.

Disjunction. In a *disjunction* we combine the two given propositions by placing an "or" between them. Thus the disjunction of the propositions mentioned above under conjunctions is

Prices are high, or people are starving.

We now run into a little trouble in deciding what tag to hang onto a disjunction, for "or" is an ambiguous word in English. It can mean that just one of the alternatives is true (*exclusive* or), or it can mean that either or both of them are true (*inclusive* or). Consider the two common statements:

Exclusive: This coin will turn up heads or it will turn up tails.

Inclusive: To qualify for this job you must be over 21 or be a veteran. (Surely veterans over 21 would qualify.)

Because of this confusion in common usage, legal documents frequently use the expression "and/or" to express the idea of the inclusive or.

When we come to define our disjunction operation, we must choose between the inclusive and the exclusive "or." Either would do, but it is more convenient to adopt the inclusive usage. We therefore set up the rules for hanging tags upon disjunctions by the following truth-table in which we denote the disjunction of p and q by $p \vee q$ (read "p or q").

p	q	$p \vee q$
T	T	T
T	F	T
F	T	T
F	F	F

DISJUNCTION

In the rare cases where the exclusive "or" is needed we use the symbolism $p \underline{\vee} q$. The construction of the truth table is left as an exercise for the reader.

Implication. An *implication* is a combined proposition of the form "If ..., then .... For example from the propositions:

People are honest

Doors can be left unlocked

we can derive two implications:

If people are honest, then doors can be left unlocked.

If doors can be left unlocked, then people are honest.

Our notation for the two implications which can be derived from the propositions p and q will be $p \rightarrow q$ (read "if p, then q") and $q \rightarrow p$ (read "if q, then p"). We shall see later that these are two distinct propositions, but for the present let us consider only $p \rightarrow q$. We must now decide what tags to hang onto it. Let us begin by stating the corresponding truth table and justify it later.

p	q	$p \rightarrow q$
T	T	T
T	F	F
F	T	T
F	F	T

IMPLICATION

The first line of this table agrees with our intuition; for by correct reasoning we believe that we can proceed from a true hypothesis (p) to a true conclusion (q). The second line of the table also is intuitive; for our reasoning must be false if it leads us from a true hypothesis to a false conclusion. The third and fourth lines are less familiar and need more discussion. What happens if we start from a false hypothesis and reason correctly; what can we say about the conclusion? As a matter of fact it may be true or false. Let us do it both ways:

Suppose we assume the false hypothesis that $2 = 5$. Then we can argue:

$$2 = 5$$
$$5 = 2$$
$$\overline{}$$
$$7 = 7$$

Thus we have reached a true conclusion although we began with a false hypothesis. This leads us to adopt the third line of the table. On the other hand we could have reasoned:

$$2 = 5$$
$$3 = 3$$
$$\overline{}$$
$$5 = 8$$

which gives us a false conclusion. Thus we can also get a false conclusion by reasoning correctly from a false hypothesis. We therefore adopt the fourth line of the table. We have always been taught the dangers of reasoning from a false hypothesis, and this illustrates what may happen: From a false hypothesis we may get either a true or a false conclusion. Since we cannot predict which will be our result, reasoning from a false hypothesis is a very futile process. Our truth table for implication expresses this intuitive observation.

At this point we should notice that there is a marked formal difference between the two operations of conjunction and disjunction and that of implication. The tags we hang onto $p \wedge q$ and those we hang onto $q \wedge p$ are the same. Hence we do not distinguish between $p \wedge q$ and $q \wedge p$ and consider these to be equivalent propositions. A similar remark holds for $p \vee q$ and $q \vee p$. But let us examine the truth table for $q \rightarrow p$, we see in particular that when p is true and q is false, then $p \rightarrow q$ is false, but $q \rightarrow p$ is true. We must therefore distinguish between $p \rightarrow q$ and $q \rightarrow p$. This is why we can combine two propositions into two distinct implications, whereas they yield only a single conjunction or disjunction.

Equivalence. We have just noted that propositions with apparently different forms may have the same tags on them. For example this is true of $p \wedge q$ and $q \wedge p$. We will formalize this idea by the notion of *equivalence*. We write $p \leftrightarrow q$ (read "p is equivalent to q") and define its truth or falsehood by the table:

p	q	$p \leftrightarrow q$
T	T	T
T	F	F
F	T	F
F	F	T

EQUIVALENCE

In this notation we can then state that $(p \wedge q) \leftrightarrow (q \wedge p)$ is true for any p and q; whereas the truth of $(p \rightarrow q) \leftrightarrow (q \rightarrow p)$ depends upon the tags on p and q.

Negation. Our final basic operation on propositions differs from those just given in that it applies to a single proposition alone. It is called *negation*. To take the negation of a given proposition we write "It is false that" in front of the given proposition. This makes a true proposition into a false one, and a false one into a true one. We use the symbol $\sim p$ (read "not p") for the negation of p, and establish the truth table:

p	$\sim p$
T	F
F	T

NEGATION

The operation of taking negations of propositions is of particular importance to us since it forms an essential step in an indirect proof. We shall have to learn how to negate various types of propositions correctly if we are to have success with this method of reasoning.

Tautology. We have seen that the truth or falsehood of a proposition derived by combining other propositions may depend upon the truth or falsehood of the various component propositions. However, we have also seen that the proposition

$$(p \wedge q) \leftrightarrow (q \wedge p)$$

is true for all propositions p and q, true or false. Such a proposition is called a *tautology*.

Definition. A proposition formed by combining other propositions p, q, r, $\cdots$ is called a *tautology* if it is true regardless of the truth or falsehood of the component propositions p, q, r, $\cdots$.

Let us proceed to establish a few useful tautologies.

Example 1. The proposition $p \vee (\sim p)$ is a tautology. In order to establish this we construct the following truth table:

p	$\sim p$	$p \vee (\sim p)$
T	F	T
F	T	T

In this table we obtain the second column from the first column by using the basic truth table for negations. Then we derive the third column from the first two by using the basic truth table for a disjunction. Since a T appears in each row of the third column we have shown that the proposed proposition is a tautology.

Example 2. The proposition $\sim [p \wedge (\sim p)]$ is a tautology. The proof is obtained from the truth table:

p	$\sim p$	$p \wedge (\sim p)$	$\sim[p \wedge (\sim p)]$
T	F	F	T
F	T	F	T

This tautology has been given the name *The Law of the Excluded Middle*. It is one of the basic relationships in our subject, and is the source of much discussion by those who wish to set up other rules of logic in which this law is replaced by other statements.

Example 3. The proposition $[(p \rightarrow q) \wedge (q \rightarrow r)] \rightarrow [p \rightarrow r]$ is a tautology.

The proof of this requires a somewhat more extensive truth table. We begin with the three columns:

p	q	r
T	T	T
T	F	T
F	T	T
F	F	T
T	T	F
T	F	F
F	T	F
F	F	F

On the basis of these we assign tags to the columns: $p \rightarrow q$, $q \rightarrow r$,

$[(p \rightarrow q) \wedge (q \rightarrow r)]$, $p \rightarrow r$. Finally we have the given proposition $[(p \rightarrow q) \wedge (q \rightarrow r)] \rightarrow [p \rightarrow r]$. Since we find that all these final tags are T's, we conclude that this proposition is a tautology.

This particular tautology is called the *Law of the Syllogism*. It is one of the most useful in deductive reasoning.

Rules of Substitution and Detachment. In carrying out a deductive process we shall often have to take two steps called respectively "substitution" and "detachment" which need justification at this stage.

The *Rule of Substitution* states that at any point we may substitute one proposition for an equivalent one.

The justification for this rule is quickly apparent when we realize that our whole process consists of hanging tags onto the various propositions which we write down. In a proposition which is found by combining others, these tags depend upon the tags of the component propositions and not on these propositions themselves. Since equivalent propositions have the same tags, we change nothing if we replace a proposition by an equivalent one.

The *Rule of Detachment* states the following: If we have established that (1) the implication $p \rightarrow q$ is true, and (2) that the proposition p is true; then we may conclude that the proposition q is true.

This follows easily from the truth table for an implication. Since we know that $p \rightarrow q$ is true, we are in lines 1, 3, or 4 of this table. Since p is true, we are in lines 1 or 2 of this table. The only line common to these two circumstances is line 1, and in this line q is true.

The Law of the Syllogism and the Rule of Detachment are frequently combined into one step in our reasoning. We have seen that a syllogism is a tautology, and consequently the condition (1) of the Rule of Detachment is automatically satisfied. In order to apply the Rule of Detachment we must also establish that $(p \rightarrow q) \wedge (q \rightarrow r)$ is also true. We can do this if we can establish the truth of both $p \rightarrow q$ and $q \rightarrow r$. Hence we make the statement:

If we can establish that (1) $p \rightarrow q$ is true

(2) $q \rightarrow r$ is true

we can conclude that $p \rightarrow r$ is true.

This process can be carried out in a repetitive fashion and gives us results such as:

If we can establish that (1) $p \rightarrow q$ is true

(2) $q \rightarrow r$ is true

(3) $r \to s$ is true

(4) $s \to t$ is true

we can conclude that $p \to t$ is true.

Finally in a case like this, we can conclude that t is true if we know that p is true. It is by reasoning along such lines that most theorems in mathematics are established.

Derived Implications. One of the more troublesome spots in deductive reasoning is that when we are given an implication $p \to q$, we can rearrange the terms of this implication in various ways and obtain new implications. Some of these derived implications turn out to be equivalent to the given implication, and others are not. Students frequently are confused on this point and use derived implications incorrectly. Let us examine the various types.

Given implication: $p \to q$

Converse: $q \to p$

Inverse: $(\sim p) \to (\sim q)$

Contrapositive: $(\sim q) \to (\sim p)$.

We have already seen that the converse is not equivalent to the given implication; and hence we cannot conclude that a converse is true just because the given implication is true. Indeed, the converse may be true or false.

Similarly let us investigate the situation for the inverse. We construct the truth table:

p	q	$p \to q$	$\sim p$	$\sim q$	$(\sim p) \to (\sim q)$	$(p \to q) \leftrightarrow [(\sim p) \to (\sim q)]$
T	T	T	F	F	T	T
T	F	F	F	T	T	F
F	T	T	T	F	F	F
F	F	T	T	T	T	T

INVERSE

Since the final column contains both T's and F's we see that the given implication and its inverse are not equivalent. It is a mistake, therefore, to substitute one for the other.

In the case of the contrapositive, the situation turns out in a different fashion; for the contrapositive is equivalent to the given implication.

To see this we again construct a truth table and notice that the last column contains nothing but T's.

p	q	$p \rightarrow q$	$\sim q$	$\sim p$	$(\sim q) \rightarrow (\sim p)$	$(p \rightarrow q) \leftrightarrow [(\sim q) \rightarrow (\sim p)]$
T	T	T	F	F	T	T
T	F	F	T	F	F	T
F	T	T	F	T	T	T
F	F	T	T	T	T	T

CONTRAPOSITIVE

We can use this information about the contrapositive in two ways: (1) If we know the given implication to be true we can infer that the contrapositive is true, and vice versa. Frequently it is easier to prove one of these than it is to prove the other, so we choose the more convenient one. (2) In any proof we can substitute a contrapositive for an implication without changing the result. This often helps us to construct a chain of syllogisms.

The catch is that the converse and inverse can not be used in this same fashion. Since they have a close resemblance to the contrapositive, this leads to confusion and errors in untrained minds.

Methods of Taking Negations. Since we shall need to take negations of compound propositions when we use the contrapositive and when we use indirect proof, we shall need to see how this is to be done.

Let us first consider the *negation of a conjunction*: $p \wedge q$. We are looking for a proposition whose tags are just the opposite of those of $p \wedge q$. A little experimenting suggests that the proper expression is $(\sim p) \vee (\sim q)$. Another way to guess this is to observe that $\sim (p \wedge q)$ is to be true when $p \wedge q$ is false, or indeed when either or both of p and q is false. This is just expressed by the statement that $(\sim p) \vee (\sim q)$ is true. To clinch the argument we construct the truth table and verify that $[\sim (p \wedge q)] \leftrightarrow [(\sim p) \vee (\sim q)]$ is a tautology.

p	q	$p \wedge q$	$\sim (p \wedge q)$	$\sim p$	$\sim q$	$(\sim p) \vee (\sim q)$	*Equivalence*
T	T	T	F	F	F	F	T
T	F	F	T	F	T	T	T
F	T	F	T	T	F	T	T
F	F	F	T	T	T	T	T

In a similar way we find that $[\sim (p \vee q)] \leftrightarrow [(\sim p) \wedge (\sim q)]$.

The *negation of an implication* is no more trouble, but the result is surprising at first sight. Consider the proposition: If it rains, I will stay home. What is your guess as to the negation? Is it: If it does not rain, I will stay home? Or is it: If it does not rain, I will not stay home? (the inverse). Or can you do better? The correct answer is: It is raining, and I am not staying home. More formally

$$[\sim(p \rightarrow q)] \leftrightarrow [p \wedge (\sim q)]$$

To prove this let us again check by a truth table:

p	q	$p \rightarrow q$	$\sim(p \rightarrow q)$	$\sim q$	$p \wedge (\sim q)$	*Equivalence*
T	T	T	F	F	F	T
T	F	F	T	T	T	T
F	T	T	F	F	F	T
F	F	T	F	T	F	T

Finally, the *negation of a negation* is easy. It is just $[\sim(\sim p)] \leftrightarrow p$.

Using these basic relations we can form the negations of more complicated propositions by going a step at a time.

Example.

$$\sim[(p \wedge q) \rightarrow (r \vee s)] \leftrightarrow (p \wedge q) \wedge [\sim(r \vee s)]$$

$$\leftrightarrow [(p \wedge q) \wedge (\sim r \wedge \sim s)]$$

Quantifiers. We conclude this section on logic by discussing the troublesome but absolutely essential topic of *quantifiers*. Undoubtedly the reader is acquainted with this to some extent in algebra where the distinction is made between identities and equations of condition. An *identity* is an algebraic expression in one or more variables which is true for all values of these variables. An easy example is

$$x^2 - y^2 = (x + y)(x - y).$$

An *equation of condition* is one which places a restriction upon the variable or variables in it, and which consequently is satisfied for only restricted values of these variables. An easy example is

$$3x + 4 = 7$$

which is satisfied only for $x = 1$.

In order to develop the idea of quantifiers properly we must begin with an extension of the idea of a proposition, namely the propositional function. A *propositional function* is a proposition which contains a symbol which is supposed to refer to an unspecified element of some set of numbers or objects or the like. For example, consider the propositional function: "x is an even integer," where x is to be chosen from the set of integers. We cannot say whether this statement is true or false, and so it is not a proposition. If we know x, however, the statement is converted into a proposition. For when $x = 3$, the corresponding proposition is false, and when $x = 4$ the corresponding proposition is true.

We can convert propositional functions into propositions by the use of quantifiers, a word which means "how many." For example we can state the proposition: *For all x (where x is an integer), x is even.* This happens to be false, but it is a perfectly good proposition. We could also have said: *For some x (where x is an integer), x is even.* This turns out to be true. These examples illustrate the two basic quantifiers which we shall use.

The *universal quantifier, all.* The symbol for this is $\forall_x[p]$, "For all x, p."

The *existential* quantifier, *some.* The symbol for this is $\exists_x[p]$, "For some x, p."

We use the word *some* in the sense that it means *at least one.* Thus the proposition

$$\exists_x[3x + 4 = 7]$$

is true, for there is an x which satisfies the equation.

A great many, indeed most, mathematical statements have quantifiers implicit in them, and for certain purposes these must be made explicit. For instance, the theorem "Circles with equal radii have equal areas" should more properly be written:

Let x be the set of circles, and let x_1 and x_2 be any two circles. Then $\forall_{x_1}\forall_{x_2}$ [if radius of x_1 = radius of x_2, then area of x_1 = area of x_2].

As another example consider the theorem: The base angles of an isosceles triangle are equal. This really means:

Let x be the set of isosceles triangles

Then $\forall_x$ [the base angles of x are equal].

Most theorems in geometry are "general" theorems, where *general* means that they apply to *all* geometric objects having a certain property. These theorems therefore imply the universal quantifier, *all.* You should beware of other kinds of statements, however, such as "It is impossible to trisect an angle with ruler and compass alone," which means that there are *some* angles which cannot be trisected in this fashion.

Since the logical behavior of propositions involving $\forall_x$ is quite different from those involving $\exists_x$, it is essential that these quantifiers be made explicit before the propositions are subjected to further logical manipulation.

These differences appear sharply when we examine the negations of propositions involving quantifiers. Consider the proposition:

Some children are troublesome.

What are your guesses as to its negation? What about the following possibilities:

Some children are not troublesome.

All children are troublesome.

All children are not troublesome.

Remember that we wish to find a statement which is false if the original one is true, and true if it is false. By thinking it over, we find that the last possibility mentioned above is the correct one to choose. We are therefore motivated to define $\sim\exists_x[p]$ by the equivalence

$$\sim\exists_x[p] \leftrightarrow \forall_x[\sim p]$$

Similarly we are led to define $\sim\forall_x[p]$ by the equivalence

$$\sim\forall_x[p] \leftrightarrow \exists_x[\sim p].$$

Before examining the applications of these definitions to mathematical statements let us try a few non-mathematical examples:

Example 1. Find the negation of: If some students are slow, then all students should be held back.

Solution. This is basically an implication of the form $p \rightarrow q$ where

p: some students are slow

q: all students should be held back.

The negation should be $p \wedge (\sim q)$. But $\sim q$ is the proposition

$\sim q$: some students should not be held back.

The correct negation of the given implication is therefore:

Some students are slow, and some students should not be held back.

Example 2. Find the negation of: All men are liars, and some women are emotional. Proceeding as above we see that the correct result is:

Some men are not liars, or all women are not emotional.

Now let us try a mathematical example.

Example 3. Suppose we are given the proposition:

If two triangles are similar, then they are congruent. (Do not mind that this is false; it is still a good example.) What is its negation?

Solution. In the first place we must rewrite the given proposition to expose the hidden quantifiers. Let x be the set of triangles. Then we have $\forall_{x_1}\forall_{x_2}$ [If x_1 and x_2 are similar, then x_1 and x_2 are congruent.] The negation of this is:

$\exists_{x_1}\exists_{x_2}[x_1$ and x_2 are similar and x_1 and x_2 are not congruent],

or more smoothly:

There exists a pair of similar triangles which are not congruent.

Let us digress a moment and consider how we might prove the given implication to be false. We may do so by proving its negation to be true. In order to do so, all that we have to do is to find a pair of similar, non-congruent triangles. Since this is easy to do, the given implication is shown to be false. This method of disproof is called the *Method of the Counterexample* to which we shall refer later on.

Example 4. Find the contrapositive of the implication:

If two lines are parallel, then the alternate interior angles cut by a transversal are equal.

Solution. First we must look for quantifiers. Let x be any pair of lines, and let y be a transversal of x. Then we have:

$\forall_x\forall_y$ [If x is parallel, then the alternate interior angles cut off from x by y are equal.]

Since the propositional function inside the quantifiers is still an implication, we take its contrapositive and write:

$\forall_x\forall_y$ [If the alternate interior angles cut from x by y are not equal, x is not parallel.]

A smoother statement of this is:

If the alternate interior angles cut from a pair of lines are not equal, then the lines are not parallel.

In this example, the quantifiers did not enter into the propositions to be negated, and so they remained unchanged. Thus the careless student would not have obtained an incorrect result if he had ignored them. It is well, however, to be aware of those quantifiers which are present, so that they may be taken into account when necessary.

METHODS OF PROOF

We have now established all the machinery which we need in order to prove theorems. There are two questions to be answered: (1) How do we conjecture theorems which we will try to prove? (2) How do we prove that certain conjectures are really theorems, and how do we prove that other conjectures are in fact false?

We shall deal only briefly with the first of these two questions. It would clearly be rather silly for us to write down every proposition that we could think of which had meaning in our deductive system. Most of them would be false and their examination would be a waste of time. We really wish to guess propositions which will turn out to be true. The process of successful guessing requires a high order of innate ability and also considerable experience. We can learn a good deal about this process, however, by observing other mathematicians in action and by reading how discoveries were made in the past. There is an excellent discussion of intuition in mathematics in the two-volume work by G. Pólya titled *Mathematics and Plausible Reasoning.*

Let us now suppose that we have at hand a guess which we hope will turn out to be true. We have two possible methods of attack: to prove it or to disprove it. It is unfortunate that mathematics textbooks are solely devoted to the proofs of true propositions and that the students never are asked to disprove a false one, for the practicing mathematician spends at least as much of his time in disproof as he spends in proof. A knowledge of the real activities of a mathematician would make any mathematical subject much more interesting to the students.

Direct Proof. The most usual method of proof is a direct one in which the mathematician arranges a chain of syllogisms from the given propositions to the desired conclusion. There is no royal road to the construction of such a proof, and the student's best procedure is to examine the proofs of similar theorems and to use these as a guide.

As an elementary example of a direct proof consider the following situation:

Example. Given: p is true

$$r \rightarrow (\sim p) \quad \text{is true}$$

$$(\sim r) \rightarrow s \quad \text{is true}$$

Prove: s is true.

Solution. We would like to have a chain of syllogisms from p to s. The first hitch is that $(\sim p)$ appears in the second given statement rather than p. Fortunately it is in the right position so that we can form the contrapositive $p \rightarrow (\sim r)$ with p as the "if" part. This ties in with the third given statement to give us the chain:

$$p \qquad \text{is true}$$

$$p \rightarrow (\sim r) \quad \text{is true}$$

$$(\sim r) \rightarrow s \quad \text{is true}$$

From our discussion of syllogisms and the Rule of Detachment, we now conclude that s is true.

Example. Discuss the nature of the reasoning in the following argument:

Given: If libraries are destroyed, the people cannot read.

If the people can read, they will be free.

Conclusion: If libraries are destroyed, the people will not be free.

Solution. Let:

p: libraries are destroyed

q: the people can read

r: the people will be free.

Then the argument can be written symbolically:

Given: $p \rightarrow (\sim q)$ is true

$q \rightarrow r$ is true

Conclusion: $p \rightarrow (\sim r)$ is true.

We would like to set up a syllogism the last term of which should be $p \rightarrow (\sim r)$. Since the first term is given as $p \rightarrow (\sim q)$, the second term must be $(\sim q) \rightarrow (\sim r)$. Unfortunately this is the inverse of the second given implication and we cannot assume that it is true. Therefore, the argument does not follow, and we have no confidence in the conclusion. Examples of this kind can be developed in great numbers to show the students the fallacies in many arguments currently before the public.

Indirect Proof. When we have been unable to find a direct proof of a proposition, we frequently turn to the much misunderstood method of indirect proof. This method relies on the fact that if $(\sim p)$ is false, then p is true. Hence to prove that p is true, we attempt to show that $(\sim p)$ is false. The best way to accomplish this is to show that $(\sim p)$ is not consistent with the given propositions. In other words, we add $(\sim p)$ to the list of given propositions and attempt to show that this augmented set of propositions leads to a contradiction. When the contradiction is reached, we know that $(\sim p)$ is not consistent with our given true propositions and hence that it is false. Hence p is true. To illustrate indirect proof let us take our first example under direct proof.

Example. Use indirect proof to establish the conclusion **in:**

Given: p is true

$r \rightarrow (\sim p)$ is true

$(\sim r) \rightarrow s$ is true

Prove: s is true.

Assume that $(\sim s)$ is true.

Then the contrapositive of the third given statement is $(\sim s) \to r$.

The second given statement is $r \to (\sim p)$.

Hence we have a chain of syllogisms which yields the conclusion that $(\sim p)$ is true. However, this contradicts the first given statement, namely that p is true. This completes the indirect proof.

The greatest single pitfall in handling an indirect proof is the first step of taking the negation of the proposition to be proved. Frequently this is an implication of the form $p \to q$. We recall that its negation is $p \wedge (\sim q)$. (See p. 84 and p. 87 Ex. 1.) If this fact is remembered, many of the difficulties of indirect proof will evaporate.

Proof by Using Contrapositive. This method can be used when the desired conclusion is an implication. We have noted that an implication is equivalent to its contrapositive, and so we should try to prove the one of these which is easier. The general methods to be used, however, are the same as those discussed under direct proofs.

The reason for mentioning this as a separate method of proof is that when the desired conclusion is an implication, the details of a contrapositive proof are almost identical with those of an indirect proof. Let us examine two examples:

Example 1. Given: $p \to q$ is true

$q \to (\sim r)$ is true

Conclusion: $r \to (\sim p)$ is true

Contrapositive proof. The contrapositive of the conclusion is $p \to (\sim r)$, but this is an easy consequence of the given implications. Hence the stated conclusion is true.

Indirect Proof. The negation of the conclusion is that $r \wedge p$ is true. Hence we assume that r is true and p is true. The given implications (as in the contrapositive proof) show that $p \to (\sim r)$ is true. Since p is true, we find (by detachment) that r is false (a contradiction).

Example 2. Given the usual theorems on triangles, prove that "If two lines are cut by a transversal so that a pair of alternate interior angles are equal, the lines are parallel."

Contrapositive Proof. The appropriate contrapositive is: "If two lines are not parallel, then the alternate interior angles obtained by cutting these lines by a transversal are not equal." (See Fig. p. 92.)

Let AB and CD be the two lines. If they are not parallel, they meet at some point O (on the right, say). Since angle 1 is then an exterior angle of triangle MNO, angle 1 is greater than angle 2, and hence these two angles are not equal. This completes the proof.

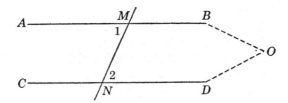

Indirect Proof. The negation of the conclusion is "Angle 1 = angle 2, and AB and CD are not parallel." From the assumption that AB and CD are not parallel we proceed as in the contrapositive proof to show that angle 1 is not equal to angle 2. This, however, is the desired contradiction.

Observe that in both examples the same essential piece of work is performed in both the contrapositive and the indirect proof. We note, however, that the contrapositive method involves fewer statements and that it appears more evident at first sight. For this reason, teachers may prefer to substitute contrapositive proofs for many of the indirect proofs which they find in their textbooks.

We should remark, however, that we cannot dispense entirely with indirect proofs. Some theorems are not stated in the form of implications, and we cannot apply the contrapositive proof to these. A famous example is the theorem of Euclid: "The number of primes is infinite." The usual proof of this is an indirect one in which we assume that the number of primes is a definite finite one, and then proceed to find another prime not in the original list. This produces a contradiction. In cases such as this the conclusion is not an implication, and so we cannot take its contrapositive.

Proof of Existence. For propositions of the form $\exists_x[p]$, we can frequently give a proof by exhibiting an example. For instance, consider the theorem:

Every equation of the form $ax + b = 0$ where $a \neq 0$ has a solution in the system of real numbers.

The proof of this theorem amounts to checking that $x = -(b/a)$ has the required property.

Although there are other forms of existence proofs, a constructive proof of this kind is considered to be of greater merit, and it is used widely in establishing the existence of solutions of various types of equations.

Proof by Enumeration. In certain simple situations we can prove a theorem involving the *all* quantifier by showing that it is true in all possible cases. Let us suppose that we have a number system consisting only of 0 and 1 and that we have assumed the addition table:

+	0	1
0	0	1
1	1	0

Consider the theorem: For all pairs of numbers (equal or unequal), a and b, in our system, the relation $a + b = b + a$ is true. To prove this theorem we have only to check the following cases:

a	b
0	0
0	1
1	0
1	1

Since our relation is verified in each case, the theorem is established. We note that, in fact, we used this method when we proved that certain propositions were tautologies by the method of truth tables.

We should add a word of caution, however, about this method. It only works when the number of cases to be examined is finite (and reasonably small). In most theorems the number of possible cases is infinite, and clearly we cannot test an infinite number of cases. Students will often make the mistake of thinking that they can prove such a theorem by verifying it in a few of these cases. It is therefore very important to stress the limitations of the method of proof by enumeration.

Disproof by Contradiction. If we have tried unsuccessfully to prove a conjectured theorem, we may well spend some time trying to disprove it. One of the usual methods of disproof is to assume that the theorem is true and then to derive consequences from this. If we succeed in arriving at a consequence which contradicts a known true theorem, we have shown that the conjectured theorem is false. This process is so similar to the method of indirect proof that no further remarks need be made about it.

Disproof by Counterexample. When we are considering a conjecture of the form $\forall_x[p]$, we may also consider its negation $\exists_x[\sim p]$. It may be possible to find a specific case in which $\sim p$ is true, and if we can do so we have shown that the negation of our conjecture is true and hence that the conjecture is false.

Example. Disprove: Let x_1 and x_2 be arbitrary odd numbers. Then $x_1 + x_2$ is an odd number. The disproof is immediate from the counter-example: $3 + 5 = 8$.

Again we must add the warning that although disproof by a single example is a valid method of procedure, we can prove theorems by considering particular cases only when the number of these cases is finite so that they can all be tested.

SOME IMPORTANT GEOMETRIC SYSTEMS

In this final section we shall discuss the logical aspects of some important geometric systems and indicate how these topics can be related to the teaching of the deductive method in a geometry course.

Euclidean Geometry. At the beginning of this chapter we hinted that there are some logical slips in the geometric system established by Euclid. Since many critical studies of Euclid's system are available (*e.g.* Felix Klein, *Elementary Mathematics from an Advanced Standpoint— Geometry* p. 188–208), we shall not enter into a systematic critique of his *Elements*, but shall discuss several points which illustrate some of the difficulties which have been encountered.

The first of these concerns the method of superposition. The standard proofs of the early theorems on the congruence of triangles use this method, but there is very considerable doubt that this method can be justified logically on the basis of the axioms of Euclid. Much scholarship has been devoted to a study of Euclid's position regarding motion in his geometry, and the only firm conclusion that can be drawn is that he was not really clear on this point. For this reason, many recent texts have abandoned the method of superposition and have taken the early theorems on congruence as axioms.

Although there is no logical objection to omitting motion from Euclidean geometry in this way, there is another point of view which makes its introduction highly desirable. Euclidean geometry is an abstraction from the world of physical experience in which motion and rigid bodies play an important role. Indeed, the possibility of rigid motion is one of the most important properties of the Euclidean plane and is one which distinguishes the geometry of the plane from that of most surfaces. In order to understand the point involved, let us consider some of these other surfaces. First, let us suppose that we have a rigid piece of material which fits snugly on the surface of a sphere (the commercial spherical protractors are good examples of this). We find that we can move this about on the sphere at will without distorting it in any way. Thus we have experimental proof that rigid motion is possible on the surface of a sphere just as it is on the plane. It is therefore just as reasonable to use the method of superposition on the sphere as it is to use it on the plane. Now consider an ellipsoid of revolution (say, the surface of a football). If again we fit a small piece of material to this surface, we cannot move it at will without some kind of deformation. Although motion without wrinkling is indeed possible along a circle perpendicular to the axis of revolution, motion in any other direction is not possible unless we wrinkle our material to keep it in contact with the surface. This wrinkling, of course, changes the lengths of curves which we may have drawn on our material, and so the motion is no longer rigid. If we try the same experiment on a general ellipsoid (with all three axes unequal), no motions at all are possible. The general ellipsoid, indeed, is typical of most surfaces, and so we see that surfaces which permit rigid motions are of a very special type.

There is a theorem in higher mathematics which states that the only surfaces which permit unrestricted rigid motions are those of constant curvature, namely surfaces like the plane, the sphere, and the pseudosphere. This result is of particular importance when we consider its three-dimensional analogue in connection with the structure of our physical space. Since our senses tell us that three-dimensional rigid motions are freely available, we are led to believe that our three-dimensional space must be of constant curvature, and thus a tremendous number of other possibilities are ruled out. This physical need, therefore, suggests that motion and superposition may be proper ideas to introduce into our treatment of plane geometry, provided that suitable axioms are developed to describe those steps of this kind that are to be allowed.

Another difficulty which often arises is the apparent need of using certain information which we can only derive from our figures. The student is taught that figures are only a crutch to his insight, but that his reasoning is supposed to be independent of the figures which he has drawn. Although this precept is of the greatest importance in the logical development of the subject, almost all textbooks are guilty of slips in which the figure is used in an essential way. It is hard to justify proofs of this kind when at the same time we caution our students about the false reasoning into which a figure can lead us and illustrate this by "proving" such statements as "all triangles are isosceles." (For a sketch of this "proof" see the previously mentioned book of Felix Klein, p. 202.)

As an illustration of how we unconsciously use figures in some proofs, let us consider the usual proof of the theorem: "An exterior angle of a triangle is greater than either remote interior angle."

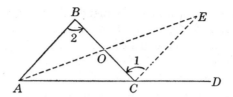

Through O, the midpoint of BC, we draw AOE so that $AO = OE$. Then we show that angle 1 = angle 2. From the figure we argue that angle DCB is greater than angle 1 and hence that angle DCB is greater than angle 2.

The difficulty with this proof is that the inequality of angles DCB and angle 1 must be inferred from the figure. Suppose that we try to establish this by supposing that angle 1 is greater than angle DCB. Then the situation must be as is illustrated in the figure below:

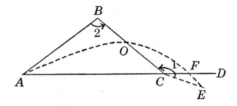

In this situation the line CD subdivides the angle BCE and apparently must intersect BE. But how do we know that this intersection is assured? We must resort to a statement to the effect that if a line subdivides an angle of a triangle it must intersect the opposite side. Obvious statements of this kind cannot be proved from the usual axioms of Euclid, and so we have arrived at an essential difficulty. It was observed by M. Pasch in 1882 that we need to assume as an axiom: "If a straight line intersects one side of a triangle, it also intersects one of the other two sides." The assumption of this axiom gets us over a whole host of difficulties and (though we omit the details) permits us to be sure that CD does intersect BE, say at F. Knowing this, we see that AE and AD have points A and F in common and hence must be the same line. This leads immediately to a contradiction, and our theorem is established.

Notice that in this argument we needed to use the axiom of plane geometry that "There is a unique straight line joining two points." In spherical geometry, however, this axiom does not hold, for two "lines" (great circles) meet in exactly two points. This makes us wonder: "Is our theorem also true in spherical geometry?" The usual proof outlined above would seem to be as valid on a sphere as on a plane, and it is only when we analyze it as we have done that we run into trouble. When we investigate spherical triangles, we find, indeed, that the theorem is false; for a spherical triangle with angles all greater than 90° serves as a counterexample.

There are two morals to this story:

(1) Because of logical difficulties with Euclid's axioms, some of the standard proofs in plane geometry make undue reliance upon the figures. These need further analysis.

(2) We can often learn a good deal about plane geometry by trying to discover whether its theorems hold in other familiar geometries, of which spherical geometry is the most convenient example. There is, in fact, much to be said for the inclusion of certain material from spherical geometry in the tenth-grade geometry course both because of the global nature of our civilization and because of the value of a comparison between plane and spherical geometries.

Finite Geometries. These very simple geometries are excellent illustrations of the deductive method and deserve consideration for inclusion in tenth-grade courses on geometry. We can give an account here of only the simplest of these and must refer the reader to the many available sources for further details. In this treatment we shall follow the approach of Veblen and Young in their *Projective Geometry* p. 1–7.

As undefined concepts we take the following and indicate the concrete notions from which these have been abstracted:

Undefined Concepts	*Concrete Illustrations*
A set of elements A, B, C, . . . belonging to a class S	*A set of points in a plane*
m-class	*A line*
"belonging to an m-class" in the usage: "A belongs to this m-class"	*A point lies on a line*

We then take the following statements as *axioms*:

I. *If A and B are distinct elements of S, there is at least one m-class to which both A and B belong.*

II. *If A and B are distinct elements of S, there is not more than one m-class to which both A and B belong.*

III. *Any two m-classes have at least one element of S in common.*

IV. *There exists at least one m-class.*

V. *Every m-class contains at least three distinct elements of S.*

VI. *All the elements of S do not belong to the same m-class.*

VII. *No m-class contains more than three distinct elements of S.*

Before we consider theorems in this system, let us first check to see that the axioms are consistent. Consider the array:

$$A \ B \ C \ D \ E \ F \ G$$
$$B \ C \ D \ E \ F \ G \ A$$
$$D \ E \ F \ G \ A \ B \ C$$

If the columns in this array represent m-classes, we see that this array satisfies all seven axioms. Hence the axioms are consistent.

Let us also consider some other interpretations of this system:

Element	m-class	Belongs to
point	line	is on
person	committee	is a member of
diameter of a sphere	great-circle of a sphere	the diameter is a diameter of the great circle
member of a lunch club	luncheon	ate at a particular luncheon

Each of these can be a concrete representation of our abstract system if they satisfy all the conditions, and the theorems stated below must then apply to each of them.

As theorems we can easily derive the following:

THEOREM 1. *Any two distinct elements of S determine one and only one m-class containing both these elements.*

THEOREM 2. *Any two m-classes have one and only one element of S in common.*

THEOREM 3. *There exist three elements of S which are not all in the same m-class.*

THEOREM 4. *Any class S satisfying these axioms contains at least seven elements.*

These theorems are not hard to prove, and give good practice in the deductive method. They are recommended to teachers and pupils alike. For further details about these geometries, you should consult some of the following books. You should be warned, however, that various authors start from different points of view, and that you will have to look carefully at the authors' axioms before examining their proofs.

BIBLIOGRAPHY

1. ALLENDOERFER, C. B., and OAKLEY, C. O. *Principles of Mathematics.* New York: McGraw-Hill Book Company, 1955.
2. ARROW, KENNETH. *Social Choice and Individual Values.* New York: John Wiley & Sons, 1951.

3. KLEIN, FELIX. *Elementary Mathematics from an Advanced Standpoint—Geometry*. New York: The Macmillan Company, 1939.

4. KLINE, MORRIS. *Mathematics in Western Culture*. New York: Oxford University Press, 1953.

5. MESERVE, BRUCE E. *Foundations of Geometry*. Cambridge, Mass.: Addison-Wesley Publishing Company, 1955.

6. O'HARA, C. W., and WARD, D. R. *Projective Geometry*. New York: Oxford University Press, 1937.

7. PÓLYA, GEORGE. *Mathematics and Plausible Reasoning*. Princeton, N. J.: Princeton University Press, 1954.

8. RICHARDSON, MOSES. *Fundamentals of Mathematics*. New York: The Macmillan Company, 1949.

9. ROSENBLOOM, PAUL. *Elements of Mathematical Logic*. New York: Dover Publications, 1950.

10. STABLER, E. R. *An Introduction to Mathematical Thought*. Cambridge, Mass.: Addison-Wesley Publishing Company, 1953.

11. TARSKI, ALFRED. *Introduction to Logic*. New York: Oxford University Press, 1946.

12. VEBLEN, O., AND YOUNG, J. W. *Projective Geometry*. Boston: Ginn & Company, 1910.

V

Algebra

SAUNDERS MacLANE

Algebra, regarded as the artful manipulation of sums and products, of radicals and exponents, of unknowns and parameters, is a venerable subject, familiar to generations of students. Less well-known is the modern algebra which has sprung up in the twentieth century, thanks to the creative efforts of German, American, and French mathematicians. This modern algebra still deals with the familiar operations of addition and multiplication, but combines these dealings with the observation that addition and multiplication apply not just to numbers, but to other objects as well. These other objects may be permutations, pairs of numbers, sets, and the like; the algebraic treatment of these objects leads to new algebraic operations as well as to better understanding of old ones.

Because algebra is carried on with new objects, it is important to make the algebraic manipulations secure. This can be done, easily and systematically, by using axioms for algebra in the same way that Euclid used axioms for geometry. The other rules of algebra then become theorems of algebra which can be proved from the axioms just as carefully as theorems of geometry. It even turns out that the proofs of algebraic theorems are neater and easier than those of geometry! Traditionally, high school geometry is said to be the subject where logic can best be learned. Algebra would be a better place!

But this will require some explanation. Usually, algebra is just a mass of rules. We have to so arrange matters that the rules can be proved systematically, once a few of them are taken as the starting point. We begin with the simplest axioms, those for equality, and then turn to those for addition and subtraction.

LAWS OF ALGEBRA

Axioms for Equality. The relation of equality between numbers is known to satisfy the basic and "obvious" laws, true for all numbers

a, b, and c:

$$\text{Reflexive:} \qquad a = a,$$

$$\text{Symmetric: If } a = b, \qquad \text{then } b = a,$$

$$\text{Transitive: If } a = b, \text{ and } b = c, \qquad \text{then } a = c$$

These axioms hold not only for equality between numbers, but also for the relation of equality between areas, between segments, or between any other sorts of mathematical objects. For the equality between numbers we have two more axioms governing the addition of equalities, as follows:

(1) $$\text{If } a = b, \qquad \text{then } a + c = b + c,$$

(2) $$\text{If } a = b, \qquad \text{then } c + a = c + b.$$

In these statements we again mean that these axioms are to hold for all choices of the numbers a, b, and c.

The axiom (1) simply provides an algebraic way of writing the familiar verbal statement: "If equal things ($a = b$) are added to the same thing (c), then the results are equal"—and how much quicker and more explicit the algebraic statement is!

Axioms (1) and (2) are almost alike; in axiom (1) the "same thing" c is added to a and b on the right, in (2) on the left. Later we want to talk about "addition" of things which may not be numbers, and for these things (see p. 131) $c + a$ may not be the same thing as $a + c$. Hence we state both laws (1) and (2) now.

Another familiar version of the rule for adding equalities runs "When equals are added to equals, the results are equal." In algebraic form, this is written:

(3) $$\text{If } a = b \text{ and } c = d, \qquad \text{then } a + c = b + d.$$

This rule, however, does not need to be taken as an additional axiom; it can be proved from the axioms (1) and (2) and from the reflexive, symmetric, and transitive laws. The proof is instructive as a first example of a formal proof. It has the following steps—note that we use both (1) and (2)

(i)	$a = b,$	given,
(ii)	$a + c = b + c,$	(i) and axiom (1),
(iii)	$c = d$	given,
(iv)	$b + c = b + d,$	(iii) and axiom (2),
(v)	$a + c = b + d,$	(ii), (iv), and transitive law.

As this example shows, algebraic proofs can be decked out with statements and reasons arranged in exactly the form often favored in geometry.

Multiplication of equalities can be treated similarly. The axioms are:

(4)   *If a = b,   then ac = bc,*

(5)   *If a = b,   then ca = cb.*

These are exact analogues of (1) and (2), with "+" replaced by "times." Again the analogue of (3) can be deduced from them.

Algebra without Numbers. Addition does not need to hold to the ordinary "addition facts." Though we may not realize it, we all are actually familiar with a system of addition where the facts are different. Sample operations in this system are:

$$8 + 7 = 3, \qquad 5 + 11 = 4, \qquad 6 + 7 = 1.$$

These examples may suffice to suggest the system at issue, but we can give some more sums:

$$3 + 5 = 8, \qquad 8 + 5 = 1, \qquad 11 + 10 = 9.$$

For that matter the corresponding subtractions also hold:

$$8 - 5 = 3, \qquad 1 - 5 = 8, \qquad 9 - 10 = 11.$$

The explanation of these curious and perverted number facts is simple: What we are adding here are not numbers, but hours. For example, the first row could be read: 7 hours after 8 o'clock is 3 o'clock; 11 hours after 5 o'clock is 4 o'clock, and 7 hours after 6 A.M. is 1 P.M. Similarly, $1 - 5$ is 8 because 5 hours prior to 1 A.M. it was 8 P.M.

These manipulations with hours masquerading as numbers are thus very familiar and easy to execute. The principle involved can be described explicitly. The essential point is that 12 hours counts again as zero hours, so that whenever ordinary addition gives an answer bigger than 12, the sum is reduced by subtracting 12. Specifically, the sum of two hours is their ordinary sum as numbers, provided that this sum does not exceed 12, but otherwise is their ordinary sum less 12. The difference between two hours is described in similar fashion: Take the ordinary difference; if it comes out negative, add 12.

This type of arithmetic operation is called addition and subtraction, *modulo* 12. It is clear that similar operations would be possible using some number other than 12 as the modulus. For example, in the Navy and on the European continent the hours are treated modulo 24.

This general idea of getting a new number system by using a suitable modulus is one of the leading constructions of modern algebra, and will be discussed in more detail below. At present we wish only to observe that addition modulo 12 still satisfies some of the same algebraic laws as does ordinary addition. For example the commutative law asserts that $a + b = b + a$. This is true for ordinary addition; hence it is also true for addition modulo 12, for if the sum of hours $a + b$ exceeds 12, both the equal numerical sums $a + b$ and $b + a$ will be reduced by the same amount in addition modulo 12, hence will give the same result, modulo 12.

Axioms for Addition and Subtraction. The laws of algebra are concise statements of experience with arithmetic. One has the facts:

$$5 + 2 = 7 \qquad 8 + 3 = 11 \qquad 6 + 9 = 15$$

$$2 + 5 = 7 \qquad 3 + 8 = 11 \qquad 9 + 6 = 15$$

The pupil may sometimes be told (thanks to a tragically mistaken pedagogical principle) that these are six different number facts—but eventually he discovers that there are really only three different number facts here. He finds that $5 + 2$ is the same as $2 + 5$, and that this observation does not depend on 5 and 2 in particular, but would be good for any two numbers in their place. He thus can understand the commutative law long before it is put in its algebraic form

$$a + b = b + a.$$

The other laws of algebra are similarly grounded in experience.

We can now systematically make a list of all the axioms for addition and subtraction. In formulating these axioms it is handy to use the special number zero, which has the useful property that the addition of zero changes nothing (see (9) below). It is also convenient to express subtraction in terms of the "negative," $- a$ in the form

(6) $$b - a = b + (-a)$$

We call $-a$ the *additive inverse* of a, because $-a$ is a number which on addition to a gives zero. The axioms are

(7) $\qquad a + b = b + a,$ $\qquad$ (*commutative law*),

(8) $\quad a + (b + c) = (a + b) + c,$ $\qquad$ (*associative law*),

(9) $\qquad a + 0 = a,$ $\qquad$ (*law for zero*),

(10) $\quad a + (-a) = 0,$ $\qquad$ (*law for additive inverse*).

They are to hold for all a, b, c.

These laws suffice to give all the usual properties of addition and subtraction. For example, we can prove

(11)   If $b + a = c + a$,   then $b = c$.

The idea is: subtract a from each side. The formal proof parallels this idea by adding $-a$ to each side with the following steps.

(i)	$b + a = c + a$,	given,
(ii)	$(b + a) + (-a) = (c + a) + (-a)$,	(i), axiom (1),
(iii)	$b + (a + (-a)) = c + (a + (-a))$,	(ii), associative law,
(iv)	$a + (-a) = 0$,	law for additive inverse,
(v)	$c + (a + (-a)) = c + 0$,	(iv), axiom (2),
(vi)	$c + 0 = c$,	law for zero,
(vii)	$b + (a + (-a)) = c$,	(iii), (v), (vi), transitive law.

A similar replacement of $a + (-a)$ by 0 will now give $b = c$, as desired.

The formal proof can be shortened by leaving out details of the use of the laws for equality (in fact we already omitted some of these details in going from (ii) to (iii) above). A similar proof will give

(12)   If $a + b = a + c$,   then $b = c$.

Alternatively, this may be proved by applying the commutative law to both sides of (11).

The axiom (10) describes the additive inverse, $-a$, as a number which gives zero when added to a. We can also establish the following very familiar property:

(13) $x + a = b$ if and only if $x = b + (-a)$,

which says in others words that the equation $x + a = b$ has one and only one solution for x; namely $b + (-a)$.

Proof. First $b + (-a)$ is a solution, for

$[b + (-a)] + a = b + [(-a) + a]$,	associative law,
$= b + [a + (-a)]$,	commutative law and (2),
$= b + 0$	law for additive inverse,
$= b$	law for zero.

Second, $b + (-a)$ is the only solution, for if x is any solution, this means that $x + a = [b + (-a)] + a$ and hence by (11) that $x = b + (-a)$.

Since we have defined subtraction in (6) as $b - a = b + (-a)$, we can summarize (13) by the statement that $x + a = b$ has the difference $b - a$ as its unique solution for x. In other words, subtraction can be defined as the solution of this equation.

As a particular case, for $b = 0$, we have

(14) $\qquad\qquad a + x = 0 \quad \textit{if and only if} \quad x = -a.$

(The "if" part here is exactly our law (10).) In other words, the additive inverse is also defined as the solution of an equation. This is why some formulations of the axioms replace our law for the additive inverse by the simple requirement that for each number a there exists a solution of the equation $a + x = 0$.

We can also derive the law that "negatives of equals are equal":

(15) $\qquad\qquad \textit{If } a = b, \quad \textit{then } -a = -b$

Proof. If $a = b$, then $a + (-b) = b + (-b)$ by the addition of equals, hence $a + (-b) = 0$ by axiom (10). This states that $-b$ is the solution x of $a + x = 0$, which means by (14) that $-b = -a$.

All the rules for the manipulations of negatives and of differences are consequences of the axioms. Such rules are, for example,

(16) $\qquad\qquad a + (b - c) = (a + b) - c,$

(17) $\qquad\qquad a - (b + c) = (a - b) - c,$

(18) $\qquad\qquad -(-a) = a.$

For example, to prove (17) we apply the associative law and then (13) twice to get

$$b + c + [(a - b) - c] = b + [c + ((a - b) - c)]$$
$$= b + (a - b)$$
$$= a.$$

The result states that $(a - b) - c$ is the solution x of the equation $(b + c) + x = a$. By proposition (13) this solution x must be $a - (b + c)$. This gives (17). The other proofs are left to the reader.

GROUPS

Abelian Groups. We pretended that the proofs we have just given dealt with numbers, their addition and subtraction. This was just pretense, for the proofs really dealt equally well with things of any sort, which could be added and subtracted, provided only that these things satisfied the basic axioms (7), (8), (9), (10). More explicitly, if we have a collection of objects a, b, c, ... which includes an object called 0, if any two of these objects have an object $a + b$ as sum and any one object a has a corresponding object $-a$ as additive inverse, and finally, if these objects, sums, and inverses, satisfy the laws of equality and the four axioms (7)—(10), then these objects also satisfy the other properties (11)—(18), proved in the last section. This is so because the proofs given used only the axioms cited, and not any other property of numbers.

For example, the 12 hours

$$0, 1, 2, 3, \cdots, 9, 10, 11,$$

form a collection of objects which can be added modulo 12 in the fashion already explained; this addition satisfies the commutative and associative laws. The hour zero, when added to any other hour a, does not change a, hence axiom (9) holds for hours. To each hour, we can assign an additive inverse

$$-0 = 0, \quad -1 = 11, \quad -2 = 10,$$
$$-3 = 9, \cdots, -10 = 2, \quad -11 = 1,$$

so that the axiom for additive inverses holds. Hence the algebra of addition and subtraction is just as valid for hours as it was for numbers.

There are many other systems of objects or elements which obey the same axioms for addition and subtraction; so many that the name *abelian group* has been used for these systems. Formally, then, an *abelian group* G is a set of mathematical objects (elements) a, b, c, $\cdots$ such that

(i) *To any two elements a, b, in G there is a sum $a + b$ in G, which satisfies the rules (1) and (2) for equality and the commutative and associative laws (7) and (8)*;

(ii) *There is an element 0 in G such that $a + 0 = a$ for all a in G*;

(iii) *To each element a in G there is an additive inverse $-a$ in G such that axiom (10) holds.*

In other words, in an abelian group G, the symbols $+$, 0, and $-$ must be defined, and the axioms (1), (2), (7), (8), (9), (10) must hold. (Actually (2) could be omitted here. Why?)

For example, the hours form an abelian group Z_{12} with twelve elements, while the whole numbers under addition modulo 6 would form an abelian group Z_6 with only six elements. Other examples of abelian groups are:

$$Z = \text{all integers } 0, 1, -1, 2, -2, \cdots$$

Q = all rational numbers m/n, m and n integers $n \neq 0$

P = all real numbers

Z_n = the integers modulo n, for any fixed positive n.

These are by no means the only examples (find some more!). In particular, what is the smallest group?

An important fact about abelian groups is that there is only one object 0 in the group; in other words, the property $a + 0 = a$ of axiom (9) characterizes 0. We state this as a theorem.

THEOREM 1. *In an abelian group, if e is an element such that $a + e = a$ for every a in the group, then $e = 0$*

Proof. We are given $a + e = a$ for every a. In particular, for $a = 0$ we have

$$0 + e = 0.$$

On the other hand, the commutative law and the law for zero give

$$0 + e = e + 0 = e$$

Combining these equations by the symmetric and transitive laws, $0 = e$, q.e.d.

Multiplicative Groups. In defining an abelian group we wrote "sum" and spoke of "addition." We could have given the same proofs with different symbols; for example, we could have replaced the ordinary sum everywhere by some other symbol, such as $a \oplus b$. We now propose more drastically to replace sum by product, $a + b$ by $a \cdot b$, or ab. This replacement is not out of line, for under it the commutative law for addition becomes the usual commutative law $ab = ba$ for multiplication, and similarly for the associative law.

More exactly, in the definition of a group we make the typographical changes

$$a + b \rightarrow a \cdot b$$

$$0 \rightarrow 1$$

$$-a \rightarrow a^{-1}$$

We then can say that a (*multiplicative*) *abelian group* G is a set of objects such that

(i) To any two objects a, b, in G, there is in G an object ab, called the *product*, which satisfies the axioms (4) and (5) and the laws

(19) $\qquad\qquad\qquad ab = ba \qquad\qquad\qquad$ *commutative law,*

(20) $\qquad\qquad\qquad a(bc) = (ab)c \qquad\qquad\qquad$ *associative law;*

(ii) *There is an object 1 in G such that, for each a in G*

(21) $a \cdot 1 = a;$

(iii) *To each $a \neq 0$ in G there is an object a^{-1} in G with*

(22) $a \cdot a^{-1} = 1.$

We call 1 the *identity element* of G; as in Theorem 1 there can be only one such element to a group. We call a^{-1} the *multiplicative inverse* of a; we know that each a has only one inverse satisfying (22).

These axioms, (4), (5), (19–22) are all true for the multiplication of ordinary numbers, with the important proviso that (22) is to be applied only to non-zero numbers. Since the product of non-zero numbers is never zero, we may then say that the non-zero real numbers form an abelian group under multiplication.

We now know that all the derived rules for addition which we found on pages 103–105 work as well for multiplication. They do not need to be proved again because the typesetter could be left to do it by the mechanical changes displayed above.

In the place of the subtraction notation, as in (6), we would instruct him to write a fraction

(23) $\dfrac{a}{b} = ab^{-1}$

and rule (13) would become the fact that a/b can be described as the unique solution of $bx = a$, for $b \neq 0$. Furthermore rule (17) becomes the familiar rule

(24) $\dfrac{\dfrac{a}{bc}}{} = \dfrac{\dfrac{a}{b}}{c}$

All the rules for manipulating double-decker fractions and the like can be obtained by similar proofs. An especially simple and basic one is

(25) $(ab)^{-1} = b^{-1}a^{-1}$

Proof.

$$(b^{-1}a^{-1})(ab) = b^{-1}a^{-1}(ab)$$
$$= b^{-1}(a^{-1}a)b = (b^{-1} \cdot 1)b = b^{-1}b = 1.$$

Hence $x = b^{-1}a^{-1}$ is a solution x of the equation $(ab)x = 1$. But rule (14) (translated from addition to multiplication) states that the equation

$(ab)x = 1$, has only one solution, $x = (ab)^{-1}$; hence $b^{-1}a^{-1} = x = (ab)^{-1}$, as asserted in (25).

Another useful rule is the *cancellation law*

(26)   *If $ba = ca$,   then $b = c$* $(a \neq 0)$.

This is a direct translation of (11) from addition to multiplication; it is useful to note explicitly that $a \neq 0$ because we cannot make 0 a part of a multiplicative group—and, for that matter, (26) would be false for $a = 0$.

The non-zero (real) numbers form, as we have already remarked, an abelian group under multiplication. This is not the only multiplicative group of numbers. For example, if we consider just the positive real numbers, we observe that the product of two positive numbers is positive, and that the identity 1 is positive, while a^{-1} is positive whenever a is. The commutative and associative laws hold automatically for the positive numbers, because they hold for all numbers. Thus the positive real numbers form an abelian group under multiplication.

The principle involved is that suitable parts of groups will themselves be groups.

DEFINITION 1. A non-empty subset S of a multiplicative group G (example: S positive reals, G reals) is called a *subgroup* of G if

(i) Whenever s and t are in S, so is the product st,

(ii) Whenever s is in S, the s^{-1} is also in S.

THEOREM 2. *Any subgroup of a group is a group.*

Proof. We must show that S is itself a group. The commutative and associative laws hold automatically for the elements of S, since they hold for all elements of G. Furthermore, S is non-empty, hence has at least one element s. By (ii), s^{-1} is also there, by (i) so is $ss^{-1} = 1$. Therefore the axioms (21) and (22) hold in S, so that S satisfies all the axioms for a group.

By picking out subgroups and using other facts we can make a list of multiplicative abelian groups, such as

(i) All non-zero complex numbers,

(ii) All non-zero real numbers,

(iii) All non-zero rational numbers,

(iv) All positive real numbers,

(v) All positive rational numbers.

This list leaves unsettled the question as to whether the objects used to make a multiplicative group need to be numbers.

Question 1. Do the non-zero integers modulo 12, under ordinary multiplication reduced modulo 12, form a group? (Note: Answers to questions are given at the end of this chapter.)

Homomorphisms. Let G and H be abelian groups under multiplication. By definition, a *homomorphism*,

$$f:G \to H \qquad \text{(read: } f \text{ maps } G \text{ into } H)$$

is a function f which assigns to each object a in G a corresponding object $f(a)$ in H in such a way that

(27) $$f(ab) = f(a)f(b)$$

for all objects a and b in G. In other words a homomorphism f must map G into H and must map products in G into products in H, as stated in (27).

Let us examine some cases to see what this definition might mean.

Example 1. Suppose that

G = multiplicative group of all non-zero real numbers,

H = multiplicative group of all positive real numbers.

Let $f(a) = |a|$ be the ordinary absolute value of the real number a, as given by the usual definition

$$|a| = a \text{ if } a \text{ is positive or zero,}$$

$$= -a \text{ if } a \text{ is negative.}$$

Then f is a homomorphism, because (27) in this case is just the familiar property

$$|ab| = |a||b|.$$

Example 2. Suppose that

$G = H$ = any multiplicative abelian group,

and let $f(a)$ be a^{-1}. Then f maps G into H (*i.e.*, into itself) and is a homomorphism because the requirement (27) is just the property

$$(ab)^{-1} = a^{-1}b^{-1}$$

which we proved in (25) above.

Example 3. Suppose that

G = multiplicative group of all positive real numbers,

H = additive group of all real numbers.

A homomorphism $L: G \to H$ must then be a function which satisfies, in place of (27), the equation

(28) $$L(ab) = L(a) + L(b),$$

which states that L maps products (in G) into sums (in H). This is a familiar equation; for the logarithm to the base 10, $L(x) = \log_{10}(x)$, is such a function, and so is the logarithm to the base e (or, for that matter, to any other base). Indeed (28) is the basic property of the logarithm.

The logarithm function, log, also has the important properties

(29) $$\log (1) = 0$$

(30) $$\log (b^{-1}) = - \log b,$$

(31) $$\log (ab^{-1}) = \log a - \log b.$$

The last of these is just the usual rule for applying logarithms to problems in division. These three properties are not new properties of the logarithm, but all follow from the simple basic property (28). We can show that this is so not just for the logarithm, but for any homomorphism L. For example, let us translate property (29) to apply to any homomorphism. Then (29) says that $L:G \to H$ carries the identity to the identity—that is, the identity element 1 of the multiplicative group G into the identity element 0 of the additive group H. The corresponding property for any two (multiplicative) groups G and H is then $f(1_G) = 1_H$, which says that f carries identity to identity, where 1_G designates the identity element of G and 1_H the identity element of H.

We now state the theorem about the properties (29) − (31):

THEOREM 3. *If $f:G \to H$ is a homomorphism of the multiplicative group G into the multiplicative group H, then*

(32) $$f(1_G) = 1_H ,$$

(33) $$f(b^{-1}) = (f(b))^{-1},$$

(34) $$f(ab^{-1}) = f(a)(f(b))^{-1},$$

are valid for all elements a and b in the group G.

Proof. Since 1_G is the identity element of G, $a = a1_G$ for any a in G. Apply the homomorphism f to this equation. By (27) we get

$$f(a) = f(a)f(1_G).$$

On the other hand 1_H is the identity element of H. By (21) this means

$$f(a) = f(a)1_H ,$$

since $f(a)$ is an element of H. By the symmetric and transitive properties of equality, these two equations give

$$f(a)f(1_G) = f(a)1_H .$$

Cancelling $f(a)$ by (26) we get $f(1_G) = 1_H$, as asserted in (32).

To get property (33) take the equation $bb^{-1} = 1_G$, which holds in G, and apply the homomorphism f. By the definition (27) of a homomorphism and by (32) this yields

$$f(b)f(b^{-1}) = f(1_G) = 1_H.$$

On the other hand, the basic property (22) for the inverse $(f(b))^{-1}$ of $f(b)$ in H states that

$$f(b)(f(b))^{-1} = 1_H.$$

Combine these equations and apply the cancellation law. This gives $f(b^{-1}) = (f(b))^{-1}$, as desired for (33).

The third part (34) of our theorem is a combination of (27) and (33):

$$f(ab^{-1}) = f(a)f(b^{-1}), \qquad\qquad \text{by (27),}$$
$$= f(a)(f(b))^{-1}, \qquad\qquad \text{by (33) and (5).}$$

Question 2. Write these proofs for the special case when $f = \log_{10}$ is the logarithm.

Isomorphisms. Let $f: G \to H$ be a homomorphism for abelian groups G and H.

We say that f is *one-one* if it maps any two distinct elements a and b of G into distinct elements $f(a)$ and $f(b)$ of H. In other words we require that $a \neq b$ in G imply $f(a) \neq f(b)$ in H—or, equivalently, that $f(a) = f(b)$ in H imply $a = b$ in G.

We say that f is *onto* H if every element of H is the map of at least one element of G. In other words we require that to each c in H there is at least one element of G with $f(a) = c$.

If f is both one-one and onto H, then to each element c of H there is exactly one element a in G with $f(a) = c$. In other words f is then a one-one correspondence of G to H, as discussed in Chapter III.

An *isomorphism* $f: G \to H$ is by definition a homomorphism which is one-one and onto. When there is such an isomorphism f we say that the two groups G and H are *isomorphic*. Since an isomorphism is a one-one correspondence which maps products into products, this means that the two groups are pretty much alike.

Example 4. If two positive real numbers x and y have the same logarithm, then $x = y$. Thus the function $\log x$ is one-one. If c is any real number, it is the logarithm of some positive number x, by the usual properties of anti-logarithms. Hence the function $\log x$ is onto the reals. Since $\log x$ is also a homomorphism, we can conclude that the logarithm is an isomorphism of the multiplicative group of positive real numbers to the additive group of all real numbers.

Example 5. Suppose that

$$G = \text{additive group of all integers}$$

$$H = \text{multiplicative group of all integral powers of 2.}$$

Then H consists of all the numbers 2^n for n a positive, negative, or zero integer. Since the products or inverses of powers of 2 are again powers of 2, it follows that H is a group; indeed, it is a subgroup of the multiplicative group of all real numbers. Consider now the function

$$(35) \qquad\qquad f(n) = 2^n$$

defined for each integer n in G and taking its values in H. The usual rule $2^{n+m} = 2^n 2^m$ for the multiplication of exponents becomes

$$f(n + m) = f(n)f(m);$$

it states that f maps sums in the first group G into products in the second group H. In other words this function f is a homomorphism of G into H. If $2^n = 2^m$, then $n = m$. This means that f is one-one. By the description of H every element of H is an integral power of 2; this means f is onto H. Therefore f is an isomorphism of the additive group G to the multiplicative group H.

These two particular groups G and H are called *infinite cyclic groups*. Specifically, to say that a group is *infinite* is to say that it has an infinite number of elements, as do both G and H. To say that a (multiplicative) group H is *cyclic* is to say that we can find a single element h in the group in such a way that every element in the group can be obtained from the single element h by repeated powers and inverses. This single element is then called a *generator* of the cyclic group. The group H in Example 5 is generated by the single element 2, because every element of H is either a product $2^n = 2 \cdots 2$ of 2's, or a product $2^{-n} = 2^{-1} \cdots 2^{-1}$ of n inverses of 2's or $1 = 2 \cdot 2^{-1}$. Hence this group H is infinite cyclic. The first group G of our example is generated (under addition and additive inverses) by 1, hence is infinite cyclic. The two infinite cyclic groups G and H are thus isomorphic; indeed, it can be proved (can you?) that any two infinite cyclic groups are isomorphic.

Question 3. Does the group H of Example 2 have any other generator?

Question 4. Find a cyclic group which is not infinite.

Question 5. Are two finite cyclic groups necessarily isomorphic?

Incidentally, Theorem 3 of the previous section applies to the homomorphism (35) above. It proves that the familiar properties

$$2^0 = 1, \qquad 2^{-n} = (2^n)^{-1}, \qquad 2^n/2^m = 2^{n-m}$$

of the exponential are all consequences of the one law $2^{n+m} = 2^n 2^m$ which states that the exponential is a homomorphism.

Example 6. An especially simple multiplicative group G is that consisting of the two numbers $+1$ and -1, with the usual rules for their multiplication

$$(36) \qquad 1 \cdot 1 = 1, \qquad 1(-1) = -1, \qquad (-1)(-1) = 1,$$

and the inverse given by $(-1)^{-1} = -1$. This group is a subgroup of the group of all non-zero real numbers under multiplication.

An additive group H with only two elements is the group of integers under addition modulo 2. In this group there are only the elements 0 and 1, added like numbers but with the answer reduced by subtracting 2 whenever needed. The addition table for these two elements is thus

$$(37) \qquad 0 + 0 = 0, \qquad 0 + 1 = 1, \qquad 1 + 1 = 0.$$

We also have the additive inverse $-1 = 1$, because $1 + 1 = 0$.

There is striking similarity between the tables (36) and (37). The element 1 in (36) is the identity element of G and acts just like the (additive) identity element 0 of (37), while (-1) in (36) acts like 1 in (37). More exactly, if we make replacements

$$1 \to 0 \qquad -1 \to 1 \qquad \cdot \to +$$

in (36) we get exactly (37). In other words the function f with $f(1) = 0$ and $f(-1) = 1$ is an isomorphism of the multiplicative group of (36) to the additive group of (37).

In this example two groups, each with two elements, turn out to be isomorphic; in other words they are not essentially different. It can be shown that any abelian group with exactly two elements is isomorphic to one of these groups and hence to both of them (can you show it?). Similarly, any two abelian groups with three elements are isomorphic. There are, however, two non-isomorphic abelian groups with four elements.

These examples may serve to illustrate how the notion of isomorphism enables us to distinguish the differences and the similarities between various additive (and multiplicative) systems.

RINGS

Systems with both Addition and Multiplication. The axioms for an abelian group deal with one operation (either addition or multiplication) at one time. For the algebra of numbers both these operations are present. Exactly one new axiom is needed to express the joint properties of these two operations. This axiom is the *distributive law*,

(38) $a(b + c) = ab + ac.$

Together with our previous axioms this law gives all the rules for removing parentheses.

For example, if we apply the commutative law of multiplication to both sides of (38) we get another form of the distributive law,

(39) $(b + c)a = ba + ca.$

From these laws we can derive more elaborate ones such as

(40) $(a + b)(c + d) = ac + bc + ad + bd.$

Proof. Start with the left-hand side of (40), apply (38), and then (39) twice to get

$$(a + b)(c + d) = (a + b)c + (a + b)d$$
$$= ac + bc + ad + bd, \qquad \text{q.e.d.}$$

Just as an abelian group was a system of objects, numbers or otherwise, which satisfied the axioms for addition and subtraction, so now we can discuss and name systems of objects, numbers, or otherwise, which satisfy the appropriate axioms for addition, subtraction, and multiplication. Such a system is called a commutative ring.

DEFINITION 2. *A commutative ring R is a set of elements a, b, c with the following operations and properties:*

(i) *Each pair of elements, a and b, in R has a sum $a + b$ in R. This sum satisfies the rules (1) and (2) and the associative and commutative laws (7) and (8);*

(ii) *There is an element 0 in R with $a + 0 = a$ for every a;*

(iii) *To each element b in R there is an element $-b$ in R with $b + (-b) = 0$;*

(iv) *Each pair of elements, a and b, in R has a product ab in R. This product satisfies rules (4) and (5) for equality as well as the associative and commutative laws (19) and (20) for multiplication;*

(v) *The distributive law (38) holds for all a, b, and c in R.*

This definition is longer than it need be. The first three properties assert simply that R is an abelian group under addition. Hence we can say that a commutative ring is a set of elements, closed under operations of addition and multiplication, which satisfy the rules for equality and are such that

(a) Under addition, R is an abelian group;

(b) Multiplication is commutative and associative;

(c) Multiplication is distributive with respect to addition.

Note that we require neither the presence of an identity element 1 for multiplication, nor the presence of multiplicative inverses for non-zero elements.

The axioms assumed for rings are all well-known properties of the algebra of numbers; hence the "numbers" form a ring. We must be more precise, and say which numbers. When we do so, we find that the following systems of numbers are commutative rings:

Z = the ring of all integers $0, \pm 1, \pm 2, \cdots$;

Q = the ring of all rational numbers m/n; m, n integers, $n \neq 0$;

P = the ring of all real numbers;

C = the ring of all complex numbers $a + b\mathrm{i}$, a, b, real.

Each of these rings is actually a *subring* of the ring C of complex numbers. This notion of a subring is just like the notion of a subgroup (see p. 109).

DEFINITION 3. A non-empty subset S of a commutative ring R is called a *subring* of R if

(i) Whenever s and t are in S, so are the sum $s + t$ and the product st;

(ii) Whenever s is in S, so is the additive inverse $-s$.

Stated more briefly the conditions (i) and (ii) require that S be "closed" under sum, product, and additive inverse.

THEOREM 4. *Any subring of a commutative ring is itself a commutative ring.*

The proof, which will be omitted, is exactly like that given for the corresponding Theorem 2 for subgroups. With this theorem and the fact that the complex numbers C form a ring, we can verify easily that the other systems of numbers listed above form rings. For example, the integers Z are contained in C, and sums, products, and negatives of integers are integers. Hence Z is closed under these operations, so is a subring of C, therefore itself is a ring.

We can use this same theorem to construct more examples of rings of numbers. One such is the set of all even integers, $0, \pm 2, \pm 4, \cdots$; this is a ring which contains no identity element for multiplication. Another is the set of all real numbers of the form $m + n\sqrt{2}$, where m and n are integers—for the set is indeed closed under the relevant operations. Can you find other rings like this?

The elements of a ring need not be numbers, since all that is required of a ring is the presence of well-behaved operations of addition, subtraction, and multiplication. For instance, we are familiar with the fact that polynomials can be added, subtracted, and multiplied, and that the usual rules of algebra apply to polynomials. Indeed, to fix the ideas, let us consider polynomials in a letter x with integer coefficients. Such polynomials are, for example,

$$3x^2 - 7x + 2, \qquad x^5 - 1, \qquad x^3 + x^2 + x + 1.$$

To add two polynomials we simply add them term by term. To multiply them, we multiply them term by term, using the laws of exponents, and then combine terms of like degree. This multiplication is associative and commutative—as we know well from experience, and as can be proved in detail by writing down the general formula for the product of two polynomials in either order. The other axioms for a commutative ring can also be verified, and indeed precisely this is done in various texts on modern algebra. In other words each polynomial with integral coefficients can be regarded as a mathematical object; these polynomials are then the elements of a commutative ring. This ring is sometimes denoted by $Z[x]$, to indicate that it consists of polynomials in x with coefficients in the ring Z of integers.

This is not the only polynomial ring. We could also manufacture the ring denoted by $Z[y]$, and consisting of polynomials in a different unknown y—it would be isomorphic to $Z[x]$; we could also consider $P[x]$, where the coefficients are all real numbers, etc.

There are also rings with only a finite number of elements. For example, take three elements u, v, and w, and define their sum and product by the tables

$+$	u	v	w		$\cdot$	u	v	w
u	u	v	w		u	u	u	u
v	v	w	u		v	u	v	w
w	w	u	v		w	u	w	v

We can verify at length that these three elements do indeed satisfy all the axioms for a ring. For example, the commutative law for addition (or for multiplication) is easily checked, because it requires only that the addition (or multiplication) table be symmetric about the diagonal leading from upper left to lower right. The other laws take more trouble to check. For example, the distributive law (38) has 27 cases, since each of the three arguments a, b, and c in (38) can take each of the three values u, v, and w, and the law must be true in all these cases. It is true, but we do not want to carry these cases all out now, because there is a much better way of showing that these tables do give a ring. The way will be described below, but perhaps the reader might discover the better way now by examining the addition table to see if he can guess its origin.

Much more could be said about rings. For example, just as with groups we could define a *ring homomorphism* $h: R \rightarrow S$ of the ring R into the ring S to be a function which assigns to each element a in R a corresponding element $h(a)$ in S in such fashion that

$$h(a + b) = h(a) + h(b), \qquad h(ab) = h(a)h(b)$$

The first condition states that a ring homomorphism preserves the sum; the second, that it preserves the product. For example, the function which assigns to any polynomial $a(x)$ its value $a(0)$ for $x = 0$ is a homomorphism of the ring $Z[x]$ of polynomials into the ring Z of integers. Can you find others?

Consider now the ring C of all complex numbers. The function h which assigns to each complex number $r + is$ its conjugate $h(r + is) = r - is$ has the two familiar properties

$$\overline{(r + is) + (u + iv)} = \overline{r + is} + \overline{u + iv}$$

$$\overline{(r + is)(u + iv)} = \overline{(r + is)}\,\overline{(u + iv)}$$

Comparing with the definition just above, we see that these properties state exactly that the function $h(r + is) = \overline{r + is}$ is a ring homomorphism $h: C \rightarrow C$ of the complex numbers to themselves. Since a homomorphism preserves both sums and products it preserves also more complicated combinations of sums and products. Suppose, in particular, that a polynomial $a_2 x^2 + a_1 x + a_0$, with real coefficients a_2, a_1, a_0, has a complex root $r + is$. This means that

$$a_2(r + is)^2 + a_1(r + is) + a_0 = 0.$$

Apply the homomorphism h to both sides of this equation. On the right side $h(0) = 0$. On the left side one gets the sum of h applied to each of the three terms separately. In each term h leaves the real coefficient a_i fixed and changes $r + is$ of $(r + is)^2$ to the conjugate; in other words

$$a_2(r - is)^2 + a_1(r - is) + a_0 = 0.$$

This states exactly that $r - is$ is a root of our polynomial; so that if $r + is$ is a root, so is its conjugate $r - is$. In other words the well-known theorem that complex roots of a real polynomial occur in conjugate pairs is a direct consequence of the fact that $r + is \rightarrow r - is$ is a homomorphism.

Number Theory. Number theory deals with the special properties of the ring Z of integers. The exact axiomatic treatment of this important ring may be left for more systematic treatises. Here we observe only that we can distinguish among the integers of Z a subset called the positive integers. These positive integers have various special properties, the most important of which is the following principle:

Every non-empty set of positive integers has a least member.

For example, the set of all even positive integers has 2 as the least member; the set of all odd prime numbers has 3 as the least member; the set of all even multiples of 7 has 14 as its least member.

This principle seems innocent, but it is powerful. It is the principle on which the process of mathematical induction may be based. We now illustrate two uses of this principle.

One use is the "division algorithm." If the integer a is divided by the positive integer b we may write the answer as:

$$\frac{a}{b} = q + \frac{r}{b}, \qquad (\text{e.g. } \tfrac{7}{3} = 2 + \tfrac{1}{3}),$$

with quotient q and remainder r; this remainder is either zero or positive and less than b. It is simpler to clear of fractions and write $a = qb + r$. The resulting "division algorithm" is then stated as follows:

THEOREM 5. *If a and b are integers, with b positive, then there are integers q and r satisfying the conditions*

(41) $$a = qb + r, \qquad 0 \leq r < b.$$

Proof. We suppose first that a is positive. The conditions (41) can be simplified. The quantity r can be determined from q by the first equation, rewritten as $r = a - qb$. Putting this value in the inequality of (41), we get

$$0 \leq a - qb < b.$$

This condition really states that the quotient q is the largest integer such that the difference $a - qb$ is not negative. Another way of putting this requirement is to say that $t = q + 1$ is the least positive integer such that $a - tb = a - (q + 1)b$ is negative.

Hence we start the proof by considering the set

$$S = [\text{all positive integers } s \text{ such that } a - sb \text{ is negative}].$$

This set is surely not empty, because (for a and b given) any sufficiently large integer s will be in S. Observe first that we get the same set S by the description

$$S = [\text{all integers such that } a - sb \text{ is negative}].$$

Indeed, since b is given positive and a is assumed positive, a negative integer s (or $s =$ zero) could not possibly make $a - sb$ negative; in other words, when $a - sb$ is negative, s must be positive.

By our principle the set S of positive integers has a least member, call it t. To say that t is the least member of S is to say that t is in S and $t - 1$ is not in S; by the second description of S this means exactly that

$$a - tb < 0, \qquad a - (t - 1)b \geqq 0.$$

Now label $t - 1$ as q. These two inequalities then are

$$a - (q + 1)b < 0 \qquad a - qb \geqq 0.$$

Next label $a - qb$ as r. The inequalities then become

$$r - b < 0 \qquad r \geqq 0;$$

these are exactly the desired conditions $0 \leqq r < b$ on the remainder r.

This completes the proof if a is positive. If a is negative, we can find a sufficiently big positive integer m so that $a' = a + mb$ will be positive. To this a' the previous proof applies, to give q' and r' with $a' = q'b + r'$. Then

$$a = a' - mb = (q' - m)b + r$$

is the desired representation of a, q.e.d.

Question 6. The integers q and r satisfying (41) of Theorem 5 are uniquely determined. This means that if we also have any q' and r' with $a = q'b + r', 0 \leqq r' < b$, then $q = q', r = r'$. Can you prove this?

The division algorithm has many uses. For example, take $b = 10$. Then by the algorithm any positive number a has the form $a = 10q_0 + r_0$ with $0 \leqq r_0 < 10$. (This inequality says exactly that r_0 is a "digit").

We can repeat by again dividing q_0 by 10 to get $q_0 = 10q_1 + r_1$, where r_1 is again a digit. The combined equation for a is

$$a = 10q_0 + r_0 = 10(10q_1 + r_1) + r_0$$
$$= 100q_1 + 10r_1 + r_0.$$

By further continuation we find $a = 10^3 q_2 + 100r_2 + 10r_1 + r_0$, and so on. This process must eventually stop when we reach a stage m at which the next quotient q_{m+1} is zero. We then have

$$a = 10^m r_m + 10^{m-1} r_{m-1} + \cdots + 10^2 r_2 + 10^2 r_1 + r_0,$$

in which each coefficient $r_m, r_{m-1}, \cdots, r_1, r_0$ is a digit. This expression of the positive integer a is in fact the usual representation of a in the scale of 10 with digits r_m, $\cdots$, r_0, ordinarily written as

$$a = r_m r_{m-1} \cdots r_0 .$$

Thus the division algorithm is at the heart of our notation for numbers—either to the base 10 or to any other base.

Another application of our basic principle is the construction of *greatest common divisors*. In particular cases this is familiar; thus 48 and 66 have 6 as the greatest common divisor—6 divides both 48 and 66; any other common divisor of 48 and 66 is a divisor of 6. These greatest common divisors come up in reducing fractions to lowest terms (in grammar school) and in many other connections. The fact that they can always be found is stated in the next theorem.

THEOREM 6. *If a and b are positive integers, then there is a positive integer d with the following three properties*

  (i) *d divides both a and b,*
 (ii) *any common divisor of a and b divides d,*
(iii) *there are integers u and v with d = ua + vb.*

Here we use the *definition*: d divides a if and only if $a = kd$ for some integer k. The property (iii) is a curious one; it does hold in the case of 48 and 66 because we can write their g.c.d. 6 as $6 = 3 \cdot 66 - 4 \cdot 48$. Furthermore, if (iii) is true, then (ii) is certainly true. For suppose that c is a common divisor, so that $a = kc$ and $b = mc$ for some integers k and m. Then $d = ua + rb = ukc + rmc = (uk + rm)c$, so that the common divisor c does divide d.

The proof of the theorem will start with the property (iii). Given a and b, consider the set S of all possible integers which can be written in the form

$$s = xa + yb$$

with arbitrary integers x and y as coefficients. For example, $a = 1 \cdot a + 0 \cdot b$ and $b = 0 \cdot a + 1 \cdot b$ are in this set S, as well as the sum $a + b$, the difference $a - b$, sums $2a + 3b$, etc., etc. Indeed, we might describe the set S as the set obtained by starting with a and b and carrying out all repeated additions and subtractions.

Let S_p be the set of all positive integers in the set S. It is surely not empty, since it contains both a and b. By our basic principle there is a least positive integer d in the set S_p; like all integers in S it can be written as

(42) $$d = ua + vb$$

for some integers u and v as coefficients.

We claim that d divides b. To prove this try to divide b by d; that is, apply the division algorithm to b and d to get a quotient q and a remainder r with

$$(43) \qquad\qquad b = qd + r, \qquad 0 \leqq r < d.$$

Solving for r and using (42), we have

$$r = b - qd = b - q(ua + vb) = (1 - qv)b - (qu)a.$$

This formula asserts that r is one of the elements of the set S. Now r is either zero or positive and less than d. If r is positive, then it is in the set S_p, but is less than the intended least positive element d of S_p. This cannot be. Hence the other alternative holds, and r is zero. By (43) this means that $b = qd$, an equation which states that d does divide b.

An exactly parallel argument will show that d divides a. We now know that d is a common divisor of a and b, as required for part (i) of the theorem. We constructed d in the form $ua + vb$, so that part (iii) is true. As noticed already, (ii) is a consequence of (iii), whence the proof is complete.

The properties of greatest common divisors are the basis of further number theory; in particular of the properties of primes. A *prime number* p is by definition a positive number, not 1, with no positive divisors other than itself and 1. The greatest common divisor of a prime number p and any other positive number is either p or 1. A number which is not prime can be broken up into factors, and these again into factors, and so on, until all the factors are prime. This shows that any positive integer can be written as a product of primes, as in $84 = 7 \cdot 3 \cdot 2^2$. No matter how this factorization is carried out experience shows that a given number always has the same prime factors. This uniqueness of factorization into primes can be proved as a theorem, using greatest common divisors.

Algebra Modulo n. We have already described the addition of integers modulo 12; now we can combine this with multiplication modulo 12. It will turn out that we get a commutative ring. We still get a ring if we replace the modulus 12 by any positive integer n as modulus.

Algebra modulo n deals with the set Z_n of residues Z_n:

$$Z_n = [0, 1, 2, \cdots, n - 2, n - 1].$$

Every integer has a unique residue $r = R_n(a)$ in Z_n. Indeed, this *residue* is defined to be the remainder upon division of a by n. According to the division algorithm this means that we write

$$(44) \qquad\qquad r = R_n(a) \quad \text{when} \quad a = qn + r, \qquad 0 \leqq r < n$$

Every element of Z_n is the residue $R_n(a)$ of at least one integer a. In fact it is the residue of many different integers a. Exactly which ones? The answer is given by

THEOREM 7. *The function $R_n : Z \rightarrow Z_n$ has $R_n(a) = R_n(b)$ if and only if $a - b$ is divisible by n.*

Proof. If $R_n(a) = R_n(b) = r$, then $a = qn + r$ and $b = q'n + r$ with a different q' but the same r. Therefore $a - b = qn - q'n = (q - q')n$. This equation states that $a - b$ is divisible by n.

Conversely, suppose that $a - b$ is divisible by n. Then $b - a = kn$ for some integer k. Hence if $a = qn + r$,

$$b = a + kn = qn + r + kn = (q + k)n + r.$$

This equation states that r is also the residue of b, q.e.d.

The rule we have used for adding two residues modulo n is

$$(45) \qquad R_n(a) + R_n(b) = R_n(a + b).$$

In words: To add the residues of the integers a and b, first add $a + b$ as integers, then take the residue modulo n. This is exactly the rule we adopted on page 102 for adding residues modulo 12. We now adopt the parallel rule for the multiplication of residues

$$(46) \qquad R_n(a)R_n(b) = R_n(ab)$$

in other words, to multiply residues, multiply the numbers and then take the residue. For example, modulo 7

$$5 \cdot 6 = 30 \quad \text{hence} \quad R_n(5) \cdot R_n(6) = 2.$$

(Subtract $4 \cdot 7 = 28$ from 30.)

THEOREM 8. *The system Z_n of integers modulo n forms a commutative ring under the addition and multiplication defined by (45) and (46).*

Proof. We must begin with the first axioms for a ring; the rules for adding and multiplying equalities. We wish to prove rule (1), which reads

If $R_n(a) = R_n(b)$, then $R_n(a) + R_n(c) = R_n(b) + R_n(c)$.

By the definition (45) the conclusion means that $R_n(a + c) = R_n(b + c)$; according to Theorem 7 this is the same as saying that

$$(a + c) - (b + c) = a - b$$

is divisible by n. But the hypothesis $R_n(a) = R_n(b)$ says exactly this. The second rule (2) for addition of equations is proved similarly.

We should also prove the rules for multiplying equalities; of these two rules (4) and (5) it will suffice to consider one:

If $R_n(a) = R_n(b)$, then $R_n(a)R_n(c) = R_n(b)R_n(c)$.

The hypothesis states that $a - b$ is divisible by n, hence that

$$a - b = kn$$

for some integer k. The conclusion states that $R_n(ac) = R_n(bc)$. By Theorem 7 this holds if and only if $ac - bc$ is divisible by n. But

$$ac - bc = (a - b)c = (kn)c = (kc)n$$

is divisible by n.

These rules for equality are the important part of the proof of this theorem. They are important because they dispose of the apparent ambiguity in the definition of our addition and multiplication. According to (45), to add two residues r and s in Z we must first pick two integers a and c with these residues, $R_n(a) = r$, $R_n(c) = s$, then add a and c, and then take the residue. There is an ambiguity in the choice of a, because we might have picked a different integer b with the same residue $R_n(b) = r$. The proofs we have just given show that we get the same residue for the final sum (or product) irrespective of the choice of a or c.

The rest of the axioms for a commutative ring are verified easily; we shall just give some samples. The identity element for addition is the residue $R_n(0)$ because the law for zero holds with this residue:

$$R_n(a) + R_n(0) = R_n(a + 0) = R_n(a).$$

Similarly, the additive inverse is $-R_n(a) = R_n(-a)$. (There cannot be any ambiguity here, because we proved long ago that the additive inverse was unique in a group.) The commutative law of multiplication is easy:

$$R_n(a)R_n(c) = R_n(ac) = R_n(ca) = R_n(c)R_n(a);$$

It comes directly from the commutative law of multiplication, $ac = ca$. The distributive law is equally easy.

Now let us pin down this ring Z_n by exhibiting what it is for $n = 3$. There are only three residues 0, 1, 2; from the definitions (45) and (46) we get the addition and multiplication tables:

+	0	1	2		$\cdot$	0	1	2
0	0	1	2		0	0	0	0
1	1	2	0		1	0	1	2
2	2	0	1		2	0	2	1

Hence we know that these tables give a ring with exactly three elements. We had such a ring before, on page 117, with elements labelled u, v, and w. With the label changes

$$u \to 0, \qquad v \to 1, \qquad w \to 2$$

it is easy to see that the tables given there are exactly like those given above. This gives the promised easy proof that those tables do really determine a ring. In other words, we have here a striking case when a general theorem, like Theorem 8 about Z_n for any n, is easier to prove than a special case, such as that on the ring with the elements u, v, and w—because the general theorem gives more insight.

This discussion has barely started the subject of rings. For example, the equations (45) and (46) used to define $+$ and times in Z_n really state that $R_n: Z \to Z_n$ is a *homomorphism* of the ring Z of integers into the ring Z_n of residues modulo n. In Theorem 7 the relation "$a - b$ is divisible by n" is sometimes written $a \equiv b \pmod{n}$ and is read "a is congruent to b modulo n." Another way of summarizing our discussion is to say that this congruence behaves just like equality.

Question 7. What are the addition and multiplication tables for a ring with four elements?

Fields. The axioms for a commutative ring covered the rules for addition, subtraction, and multiplication. If we adjoin rules for division, we get a system called a *field*.

DEFINITION 4. A *field* F is a commutative ring in which the non-zero elements form an abelian group under multiplication.

This covers the ground, but we can also spell out what it means. A field F is closed under four operations:

$$a + b, \qquad ab, \qquad -a, \qquad a^{-1} \quad \text{(the latter only for } a \neq 0\text{)}.$$

There are two special elements, 0 and 1. The rules for equality must hold. The rest of the axioms are

$$a + b = b + a \qquad\qquad ab = ba$$
$$a + (b + c) = (a + b) + c \qquad a(bc) = (ab)c,$$
$$a + 0 = a \qquad\qquad a \cdot 1 = a,$$
$$a + (-a) = 0 \qquad\qquad a \cdot a^{-1} = 1 \qquad (a \neq 0),$$
$$a(b + c) = ab + ac.$$

The beauty of this list of postulates is that all the rules for rational operations in algebra follow from these laws.

To illustrate this, let us prove the familiar but curious rule $(-1)(-1) = 1$. We must first prove that $a \cdot 0 = 0$. Indeed, $0 + 0 = 0$, hence multiplying by a and using the distributive law $a \cdot 0 + a \cdot 0 = a \cdot 0$. Subtracting $a \cdot 0$ from both sides gives $a \cdot 0 = 0$. Next, by the law for additive inverses, $1 + (-1) = 0$. Multiply this equation once on the right by 1 and once on the left by -1, and use the distributive law. The two results are

$$1 \cdot 1 + (-1) \cdot 1 = 0 \cdot 1 = 0, \qquad (-1) \cdot 1 + (-1)(-1) = (-1)0 = 0.$$

Therefore, by the rules for equality,

$$1 \cdot 1 + (-1) \cdot 1 = (-1) \cdot 1 + (-1)(-1).$$

Subtracting $(-1)1$ from both sides—or, more exactly, adding $-(-1)1$ to both sides—we get

$$1 \cdot 1 = (-1)(-1).$$

Since $1 \cdot 1 = 1$, this is the desired rule.

Question 8. Prove that $(-a)(-b) = ab$.

There are many different fields of numbers, such as

$$Q = \text{the field of all rational numbers}$$

$$P = \text{the field of all real numbers}$$

$$C = \text{the field of all complex numbers.}$$

Note that C is both a ring and a field. However, Z, the system of all integers, is a ring but not a field—not a field because the integer 2 has no multiplicative inverse in Z, so that the non-zero elements do not form a group under multiplication.

As with subrings, we may also consider *subfields*, with the definition that a *subfield* S of a field F is the subset of F closed under the four rational operations of addition, subtraction, multiplication, and division (except by 0). For example, the set of all real numbers of the form $a + b\sqrt{2}$ with a and b rational is a subfield of the field P of real numbers. The essential part of the proof that this is a subfield is the ordinary "rationalization of the denominator,"

$$\frac{1}{a + b\sqrt{2}} = \frac{a - b\sqrt{2}}{(a + b\sqrt{2})(a - b\sqrt{2})} = \frac{a - b\sqrt{2}}{a^2 - 2b^2}$$

$$= \frac{a}{a - 2b^2} - \frac{b}{a^2 - 2b^2}\sqrt{2}$$

This calculation is precisely a proof that any $a + b\sqrt{2}$ has an inverse of the same form (question: why is $a^2 - 2b^2 \neq 0$?).

There are many such fields of numbers. However, a field need not have numbers as its elements. A simple example is the ring Z_3 of residues modulo 3. This is a field because the two non-zero elements 1 and 2 form a group under multiplication, 1 is the identity, and 2 has inverse $2^{-1} = 2$. For that matter the ring Z_2 of integers modulo 2 is also a field. However, the ring Z_n of residues modulo n is not always a field. For example, if $n = 4$, Z_4 is not a field, because in Z_4 we have the multiplication rule $2 \cdot 2 = 0$. If 2 had an inverse we would get $2 \cdot 2 \cdot 2^{-1} = 0 \cdot 2^{-1}$, hence $2 = 0$, a contradiction.

Question 9. If n is a prime number, show that Z_n must be a field (this is harder).

Question 10. If n is not a prime number, show that Z_n cannot be a field.

There are, in fact, many finite fields. One of the most beautiful theorems of algebra states that the number of elements in any finite field is always a power of some prime number, p. Conversely, for each power p^k of a prime number p, there is a field with exactly p^k elements and any two fields with p^k elements are necessarily isomorphic.

APPLICATIONS TO GEOMETRY

Transformations. It is hard to keep geometry and algebra apart, because geometric objects continually crop up and fulfill the laws of algebra. One simple example is the algebra of vectors. Vectors are originally geometric objects, but under addition they form an abelian group, as explained in Chapter VI. Another example of algebra-in-geometry is the algebra of transformations which we now consider.

DEFINITION 5. A *transformation* $T : X \rightarrow X$ of the set X is a function which assigns to each element a in X a unique element $T(a)$ in X.

For example, the logarithm is a transformation of the set P of real numbers. Indeed, the definition simply states that a transformation on X is the same thing as a function on X. In the definition, the statement that $T(a)$ is unique has the usual meaning:

$$\text{If } a = b, \quad \text{then} \quad T(a) = T(b).$$

The best examples of transformations are geometric, for instance, when X is the set of points in the plane. A rotation R of the plane through 60° about some fixed point is a transformation of X; it carries each point p into the point $R(p)$ rotated 60° from p. Similarly, a translation of the plane up three inches is a transformation T which carries each point p of X into the point $T(p)$ three inches above p. A reflection of the plane in a line is also a transformation of the set X of points of the plane.

There are also non-geometric examples. If G is an abelian group under multiplication, then the function $T(a) = a^{-1}$ carrying each element of G into its multiplicative inverse is a transformation $T:G \to G$. This particular transformation, applied twice, is the identity, since

$$T(T(a)) = (a^{-1})^{-1} = a.$$

We therefore say that this transformation T is its own inverse. More generally we can define the inverse S of any transformation T by stating that S undoes what T does, and vice versa:

DEFINITION 6. A transformation $S:X \to X$ is said to be an *inverse* of a transformation $T:X \to X$, if for every a in X we have

(47) $T(S(a)) = a, \qquad S(T(a)) = a.$

A transformation T is said to be *invertible* if it has an inverse.

All the examples of transformations which we have given above (except for the logarithm) are actually invertible transformations. There is, in fact, a general theorem which says that a transformation is invertible if it is one-one (condition (i) below) and onto (condition (ii)).

THEOREM 9. *A transformation* $T:X \to X$ *is invertible if and only if* T *satisfies the two conditions*

  (i) *If* $T(a) = T(b)$, *then* $a = b$,
  (ii) *Each element* c *in* X *has the form* $c = T(a)$ *for some* a.
When T *satisfies these conditions it has exactly one inverse transformation* S.

Proof. Suppose first that T satisfies conditions (i) and (ii). Then we can define a transformation S for any element c in X by specifying that $S(c)$ is an element a with $T(a) = c$. Condition (ii) assures us that there is such an element, condition (i) assures us that there is only one such element. In symbols, the definition of S reads

(48) $S(c) = a$ if and only if $c = T(a)$.

We now show that S has the properties required in (47) for an inverse. With elements c and a chosen, as in (48), $T(S(c)) = T(a) = c$, $S(T(a)) = S(c) = a$. Hence (47) indeed holds.

Conversely, suppose that T has an inverse S as in the definition (47). We are to prove conditions (i) and (ii). As for (i), if $T(a) = T(b)$, then $S(T(a)) = S(T(b))$. Apply the second part of (47) to both sides; then $a = b$. As for (ii), the first part of (47) asserts that $a = T(S(a))$, hence that $a = T(c)$, for $c = S(a)$.

Finally, suppose that T had two inverses S and S'. The first equations of (47) applied to both these inverses would give

$$T(S(a)) = a = T(S'(a)).$$

Apply S to both sides; then $S(T(S(a))) = S(T(S'(a)))$. But, by the second equation of (47), $S(T(b)) = b$ for any b, be it $b = S(a)$ or $b = S'(a)$. Hence

(49) $S(a) = S'(a)$ for any a in X.

This states that S and S' have the same effect on every element a in X, and hence that the transformations S and S' are equal. Indeed, the *equality $S = S'$ of two transformations* S and S' on the set X is to be defined by the statement (49).

Since we now know that an invertible transformation T has a unique inverse S, it is more natural to write the inverse as $S = T^{-1}$. Its definition, as given in (48), now reads:

(48') $T^{-1}(c) = a$ if and only if $c = T^{-1}(a)$.

Example 7. We first give a number of examples of transformations of the set Z of integers:

(i) The process of adding a fixed integer to each integer z gives a transformation

$$T_k(z) = z + k.$$

Thus, for instance, T_3 carries 0 to 3, 1 to 4, 2 to 5, and so on. The transformation T_k is invertible; in fact its inverse is the transformation obtained by subtracting k. In symbols this means that $(T_k)^{-1} = T_{(-k)}$.

(ii) The process of multiplication by 2 is transformation on Z,

$$M(z) = 2z.$$

It is one-one because if $2z = 2z'$ then $z = z'$. However, it is not onto Z, because not every integer is even (*i.e.*, not every integer can be written in the form $2z$). Hence condition (ii) in Theorem 9 fails, and M has no inverse.

(iii) Another transformation N of the set Z is defined by

$$N(z) = z/2, \qquad \text{if } z \text{ is even,}$$

$$= z \qquad \text{if } z \text{ is odd.}$$

Clearly it has the property that $N(M(z)) = z$. This shows that every integer z is $N(z)$ for some x, namely $x = M(z)$, hence N transforms Z onto Z. However, both $N(6) = 3$ and $N(3) = 3$, so that N is not one-one. Condition (i) of Theorem 9 fails, so that N is not invertible.

Since $N(M(z)) = z$, we may say that N is a "left" inverse of M, and M a "right" inverse of N. However, it can be shown that N has no left inverse, and that M has no right inverse.

Example 8. To have some geometric examples of transformations on a set X, let X be the set of points on the line. By using a zero point as origin and a unit of measure on the line we can also regard X as the set of all real numbers. Some transformations of X are as follows:

(i) Translation of the line by k units to the right is the transformation T_k given for each real number x as

$$(50) \qquad\qquad T_k(x) = x + k.$$

This transformation is invertible, with inverse T_{-k}, for translation by k units and then back by k units (or vice versa) gives no final change in any point.

(ii) The transformation $F(x) = |x|$, the absolute value of x, folds the line over into the positive part (see the discussion of absolute values in Example 1, p. 110). This transformation is not invertible because it is not one-one. Thus, for example $|-2| = |2|$, so that condition (i) of Theorem 9 fails. For that matter, condition (ii) of the Theorem also fails.

(iii) The transformation

$$(51) \qquad\qquad R(x) = -x$$

reflects the line on the origin—right half to left half, and vice versa. Since $R(R(x)) = -(-x) = x$, this transformation is its own inverse.

(iv) The transformation

$$S_3(x) = 3x$$

leaves the origin fixed, but moves each other point three times as far away from the origin as it was. This transformation is clearly invertible; its inverse is the transformation $S_{1/3}(x) = (1/3)x$ which compresses each point x toward the origin by a factor of 3.

(v) The transformation S_{-3}, defined by $S_{-3}(x) = -3x$, moves each point three times as far from the origin and reflects the point in the origin. In other words this transformation S_{-3} can be had by first applying the transformation S_3, then applying a reflection:

$$(52) \qquad\qquad S_{-3}(x) = R(S_3(x)).$$

(vi) For any constant a we have a transformation S_a defined by

$$(53) \qquad\qquad S_a(x) = ax;$$

it multiplies the distance of points from the origin by a. This definition includes S_3 for $a = 3$, R for $a = -1$, and so on. If $a > 1$, S_a is an expansion; if a is positive but less than 1, S_a is a compression toward the origin. If $a \neq 0$, S_a is invertible.

(vii) For any constants a and k we have a transformation $L_{a,k}$ given by

$$(54) \qquad\qquad L_{a,k}(x) = ax + k, \qquad\qquad a \neq 0.$$

Geometrically, this transformation means first expand by the factor a, then translate by the factor k. In other words, in the notation of (50) and (53)

$$(55) \qquad\qquad L_{a,k}(x) = T_k S_a(x).$$

Notice that we first expand, then translate. If we do the operations in the opposite order, we get

$$S_a T_k(x) = S_a(x + k) = a(x + k) = ax + ak,$$

so that $S_a T_k$ in that order has the same effect as $L_{a,\,ak}$. In other words,

$$(56) \qquad\qquad S_a T_k(x) = T_{ak} S_a(x).$$

Composition of Transformations. We have had occasion to apply one transformation after another. This is a natural process which leads to a new transformation called the *composite*.

DEFINITION 7. If $S: X \to X$ and $T: X \to X$ are two transformations on the same set X, the *composite* transformation $S \circ T$ is the transformation obtained by first applying T, then S. In symbols

$$(57) \qquad\qquad (S \circ T)(x) = S(T(x)).$$

We always write the transformation (like a function) in front of the objects x which it is transforming. This is why the composite $S \circ T$, as written in (57), means *first* apply T, *then* S.

We have already found examples of this process in the last section. Thus (52) means that $S_{-3} = R \circ S_3$, and (55) that $L_{a,k} = T_k \circ S_a$. In (56) we have the fact that

$$S_a \circ T_k = T_{ak} \circ S_a.$$

The order matters; the composition of transformations is not commutative.

Question 11. Compound $S_a \circ S_b$, $L_{3,1} \circ L_{2,2}$.

The commutative law may fail, but otherwise the composition of transformations behaves like multiplication. Let us examine some algebraic properties. First we have

$$(58) \qquad\qquad \text{If } S = S', \quad \text{then} \quad S \circ T = S' \circ T.$$

Indeed $(S \circ T)(x) = S(T(x)) = S'(T(x)) = (S' \circ T)(x)$ for any x; according to our definition for the equality of transformations this means that $S \circ T = S' \circ T$. The other rule for the multiplication of equations,

(59)   If $T = T'$,   then $S \circ T = S \circ T'$

is also valid. Proof?

Let us test the associative law for the operation of composition. Given three transformations R, S, and T on the same set X, the composite $R \circ (S \circ T)$ applied to an element x of X gives

$$[R \circ (S \circ T)(x) = R[(S \circ T)(x)] = R[S(T(x))],$$

while in the other association one has

$$[(R \circ S) \circ T](x) = [R \circ S](T(x)) = R[S(T(x))].$$

The results are equal, hence we have

(60) $R \circ (S \circ T) = (R \circ S) \circ T$

For any set X we have an *identity* transformation $I : X \to X$ which simply leaves every element put:

$$I(x) = x, \qquad\qquad x \text{ in } X.$$

Clearly composition with I changes nothing:

(61) $T \circ I = T = I \circ T.$

The identity transformation is invertible; indeed, it is its own inverse, since $I \circ I = I$.

Finally, the definition (47) of the inverse $S = T^{-1}$ of an invertible transformation T requires precisely that

(62) $T \circ T^{-1} = I = T^{-1} \circ T.$

Groups of Transformations. The algebra of transformations has led us back to multiplicative groups—without the commutative law.

DEFINITION 8. A *group of transformations* on a set X is a non-empty set $G = (R, S, T, \cdots)$ of invertible transformations $T : X \to X$, such that

(i) *If T is in G, so is its inverse T^{-1},*

(ii) *If S and T are in G, so is their composite $S \circ T$.*

These conditions insure also that the identity transformation I on X must be in G, because $I = T^{-1} \circ T$.

We should like to assert that every group of transformations is an abelian group. We cannot because in an abelian group the composition must be commutative. This indicates that we should enlarge our conception of algebra in general and of groups in particular by allowing a non-commutative multiplication.

DEFINITION 9. A (multiplicative) *group* G is a set of elements a, b, $\cdots$ such that

(i) *To any two elements a, b in G there is a product ab in G which satisfies the rules (4) and (5) for multiplication of equalities;*

(ii) *For a, b, c, in G, $a(bc) = (ab)c$;*

(iii) *There is an element 1 in G such that*

$$(63) \qquad\qquad 1a = a = a1, \qquad\qquad \text{all } a \text{ in } G;$$

(iv) *To each element a in G there is in G an element a^{-1} such that*

$$(64) \qquad\qquad a^{-1}a = 1 = aa^{-1}.$$

These are the axioms of group theory. (Sometimes these axioms are deduced from a weaker set, involving only half of each of (63) and (64); this need not concern us here.) In the previous section we have shown that the composition of transformations when regarded as a "product" satisfies all these axioms; we have established axiom (i) in (58) and (59), (ii) in (60), (iii) in (61) and (iv) in (62). In other words we have proved the following:

THEOREM 10. *A group of transformations is a group.*

This theorem sounds tautological (a group is a group is a group), but it really is not. If we look back at the definition of the phrase "group of transformations" just above, we see that the theorem really says that a set of transformations with the properties (i) and (ii) is a group. Because the theorem is true, the terminology is so chosen that a set of transformations with these two properties is called a *group of transformations.*

The generality of the notion of group is now apparent; it includes both groups of numbers, in which case the group multiplication is the usual product of numbers, and groups of transformations, in which case the group multiplication is the composition of transformations.

Examples of groups of transformations arise from the examples of transformations given on pages 127–31.

Example 8 (continued). Groups on the set X of real numbers.

(i) Translations of the line, T_h and T_k by h and k units respectively, have a composite given for any real number x by

$$(T_h \circ T_k)(x) = T_h(T_k(x)) = T_h(x + k) = x + k + h = T_{h+k}(x).$$

This equation means that

$$(65) \qquad\qquad T_h \circ T_k = T_{h+k}.$$

This formula shows that the set G of all translations of the line is closed under composition. Since the inverse T_h^{-1} is known to be T_{-h}, G is also closed under inverses, hence is a group of transformations. We call it the *translation group of the line*. There is exactly one translation T_h for each real number h, and the formula (65) shows that the one-one correspondence $h \rightarrow T_h$ is a homomorphism. In other words, the additive group of real numbers is isomorphic to the translation group of the line, under the isomorphism $f(h) = T_h$.

(vi) (continued) The transformations $S_a : X \rightarrow X$ with $S_a(x) = ax$ have a composite given for any real number x by

$$(S_a \circ S_b)(x) = S_a(S_b(x)) = S_a(bx) = (ab)x = S_{ab}(x);$$

hence $S_a \circ S_b = S_{ab}$. This formula also shows that $S_a \circ S_{a^{-1}} = S_1 = I$, if $a \neq 0$. Hence the set H of all these transformations S_a with $a \neq 0$ is a group of transformations. The formulas above show that the correspondence $a \rightarrow S_a$ of numbers a to transformations S_a preserves products and hence is a homomorphism. In other words, the multiplicative group of non-zero real numbers is isomorphic to the group H of all the transformations S_a.

(vii) (continued) The set of all transformations $L_{a,k}(x) = ax + k$ with $a \neq 0$, described as in (54), also is a group of transformations. To see this we calculate the composite of two such transformations $L_{a,k}$ and $L_{b,h}$ as

$$(L_{a,k} \circ L_{b,h})(x) = L_{a,k}(L_{b,h}(x)) = L_{a,k}(bx + h)$$

$$= a(bx + h) + k = (ab)x + (ah + k)$$

$$= L_{ab,ah+k}(x).$$

This gives the composition formula

(66) $L_{a,k} \circ L_{b,h} = L_{ab,ah+k},$ $a, b \neq 0.$

In particular, this formula shows that

$$L_{a,k} \circ L_{a^{-1},-a^{-1}k} = I = L_{a^{-1},-a^{-1}k} \circ L_{a,k}.$$

Hence the inverse of any one transformation $L_{a,k}$ is another such. The set of all transformations $L_{a,k}$, for $a \neq 0$, is therefore a transformation group. It is known as the *affine* group of the line. The essential feature of the affine group is that each transformation L is given by a linear expression $ax + b$. By using similar linear expressions in two coordinates x and y one can construct the affine group of the plane.

The affine group contains as a subgroup the group of all translations

$L_{1,k}$, for $L_{1,k}(x) = x + k = T_k(x)$, hence $L_{1,k} = T_k$. The affine group also contains the compressions S_a as a subgroup. If we neglect the geometric meaning of the affine group and attend only to the formula (66) for the composition of two affine transformations, we see that the transformation $L_{a,k}$ is completely determined by a pair of real numbers (a, k) with $a \neq 0$, with the product of two such pairs defined by a formula

$$(a, k)(b, h) = (ab, ah + k)$$

written to be parallel to (66). This direct numerical formula has the advantage that we can see at once that this product is not commutative, for the product in the other order

$$(b, h)(a, k) = (ba, bk + h)$$

will be equal to the previous product only for very special choices of b, a, h, and k.

Geometric Groups. The affine group of the line is but one example of the various groups of geometric transformations which are possible. Indeed, different systems of geometry can be regarded as the study of the behavior of various groups of transformations, as is set forth later in Chapter IX. As an example, we note that the transformations which belong to ordinary Euclidean geometry are the "rigid motions."

A *rigid motion* M of the plane X is an invertible transformation $M:X \to X$ which leaves fixed the distance between points. This means that the distance between any two points x and y must be equal to the distance between the transformed points Mx and My. As samples of such transformations we mention *translation* in a fixed direction by a fixed amount, *rotation* by a given angle about a fixed point, and *reflection* in a fixed line. In fact it can be shown that any rigid motion can be written as a translation followed by a rotation, perhaps followed by a reflection; in other words as a composite of translations, rotations, and reflections.

The set E of all rigid motions of the plane is a group of transformations. To prove this we must show that E satisfies the two conditions (i) and (ii) in definition 8. As for the first condition, if M is a rigid motion, it leaves distances fixed, hence so does its inverse M^{-1}, so that M^{-1} is also a rigid motion. As for the second condition, suppose that M and N are both rigid motions. Then, for any points x and y, the distance from x to y equals the distance from $N(x)$ to $N(y)$, because N is rigid and in turn equals the distance from $(M \circ N)(x)$ to $(M \circ N)(y)$, because M is rigid. Hence $M \circ N$ is also a rigid motion, and our set E is a transformation group. It is often called the *Euclidean group*. However, it is not the only group which appears in the ordinary study of plane geometry. Similar triangles clearly involve expansions and contractions, so that similarity involves such additional transformations (the group needed is actually the group of affine transformations of the plane).

Symmetry is best studied by groups of transformations. This means that the symmetry of an object can be measured by the set of transformations which carry the object onto itself. The idea appears in quantum mechanics where the applications of group theory depend upon the group of symmetry of a system of atoms. We can illustrate the geometric idea by finding the group of symmetries of a linear ornament, such as the arrowhead pattern imagined as extending to infinity in both directions.

FIG. 1

We define a *symmetry* of this pattern to be a rigid motion of the plane of the pattern which so moves the ornament that it covers itself exactly. Two examples of such symmetries are the reflection R in the horizontal axis of the pattern, and the translation $T = T_1$ to the right by one unit. By compounding T_1 with itself we can get other translational symmetries as "powers" of T

$$T^n = T \cdots T = T_n \qquad (n \text{ factors}).$$

In fact any symmetry of the pattern must have one of the forms T^n or $R \circ T^n$, for n a positive, zero, or negative integer—for any symmetry must carry the arrowhead labelled 0 into some other arrowhead, n units to the right or to the left; thereafter the symmetry can be either nothing more than T_n or a reflection following T_n.

The set of all symmetries of this pattern is a group, because if S and S' are any two symmetries, each must be a rigid motion which carries the ornament to cover itself, so that the composite $S \circ S'$ and the inverse S^{-1} must also be rigid motions which carry the ornament so as to cover itself. We have just seen that this group is generated by two particular symmetries R and T. One can see explicitly how the composites of these symmetries behave:

$$R \circ R = I, \qquad R \circ T = T \circ R.$$

The symmetries R and T thus commute. This means also that R commutes with any other power of T; since any two powers of T commute, the group of symmetries is in fact an abelian group.

Another ornament is

This has a bigger group of symmetries—we again have the symmetries T and R above, but there is an additional symmetry, the reflection V in the vertical axis through 0. This group is not abelian, because $T \circ V$ is not the same as $V \circ T$; for instance $V \circ T$ carries 0 to 1 and thence, by the reflection V, to -1, while $T \circ V$ carries 0 to 0 and thence to 1:

$$(V \circ T)(0) = -1 \neq 1 = (T \circ V)(0).$$

These three symmetries, T, R, and V suffice to generate the whole group of symmetries of this pattern. This may seem mysterious, because they do not automatically include such other symmetries as the reflection V_2 in the vertical axis through the point 2. We claim, however, that

(67) $V_2 = T^2 \circ V \circ T^{-2}$

Indeed V_2 takes the point x into the point $4 - x$ (check that this is reflection in $x = 2$), while the composite indicated in (67) is

$$(T^2 \circ V \circ T^{-2})(x) = (T^2 \circ V)(x - 2) = T^2(-x + 2) = -x + 4,$$

with the same result.

The equation (67) clearly displays the non-abelian character of this group of symmetries.

Question 12. Describe the group of symmetries of the following three ornaments (each imagined as extending to infinity in both directions).

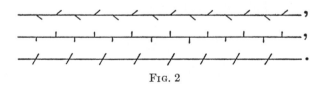

FIG. 2

ABSTRACT GROUPS

Permutation Groups. Let X be any set of objects.

THEOREM 11. *The set G of all invertible transformations $T:X \to X$ is a group of transformations.*

Proof. According to the definition of a group of transformations given on page 132 we must first show that the composite of any two elements of G is in G. Let S and T be invertible transformations, with the inverses S^{-1} and T^{-1}, so that by the definition (62) of an inverse

$$(68) \qquad S \circ S^{-1} = I = S^{-1} \circ S, \qquad T \circ T^{-1} = I = T^{-1} \circ T.$$

Then $(S \circ T) \circ (T^{-1} \circ S^{-1}) = S \circ (T \circ T^{-1}) \circ S^{-1} = S \circ I \circ S^{-1} = I,$
while by a similar argument

$$(T^{-1} \circ S^{-1}) \circ (S \circ T) = I.$$

Hence we have $S \circ T$ invertible, with inverse

$$(69) \qquad\qquad (S \circ T)^{-1} = T^{-1} \circ S^{-1} \qquad\qquad \text{(note the order).}$$

Secondly, we must show that if T is in G, so is T^{-1}. But equation (68) shows that T is the inverse of T^{-1}, so that

$$(70) \qquad\qquad\qquad (T^{-1})^{-1} = T,$$

and T^{-1} is thus invertible, hence is in G.

In particular, when the set X is finite, an invertible transformation T of X is commonly called a *permutation* of the set X. In this case our theorem means that the set of all permutations of a finite set X is a group. This group is called the *symmetric group*. If X has n elements, the symmetric group has exactly $n!$ elements. For, taking any convenient order of the n elements, a permutation can carry the first element into any one of the n elements, the second into any one of the remaining $n - 1$ elements, and so on, giving $n(n - 1) \cdots 2 \cdot 1$ choices for the permutation.

To fix the ideas, consider the symmetric group on the set X consisting of the three letters a, b, and c. A permutation T can now be completely described by giving the three transforms Ta, Tb, and Tc; it is convenient to write this description of T in the form

$$T = \begin{pmatrix} a & b & c \\ Ta & Tb & Tc \end{pmatrix}$$

The transformation T is invertible if and only if it is one-one and onto. This condition means precisely that the three letters Ta, Tb, Tc, in the second row of this symbol must be all distinct, for then they will include all of a, b, c in some order. This lets us list all the possible permutations of a, b, and c, as

$$I = \begin{pmatrix} a & b & c \\ a & b & c \end{pmatrix}, \qquad L = \begin{pmatrix} a & b & c \\ b & c & a \end{pmatrix}, \qquad M = \begin{pmatrix} a & b & c \\ c & a & b \end{pmatrix},$$

$$U = \begin{pmatrix} a & b & c \\ a & c & b \end{pmatrix}, \qquad V = \begin{pmatrix} a & b & c \\ c & b & a \end{pmatrix}, \qquad W = \begin{pmatrix} a & b & c \\ b & a & c \end{pmatrix}.$$

This permutation group thus consists of the six elements, I, L, M, U V, W. There are 6×6 "multiplication facts" for this group.

Let us compute a few. The composite $L \circ L$ is a permutation which carries $a \to b \to c$, $b \to c \to a$, $c \to a \to b$. Hence $L \circ L = M$. Similarly

$$L \circ L \circ L = I, \qquad L \circ U = W, \qquad V \circ L = W, \qquad U \circ U = I.$$

If we write $L \circ L$ as L^2, $L \circ L \circ L$ as L^3, and so on, the elements of this group are

(71) $$I, L, M = L^2, U, V = U \circ L, W = U \circ L^2,$$

or more briefly, I, L, L^2, U, UL, UL^2. Thus the whole group is generated by L and U. Furthermore

(72) $$L^3 = I, \qquad U^2 = I, \qquad L \circ U = U \circ L^2.$$

If we assume that the ordinary rules for exponents work properly, the whole multiplication table (all 36 facts) can be computed from these two lists (71) and (72). This we leave to the reader.

Abstract Groups. The development of transformation groups now calls for further study of the algebraic properties of arbitrary multiplicative groups G—abelian or not. We start with the axioms for a group as given on page 133, and deduce first some properties of inverses.

THEOREM 12. *The inverse a^{-1} is the only solution of either of two equations*

$$ax = 1, \qquad xa = 1.$$

Proof. The axiom (64) says that a^{-1} is a solution of these equations. If x is any solution of the first equation, multiply by a^{-1} on the left to get

$$a^{-1}(ax) = a^{-1}; \qquad (a^{-1}a)x = a^{-1}; \qquad x = a^{-1}.$$

This says that the solution x is indeed a^{-1}. The second equation $xa = 1$ is treated similarly.

THEOREM 13. *The inverse of a product ab is given by*

(73) $$(ab)^{-1} = b^{-1}a^{-1} \qquad \text{(note the order).}$$

We have just seen in (69) that this formula holds in the special case of a group of transformations. To prove it for a general group, calculate

$$(ab)b^{-1}a^{-1} = a(bb^{-1})a^{-1} = a1a^{-1} = aa^{-1} = 1.$$

This shows that $b^{-1}a^{-1}$ is a solution of the equation $(ab)x = 1$. By the previous theorem, the only solution of this equation is $(ab)^{-1}$, q.e.d.

As the examples in the previous section may have indicated, we can define integral powers of an element a in any group, using the standard recursive formulas

(74) $a^0 = 1, \qquad a^{n+1} = a^n a$ (n non-negative)

(75) $a^{-n} = (a^n)^{-1}$ (n negative)

The formula (74) means in effect that $a^{n+1} = a \cdot a \cdots a$ to $n + 1$ factors. We also have the familiar rule for multiplying two powers of a:

THEOREM 14. *For a in a group G and any integers n and m,*

(76) $a^{m+n} = a^n a^m.$

Proof. The proof subdivides into cases, according to the case subdivisions in the definitions (74) and (75) of the exponent.

Case 1. Either n or m is zero. In this case the result is immediate, by virtue of properties of the identity element $a^0 = 1$ of our group.

Case 2. Both n and m are positive. We proceed by mathematical induction on m, the first case $m = 0$ being known. If the equation (76) is true for m, then, for $m + 1$, we compute each side of (76) as

$$a^{n+(m+1)} = a^{(n+m)+1} = a^{n+m}a, \qquad \text{by (74)}$$

$$a^n a^{m+1} = a^n(a^m a) = (a^n a^m)a = a^{n+m}a,$$

by (74) and the induction assumption. Since the results are equal, this completes the induction.

Case 3. Positive n, negative m. Write $k = -m$ and subdivide further as follows:

Case 3a. $n > k$. Then $a^{n+m} = a^{n-k}$, and

$$a^n a^m = a^n a^{-k} = a^n(a^k)^{-1}$$

We can write $a^n = a^{n-k}a^k$ by case 2, since $n - k$ and k are both positive. Then

$$a^n a^m = a^{n-k}a^k(a^k)^{-1} = a^{n-k} = a^{n+m}, \qquad \text{q.e.d.}$$

Case 3b. $n = k$. Here $a^{n+m} = a^{n-k} = a^0 = 1$, while, by the definition (75),

$$a^n a^m = a^n a^{-k} = a^n(a^k)^{-1} = a^n(a^n)^{-1} = 1, \qquad \text{q.e.d.}$$

Case 3c. $n < k$. Then

$$a^n a^m = a^n a^{-k} = a^n(a^k)^{-1} = a^n(a^{k-n}a^n)^{-1},$$

where we have used case 2 to split a^k. This inverse can be computed by

Theorem 13; we get

$$a^n a^m = a^n (a^n)^{-1} (a^{k-n})^{-1} = (a^{k-n})^{-1},$$

By (75), the result is $a^{n-k} = a^{n+m}$, as desired.

Case 4. Positive m, negative n. Similar proof or use commutative laws on case 3.

Case 5. Both m and n negative. Write $n = -t$, $m = -k$. Then

$$a^{n+m} = a^{-t-k} = (a^{t+k})^{-1} = (a^k a^t)^{-1},$$

$$a^n a^m = a^{-t} a^{-k} = (a^t)^{-1} (a^k)^{-1}.$$

The results are equal, again by formula (73).

This proof, though complicated, is really not new—it is exactly the same proof that we could use in elementary algebra to deduce the usual law (76) for exponents.

This rule (76) looks like the definition of a homomorphism (compare Example 5, p. 113). In fact, when we recall the definition of a homomorphism, we see that (76) says exactly that the function $f(n) = a^n$ is a homomorphism $f: Z \to G$ of the additive group Z of integers into the multiplicative group G. Then Theorem 3 will also apply here (changed so that the first group in that theorem is an additive group). It states that $f(-m) = f(m)^{-1}, f(n - m) = f(n)f(m)^{-1}$; in other words, that

(77) $$a^{-m} = (a^m)^{-1}, \qquad a^{n-m} = a^n (a^m)^{-1}.$$

In other words we get these two laws of exponents free of charge from the first law (76) and the general theorem about homomorphisms.

We have repeatedly found group elements $a \neq 1$ with $a^2 = 1$; we say that such an element a has *order* 2. In the symmetric group in three letters in the previous section we had an element $L \neq 1$, with $L^2 \neq 1$, but $L^3 = 1$. We say that the element L has *order* 3.

DEFINITION 10. The *order* of an element b in a group G is the least positive integer m, such that $b^m = 1$.

There may be no such positive integer m; in this case we say that b has *infinite order*. However, if there is any positive integer h with $b^h = 1$, there is a least such positive integer by the principle enunciated on page 119 on the least elements of sets of positive integers.

THEOREM 15. *If the element b has finite order m in the group G, then $b^s = b^t$ for integers s and t if and only if $s - t$ is divisible by m.*

Proof. If $s - t$ is divisible by m, then $s - t = km$ for some integer k, or $s = t + km$. By the laws of exponents

$$b^s = b^{t+km} = b^t b^{km} = b^t (b^m)^k = b^t 1^k = b^t,$$

as desired.

Conversely, suppose that $b^s = b^t$. Multiply both sides by b^{-t}; then $b^{s-t} = 1$. Divide $s - t$ by m, getting $s - t = qm + r$, as in the division algorithm. Then

$$1 = b^{s-t} = b^{qm+r} = (b^m)^q b^r = 1 \cdot b^r = b^r.$$

But the remainder r is so chosen that $0 \leq r < m$. Hence $b^r = 1$; but m was, by its definition as the order of b, the least positive integer with $b^m = 1$. Thus r has no option but to be 0, so that $s - t = qm + 0$, and $s - t$ is divisible by m, as desired.

In this proof we also used the rule $b^{qm} = (b^m)^q$ for exponents; the proof of this we leave to the reader (the proof for b in a group G is again just like the usual proof of the law of exponents).

As we have already indicated, a *cyclic* group G is one which is composed of a single element b and its powers. If the order of b is infinite, then the cyclic group G consists of the infinitely many elements

$$\cdots b^{-2}, b^{-1}, 1, b, b^2, \cdots ;$$

no two are equal. The correspondence $n \rightarrow b^n$ is an isomorphism of the additive group Z of integers to this cyclic group.

If G is generated by a single element b of order m, the theorem we have just proved shows that G consists of exactly m distinct elements

$$1, b, b^2, \cdots, b^{m-1}.$$

These elements multiply according to the rule (76). The correspondence $s \rightarrow b^s$ carries residues modulo m into elements of the group G. The residues of s and t, modulo m, are equal if and only if $s - t$ is divisible by m (Theorem 7). On the other hand $b^s = b^t$ if and only if $s - t$ is divisible by m, by our most recent theorem. Hence the correspondence $s \rightarrow b^s$ is an isomorphism of the additive group of integers modulo m to the arbitrary cyclic group G with generator b of order m. In other words, there is only one cyclic group of order m—all others must be isomorphic copies of it.

This cyclic group of order m is actually familiar from the studies of mth roots of unity. Using complex numbers we can find a primitive complex mth root of unity, say w. We know then that all the complex mth roots of unity are

$$1, w, w^2, w^3, \cdots, w^{m-1}.$$

(Picture them spread out on the unit circle in the complex plane.) The complex mth roots of unity thus form a group under multiplication, and this group is simply a cyclic group of order m with generator a primitive root w. (Try this on the fourth roots of unity, 1, i, -1, $-i$).

We have only skimmed the surface of group theory and algebra. There are many more groups; they may be studied in detail, using homomorphisms as one of the basic tools. There are many other sorts of algebraic systems besides groups, abelian groups, rings, and fields. They go far, but they come from the elementary axioms for ordinary algebraic operations.

Answers to Questions

1. No, 2 has no multiplication inverse, modulo 12, and the same applies to 3, 4, 6, 8, 9, and 10. But the remaining residues 1, 5, 7, 11 do form a group under multiplication. Explanation? Generalization?

2. Refer this to your typographer! It's mechanical.

3. Yes, 2^{-1} is the only other generator.

4. Z_n as described on page 122. Are there any others?

5. No. Z_{12} (the hours) is not isomorphic to Z_{24} (the continental hours).

6. Suppose $a = qb + r = q'b + r'$ with both r and r' between 0 and b. If $r' > r$, then $r' - r = (q - q')b$ is a positive multiple of b less than b; impossible. Likewise $r' < r$. Hence $r = r'$, and then easily $q = q'$.

7. Z_4 is one such ring; another one has 4 elements 0, 1, a, b, and

+	0	1	a	b		·	0	1	a	b
0	0	1	a	b		0	0	0	0	0
1	1	0	b	a		1	0	1	a	b
a	a	b	0	1		a	0	a	b	1
b	b	a	1	0		b	0	b	1	a

The second one is a field. Can you prove it?

8. $a \cdot 0 = a(b + (-b)) = ab + a(-b)$. Hence, by (14), $a(-b) = -ab$. By similar proof $(-a)b = -ab$. Then $ab = -(-ab) = -a(-b) = (-a)(-b)$. Would it be shorter to prove $(-1)a = -a$ first?

9. If r is a residue of the prime n, r and n have no common divisor other than 1, hence by Theorem 6, $1 = ur + vn$. Then $R_n(1) = R_n(u)R_n(r)$, and $R_n(u)$ is the desired inverse of r.

10. If $n = km$ is a factorization of n, $R_n(k)$ can have no inverse, for $R_n(k)R_n(m) = 0$.

11. $S_a \circ S_b = S_{ab}$; $L_{3,1} \circ L_{2,2} = L_{6,7}$.

12. (a) Generated by V, T;   (b) by V, T, and H;   (c) by T.

BIBLIOGRAPHY

1. Birkhoff, Garret, and MacLane, Saunders. *A Survey of Modern Algebra.* Revised edition. New York: The Macmillan Company, 1953.

2. Davenport, Harold. *The Higher Arithmetic.* London: Hutchinson's University Library, 1952.

3. JONES, BURTON W. *The Theory of Numbers*. New York: Rinehart & Company, 1955.

4. ORE, OYSTEIN. *Number Theory and Its History*. New York: McGraw-Hill Book Company, 1948.

5. USPENSKY, J. V., AND HEASLET, M. A. *Elementary Number Theory*. New York: McGraw-Hill Book Company, 1938.

6. WEISS, M. J. *Higher Algebra for the Undergraduate*. New York: John Wiley & Sons, 1949.

VI

Geometric Vector Analysis and the Concept of Vector Space

WALTER PRENOWITZ

VECTOR analysis had its origin in the desire of 19th-century physicists and mathematicians to deal with quantities like forces and velocities whose specification involves spatial orientation, and hence cannot be given by means of a single number. It was found that such physical quantities could be represented geometrically by "arrows" or by directed segments, and that the processes for combining the physical quantities could be reduced to geometrical operations applied to the corresponding geometrical entities. Thus there arose an algebra of geometrical entities, called vectors, which had many close analogies to the familiar algebra of numbers. This *vector algebra* or *vector analysis* not only had important applications to physics and engineering but became a powerful method for the study of geometry, since its laws were general algebraic principles easy to remember and apply, and its operations dealt with geometrical material directly, rather than with algebraic counterparts based on coordinate axes as in cartesian analytic geometry.

When the purely algebraic properties of vectors are abstracted from the geometrical framework in which they arose, the concept of *vector space* is formed. This is a central idea in modern mathematics, having applications in and serving to interrelate algebra, geometry, and analysis. Part of the importance of the vector space concept is due to the fact that it permits the formulation of a general theory of dimensionality, which can be applied to "spaces" of arbitrary finite dimension or even of infinite dimension.

First, we formulate the vector concept geometrically, indicating some of its applications. Then, we introduce three basic operations on vectors, and study their algebraic properties and geometrical significance before, finally, giving an introduction to the theory of vector spaces.

145

THE CONCEPT OF GEOMETRIC VECTOR

Force. Before beginning the formal development of vector analysis let us consider some of the concrete situations which motivate it. These involve entities ("vector quantities") which may be described somewhat crudely as oriented or directed quantities. Consider the important physical notion of force—"push or pull applied to a body." A force is considered specified if we know three basic characteristics: (1) its magnitude, say 10 lbs.; (2) its line of action in space; (3) its sense of action along this line, upward or downward, rightward or leftward. A *directed segment* (A, B), conceived as issuing from point A and terminating at point B, has three corresponding characteristics: (1) its *length* or *magnitude* which is the distance between A and B; (2) its *direction*, which is the direction of line AB or of any line parallel to AB; (3) its *sense* of description, which is from A to B along line AB.[1] Directed segments clearly are well suited for the geometrical representation of forces. Thus we represent a force F by a directed segment (A, B) chosen such that: (1) the *length* of (A, B) is a measure of the magnitude of F; (2) the *direction* of (A, B) is that of the line of action of F; (3) the *sense* of (A, B) is the sense of action of F.

If two forces F_1, F_2 are applied at the same point, it is a physical fact that there exists a third force F_3, whose effect acting alone is equivalent to that of F_1 and F_2 acting together. F_3 is called the *resultant* or *sum* of F_1 and F_2. By a law of physics the resultant of two forces can be characterized geometrically by the Parallelogram Law.

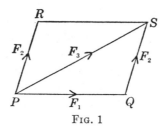

FIG. 1

Parallelogram Law. Let F_1, F_2 be represented respectively by directed segments (P, Q), (P, R); then their resultant F_3 is represented by (P, S) where PQSR is a parallelogram (Fig. 1).

---

[1] "Direction," as commonly used, is ambiguous: (1) parallel or coincident lines are said to have the same direction; (2) a line may be traversed in either of two directions. Here *direction* is used strictly in the meaning of (1); *sense* conveys the meaning of (2). Thus to each direction there correspond two senses.

We can also determine (P, S) by a simple *Triangle Law*. For we can represent F_2 in Figure 1 just as well by directed segment (Q, S), which has the same magnitude, sense and direction as (P, R) but issues from Q, the point where (P, Q) terminates.

Velocity. Another important physical concept which is vectorial in character is velocity. To determine the velocity of a particle we must specify: (1) its speed or the magnitude of the velocity, say 30 mi/hr; (2) its direction; (3) its sense of motion in this direction. Thus velocities are represented by directed segments essentially as forces are. The Parallelogram Law applies to velocities also. For example, suppose particle P moves at the rate of 30 mi/hr in a northeast-southwest line, and in a northerly sense. Then we can associate to P (Fig. 2) *component velocities* V_x, V_y of magnitude $30/\sqrt{2}$ mi/hr in the east-west and north-south directions respectively, which indicate how fast and in which sense P is moving in these directions. These *component velocities* V_x, V_y when "combined" by the parallelogram principle yield as a *resultant* or *sum*, the velocity V of P.

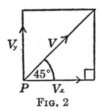

FIG. 2

Location. A geometrical situation which suggests the vector concept is the elementary problem of determining the *position* or *location* of point P *relative to* point O. This can be done by specifying rectangular coordinates of P relative to O as origin, as when we say P is 4 miles east and 3 miles north of O. Polar coordinates suggest another method: Specify the direction of OP, the sense from O to P and the distance from O to P. This is used in gunnery, where a target P is located from an observation post O, by specifying: (1) the distance of P from O; (2) the angular rotation from a due north line ON to the line of sight OP. Clearly the location of P relative to O is determined if we know: (1) the distance from O to P; (2) the direction of line OP; (3) the sense from O to P along line OP. Thus we may consider the location of P relative to O to be determined by the specification of a directed segment (A, B), such that: (1) the length of (A, B) represents distance OP; (2) the direction of (A, B) is that of line OP; (3) the sense of (A, B) is the same as the sense from O to P on line OP. Note that A, the "initial point" of (A, B), is chosen arbitrarily and need not coincide with O. Thus we see that the concept of directed segment is implicit in, and underlies, the practical methods used to determine positions.

Now suppose three points O, P, Q are given. Let us consider the locations of P relative to O and of Q relative to P. Let these locations be determined by directed segments (A, B) and (B, C) respectively (Fig. 3).

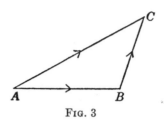

FIG. 3

Then we see that the location of Q relative to O is determined by (A, C), the directed segment which "runs" from A, the "initial point" of (A, B) to C, the "terminal point" of (B, C). Thus to obtain the "location" of Q relative to O we merely "combine" (A, B) and (B, C) by suppressing B to get (A, C). This *Triangle Law* for combining directed segments is related to the Parallelogram Law and is very important.

Summary. Our informal discussion may be summarized as follows: (1) there exist "vector quantities" of several types, both physical and geometrical, which have the essential characteristics of *magnitude*, *direction* and *sense*; (2) consequently they are representable in a natural way by *directed segments*; (3) two vector quantities (of a given type) may be combined to produce a third, by a simple "parallelogram" or "triangle" law, involving the directed segments which represent them.

Directed Segments. Having introduced the notion, directed segment, and shown its utility in dealing with vector quantities, let us examine it more closely. A directed segment (A, B) is merely the line segment AB, conceived as issuing from A, its *initial* point, and terminating at B, its *terminal* point. Consider directed segment (B, A), with initial point B and terminal point A. This has the same length and direction as (A, B), but *opposite* sense. (B, A) is no more to be identified with (A, B) than -1 with $+1$.

We have implicitly assumed that A, B are distinct points, that is $A \neq B$. Situations like zero velocity, or the location of P relative to O when $P = O$, suggest the idea of a directed segment whose initial and terminal points coincide. Thus we extend the concept directed segment, to include the case of a *null* directed segment (A, A), whose initial and terminal points are the same. Naturally we assign to (A, A) length 0, and no direction or sense. The idea of a null directed segment is as important and useful in vector analysis as the number 0 in algebra. It helps us to develop a geometrical theory of vectors, based on general principles which cover all limiting and degenerate cases, without requiring frequent consideration of special cases as in classical geometry.

Equivalent Directed Segments. Glancing back at the representation of a force, a velocity or a "location" by a directed segment (A, B), we recall that (A, B) in each case was specified in terms of its *length*, *direction* and *sense*. But these three characteristics do not completely determine (A, B), since they do not tell us where its initial point is. Thus there are many directed segments which equally well represent a given vector quantity, and all agree in the three basic characteristics of *length, direction, sense*. Two directed segments which do agree in the three basic characteristics, but do not necessarily have the same initial point are called *equivalent directed segments*.

Precisely, (A, B) is *equivalent* to (C, D), written $(A, B) \equiv (C, D)$, if (1) (A, B) and (C, D) have the same length; (2) lines AB, CD have the same direction that is, are parallel or coincide; (3) the senses of description of (A, B), (C, D) along these lines are the same. Since null directed segments have no assigned direction or sense but have the same length, 0, we interpret the definition to mean that any two null directed segments are equivalent; that is, $(A, A) \equiv (B, B)$ for any points A, B.

Observe that in the general case (Fig. 4) when A, B, C, D do not lie on a line, $(A, B) \equiv (C, D)$ has the simple graphical meaning: AB and CD are opposite sides of a parallelogram, namely $ABDC$.

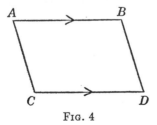

FIG. 4

Properties of Equivalence of Directed Segments. In the idea *directed segment* and the relation of *equivalence* between directed segments, we have the foundation on which rests the whole of geometric vector analysis. Equivalence of directed segments is as important here as congruence of triangles in school geometry. Let us examine its properties. First we observe that equivalence of directed segments satisfies the three basic laws of equality:

Reflexive: $(A, B) \equiv (A, B)$;

Symmetric: If $(A, B) \equiv (C, D)$ then $(C, D) \equiv (A, B)$;

Transitive: If $(A, B) \equiv (C, D)$ and $(C, D) \equiv (E, F)$ then $(A, B) \equiv (E, F)$.

Since an equivalence relation between directed segments carries three items of information (length, direction, sense), we can use equivalence to express certain geometric relations neatly and compactly. Thus we can easily justify:

(i) *M is the midpoint of segment* $\overline{AB}$ (Fig. 5) *if and only if* $(A, M) \equiv (M, B)$.

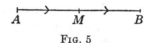

FIG. 5

(ii) *Quadrilateral ABDC* (Fig. 4) *is a parallelogram if and only if* $(A, B) \equiv (C, D)$.

Consider now the important

Alternation Principle for Directed Segments. If $(A, B) \equiv (C, D)$ *then* $(A, C) \equiv (B, D)$. This is easily verified by considering two cases: (a) A, B, C, D (Fig. 6) do not lie on a line (and hence determine a parallelo-

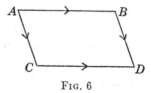

FIG. 6

gram); (b) A, B, C, D (Fig. 7) lie on a line. Do not fail to observe that (b) covers degenerate cases like $A = B$ or $A = C$ or even $A = B = C = D$,

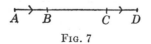

FIG. 7

which must be considered and for which the principle holds. The Alternation Principle enables us to manipulate four points in an equivalence relation abstractly, without recourse to diagrams or the need to study special cases. This is how we manipulate four numbers in a proportion, $a/b = c/d$, to obtain $a/c = b/d$. This foreshadows the success of Geometric Vector Analysis in studying geometry by algebraic operations applied to geometric objects.

Finally we have the fundamental

Representation Principle for Directed Segments. Let (A, B) and point P be given (Fig. 8). *Then there exists one and only one point Q such that*

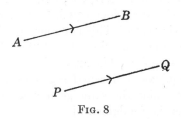

Fig. 8

$(A, B) \equiv (P, Q)$. In effect this says we can uniquely "represent" (A, B) by an equivalent directed segment (P, Q) with the arbitrarily chosen point P as initial point. We "slide" (A, B), so to speak, so that A takes the position of P and obtain the equivalent directed segment (P, Q).

Geometric Vectors. We have seen that equivalent directed segments play the same role in the representation of "vector quantities" by directed segments. Otherwise put, there is no natural way to single out a particular directed segment from a whole class of equivalent directed segments to represent a "vector quantity." Thus the important issue in our study will be the properties *length, direction, sense* of a directed segment—not its initial point. This suggests that we abstract from the idea of directed segment these three properties, divorced from the idea of an initial point.

We are thus led to introduce the notion of a *geometric vector*, which may be described informally as having the three basic characteristics of a directed segment but no specific initial point. Expressed otherwise, a geometric vector is an abstract or "ideal" entity determined by and associated with a directed segment, such that: *Equivalent directed segments determine the same geometric vector* and *inequivalent ones determine distinct geometric vectors.* Thus a geometric vector is an abstract entity uniquely associated with a whole class of equivalent directed segments, each of which determines it, and each of which in a sense is a concrete representative of it. Similarly the cardinal number 2 is an abstraction associated with the class of all couples of objects, and each couple of objects is a concrete instance of the abstract number 2.

Notation. In order to deal successfully with geometric vectors, we must introduce a convenient notation. Certainly we do not want to have to repeat constantly the phrase "geometric vector determined by a directed segment." Thus given directed segment (P, Q), we adopt the "shorthand" notation $\overrightarrow{PQ}$, to stand for the *geometric vector determined by* (P, Q). We may picture $\overrightarrow{PQ}$ as having the *length*, *direction* and *sense* of (P, Q), but no specific initial point.

Conversion Principle. We can now write down many expressions for geometric vectors, for example: $\overrightarrow{AB}$, $\overrightarrow{CD}$, $\overrightarrow{PQ}$, $\cdots$. You may naturally ask: When are two such expressions equal, that is, when do they represent the same geometric vector? This is important since we might have $\overrightarrow{AB} = \overrightarrow{CD}$ even though the expressions do not "look" equal—that is they involve different pairs of points. The answer is contained in the principle stated above, that equivalent directed segments and only equivalent directed segments determine the same geometric vector (p. 151). This we state in symbols as the fundamental

Conversion Principle. $\overrightarrow{AB} = \overrightarrow{CD}$ *if and only if* $(A, B) \equiv (C, D)$. This principle is very important, since it enables us automatically to convert or "translate" any statement about equivalent directed segments into a corresponding statement about equal geometric vectors. For example, "translating" the Alternation Principle for Directed Segments (p. 150), we get the

Alternation Principle (for geometric vectors). *If* $\overrightarrow{AB} = \overrightarrow{CD}$ *then* $\overrightarrow{AC} = \overrightarrow{BD}$.

Similarly from the Representation Principle for Directed Segments (p. 151), we obtain the

Representation Principle (for geometric vectors). *Let* $\overrightarrow{AB}$ *and point P be given. Then there exists one and only one point Q such that* $\overrightarrow{AB} = \overrightarrow{PQ}$. This enables us to represent any geometric vector $\overrightarrow{AB}$ in the notation $\overrightarrow{PQ}$, where P is arbitrarily chosen beforehand. And $\overrightarrow{AB}$ is representable in the form $\overrightarrow{PQ}$ in only one way, since Q is uniquely determined. We call the process of representing $\overrightarrow{AB}$ in the form $\overrightarrow{PQ}$, *localizing* $\overrightarrow{AB}$ at P. Thus we can *localize* our geometric vector at any convenient point, and even consider several different "*localizations*" if necessary.

The Nature of Geometric Vectors. Although we have now set the basis for our treatment of vector analysis, you may not feel wholly satisfied with our explanation of the concept, geometric vector. You may say that directed segments are definite geometric objects—but what can geometric vectors be as abstract entities? We can answer this very explicitly.

To be specific, consider $\overrightarrow{PQ}$. Whatever $\overrightarrow{PQ}$ is, it is determined by (P, Q) or equally well by any directed segment which is equivalent to (P, Q). Moreover any directed segment not equivalent to (P, Q) determines a geometric vector distinct from $\overrightarrow{PQ}$. Thus $\overrightarrow{PQ}$ is uniquely associated with the class of all directed segments equivalent to (P, Q). Clearly the simplest way to get something uniquely associated with this class is to take the class itself. Thus we define formally: *Geometric vector* $\overrightarrow{PQ}$ is the class of all directed segments equivalent to (P, Q). We say (P, Q) *determines* $\overrightarrow{PQ}$. The *magnitude (direction, sense)* of $\overrightarrow{PQ}$ is that of (P, Q) or of any directed segment equivalent to (P, Q).

This definition of geometric vector may seem more complicated than you expected—but observe it has the great advantage of making geometric vectors as definite and concrete as directed segments. In other words our abstraction turns out to be rooted in specific things, which should not disappoint us at all. (A similar type of reasoning led Bertrand Russell to define the cardinal number 2 as the class of all couples.)

If our definition of geometric vector as a class of equivalent directed segments seems too complicated and artificial, you can disregard it in what follows. (You will probably be better prepared to accept this kind of definition after having studied vector analysis.) For the essential connection between directed segments and geometric vectors is given in the Conversion Principle; and you can have in mind any intuitive conception of geometric vector you please, so long as it conforms to the Conversion Principle. In the last analysis everything we say about geometric vectors can be expressed in terms of two ideas: directed segments, and their equivalence.

THE ADDITIVE ALGEBRA OF GEOMETRIC VECTORS

In this part we introduce addition of geometric vectors, the basic operation in vector analysis, and derive its essential properties. They are remarkably similar to those of addition in school algebra. For the present we refer to geometric vectors simply as vectors, since no ambiguity is involved.

Addition of Vectors. For many purposes it is not necessary to represent vectors *geometrically*, as $\overrightarrow{AB}$, $\overrightarrow{CD}$ $\cdots$. Thus we shall often represent vectors abstractly, by small letters: **a**, **b**, **c**, $\cdots$. The two notations are tied together by the Representation Principle: **a** is uniquely representable in the form $\overrightarrow{PQ}$, when P is given.

The Triangle Law for combination of vector quantities suggests

DEFINITION 1. Let **a**, **b** be arbitrary vectors. Express **a** as $\overrightarrow{PQ}$, **b** as $\overrightarrow{QR}$. Then **a** $+$ **b** the *sum* or *resultant* of **a** and **b** is defined to be $\overrightarrow{PR}$

We base our definition on the Triangle Law rather than the Parallelogram Law, because it is simpler and applies to all cases. Definition 1 applies equally well whether or not P, Q, R lie on a line. However there is an element of ambiguity in Definition 1, since the location of P is not specified, and we might conceivably get different results depending on the position of P. This is clarified in

THEOREM 1. (Closure Law.) *The sum of two vectors is a uniquely determined vector.*

Proof. Let $\mathbf{a}$, $\mathbf{b}$ be given. Then by the Representation Principle we can choose P arbitrarily, and determine Q, R such that $\mathbf{a} = \overrightarrow{PQ}$, $\mathbf{b} = \overrightarrow{QR}$ (Fig. 9). Hence there exists at least one "value" of the sum $\mathbf{a} + \mathbf{b}$,

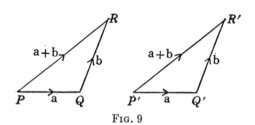

FIG. 9

namely $\overrightarrow{PR}$. Let us choose a new point, say P'. Then by the Representation Principle we can express $\mathbf{a}$ as $\overrightarrow{P'Q'}$, $\mathbf{b}$ as $\overrightarrow{Q'R'}$. Thus we can just as well take $\overrightarrow{P'R'}$ as a "value" of $\mathbf{a} + \mathbf{b}$. Hence to prove $\mathbf{a} + \mathbf{b}$ has a unique value, we must show $\overrightarrow{PR} = \overrightarrow{P'R'}$. This is not hard to do. For we have

$$\overrightarrow{PQ} = \overrightarrow{P'Q'}, \qquad \overrightarrow{QR} = \overrightarrow{Q'R'}.$$

The Alternation Principle applied to each of these, yields

$$\overrightarrow{PP'} = \overrightarrow{QQ'}, \qquad \overrightarrow{QQ'} = \overrightarrow{RR'}.$$

Hence $\overrightarrow{PP'} = \overrightarrow{RR'}$. To this we apply the Alternation Principle, getting $\overrightarrow{PR} = \overrightarrow{P'R'}$, and the proof is complete.

Several observations are in order. First, do not be misled by the seemingly "obvious" nature of the theorem as an algebraic principle. It states an important property of vectors, and has important geometric content. It implies for example, that two triangles lying in the same plane or in parallel planes are congruent provided two pairs of corresponding sides are equal, parallel and similarly sensed.

Second, the proof does not depend on the diagram (Fig. 9). The diagram presents a "typical" case—it illustrates but does not justify the argument. The statements in the proof are justified by general principles and definitions which apply to all cases without restriction. Hence the theorem applies to all cases without restriction. For example P, Q, R may be on a line, or we may have $P = Q$ or $Q = R$. Thus our apparently "trivial" algebraic principle enables us to deal with a variety of geometrical information, abstractly, without requiring us to have a particular geometric picture in mind.

Third, note that in the proof we have informally used familiar properties of equality. The basic properties of equality of vectors are easily justified and will be used when needed without explicit reference:

Reflexive: $\mathbf{a} = \mathbf{a}$;

Symmetric: If $\mathbf{a} = \mathbf{b}$ *then* $\mathbf{b} = \mathbf{a}$;

Transitive: If $\mathbf{a} = \mathbf{b}$ *and* $\mathbf{b} = \mathbf{c}$ *then* $\mathbf{a} = \mathbf{c}$;

Additive: If $\mathbf{a} = \mathbf{b}$ *and* $\mathbf{c} = \mathbf{d}$ *then* $\mathbf{a} + \mathbf{c} = \mathbf{b} + \mathbf{d}$.

The Additive Property merely says: "If equal vectors are added to equal vectors, the sums are equal." For a discussion of the properties of equality in algebra see pages 100–102.)

Suppose now $\mathbf{a}$ is given *geometrically* as $\vec{PQ}$ and $\mathbf{b}$ as $\vec{QR}$. Clearly their sum is $\vec{PR}$. Thus we have the very useful principle

$$(1) \qquad\qquad \vec{PQ} + \vec{QR} = \vec{PR}.$$

Further Properties of Addition of Vectors. The basic properties of addition of numbers also hold for addition of vectors. Thus the sum of three vectors $\mathbf{a}$, $\mathbf{b}$, $\mathbf{c}$ is independent of how they are "associated" in pairs:

THEOREM 2. (Associative Law.) $(\mathbf{a} + \mathbf{b}) + \mathbf{c} = \mathbf{a} + (\mathbf{b} + \mathbf{c})$.

Proof. As we have seen the sum of two vectors is independent of the representation of the first vector in the application of Definition 1. Hence we may choose any convenient representation. Thus in this case we take $\mathbf{a} = \vec{PQ}$, $\mathbf{b} = \vec{QR}$, $\mathbf{c} = \vec{RS}$. Then applying (1) several times we have

$$(\mathbf{a} + \mathbf{b}) + \mathbf{c} = (\vec{PQ} + \vec{QR}) + \vec{RS} = \vec{PR} + \vec{RS} = \vec{PS},$$

$$\mathbf{a} + (\mathbf{b} + \mathbf{c}) = \vec{PQ} + (\vec{QR} + \vec{RS}) = \vec{PQ} + \vec{QS} = \vec{PS}$$

and the conclusion is immediate.

The sum of two vectors does not depend on the order in which they are added:

THEOREM 3. (Commutative Law.) $\mathbf{a} + \mathbf{b} = \mathbf{b} + \mathbf{a}$.

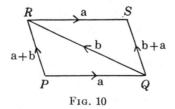

FIG. 10

Proof. Let $\mathbf{a} = \overrightarrow{PQ}$, $\mathbf{b} = \overrightarrow{QR}$ (Fig. 10). Then we have

(2) $$\mathbf{a} + \mathbf{b} = \overrightarrow{PQ} + \overrightarrow{QR} = \overrightarrow{PR}.$$

To obtain $\mathbf{b} + \mathbf{a}$ we represent $\mathbf{a}$ as $\overrightarrow{RS}$. Then we have

(3) $$\mathbf{b} + \mathbf{a} = \overrightarrow{QR} + \overrightarrow{RS} = \overrightarrow{QS}.$$

But $\mathbf{a} = \overrightarrow{PQ} = \overrightarrow{RS}$. Hence by the Alternation Principle $\overrightarrow{PR} = \overrightarrow{QS}$. Thus we have by (2), (3)

$$\mathbf{a} + \mathbf{b} = \overrightarrow{PR} = \overrightarrow{QS} = \mathbf{b} + \mathbf{a}$$

and the proof is complete.

The Null or Zero Vector. Is there a vector which behaves like the number 0? This calls to mind the idea of null segment. All null segments are equivalent, that is $(A, A) \equiv (B, B) \equiv (C, C) \equiv \cdots$. Hence they all determine the same vector, which we call the *null vector* or *zero vector* and denote $\mathbf{0}$. Thus $\mathbf{0} = \overrightarrow{AA} = \overrightarrow{BB} = \overrightarrow{CC} = \cdots$. The magnitude of $\mathbf{0}$ is 0, it has no direction or sense. We show $\mathbf{0}$ does behave like the number 0.

THEOREM 4. (Identity Law.) $\mathbf{a} + \mathbf{0} = \mathbf{a}$ *for each vector* $\mathbf{a}$.

Proof. Let $\mathbf{a} = \overrightarrow{PQ}$. Then we can represent $\mathbf{0}$ as $\overrightarrow{QQ}$. Hence

$$\mathbf{a} + \mathbf{0} = \overrightarrow{PQ} + \overrightarrow{QQ} = \overrightarrow{PQ} = \mathbf{a}.$$

The Negative of a Vector or Opposite of a Vector. Every number is paired with an "opposite number" which when added to it produces 0. For example, the "opposite" of $+5$ is -5, of -5 is $+5$. A similar principle holds for vectors.

THEOREM 5. (Existence of Additive Inverse.) *For each vector* $\mathbf{a}$, *the equation* $\mathbf{a} + \mathbf{x} = \mathbf{0}$ *has one and only one solution* $\mathbf{x}$.

Proof. Let $\mathbf{a} = \overrightarrow{PQ}$. Take $\mathbf{x} = \overrightarrow{QP}$. Then $\mathbf{a} + \mathbf{x} = \overrightarrow{PQ} + \overrightarrow{QP} = \overrightarrow{PP} = \mathbf{0}$. Hence the equation has *at least* one solution, namely $\overrightarrow{QP}$. To show it has only one solution, let $\mathbf{y}$ be any solution of the equation. Then $\mathbf{a} + \mathbf{y} = \mathbf{0}$. Represent $\mathbf{y}$ as $\overrightarrow{QR}$. Thus

$$\mathbf{0} = \mathbf{a} + \mathbf{y} = \overrightarrow{PQ} + \overrightarrow{QR} = \overrightarrow{PR}.$$

But we know $\mathbf{0} = \overrightarrow{PP}$. Thus we have two representations of $\mathbf{0}$, namely $\overrightarrow{PR}$, $\overrightarrow{PP}$. By the uniqueness part of the Representation Principle $R = P$. Then $\mathbf{y} = \overrightarrow{QR} = \overrightarrow{QP} = \mathbf{x}$, and the equation has a unique solution, namely $\overrightarrow{QP}$.

The basic algebraic properties of vector addition[2] have now been established in Theorems 1, $\cdots$, 5. All further algebraic properties in

---

[2] By these properties geometric vectors form an abelian group (see pages 105–109) and the theory of abelian groups is applicable to them.

this section will be derived algebraically from them without recourse to geometric representation of vectors.

Theorem 5 associates with each **a** a definite vector **x**. This new vector deserves a name and a special notation:

DEFINITION 2. The unique solution **x** of **a** + **x** = **0** is called the *opposite* or *negative* of **a** and is denoted −**a**. Thus **a** + (−**a**) = **0**.

As a direct consequence of Theorem 5 and Definition 2, we have

COROLLARY 1. **a** + **x** = **0** *if and only if* **x** = −**a**.

Further, "the opposite of the opposite of a vector is the original vector."

COROLLARY 2. −(−**a**) = **a**.

Proof. **a** + (−**a**) = **0**. Thus (−**a**) + **a** = **0** and Corollary 1 implies **a** = −(−**a**).

In proving Theorem 5 we showed **x** = $\overrightarrow{QP}$. Thus we have

COROLLARY 3. *If* **a** = $\overrightarrow{PQ}$ *then* −**a** = $\overrightarrow{QP}$.

Note the basic properties of −**a**: (1) algebraically, it is the "neutralizer" of **a**, since when added to **a** it yields **0**; (2) geometrically, it is truly the "opposite" of **a**, since it has the same magnitude and direction as **a** but has opposite sense.

Subtraction of Vectors. When applied to vectors the process of subtraction has the same significance as when applied to ordinary numbers.

DEFINITION 3. **a** − **b**, the *difference* of **a** and **b**, denotes the vector **a** + (−**b**). Thus to subtract **b** from **a** we "change" **b** to its "opposite" −**b** and add to **a**. Just as in algebra we get +6 − (−2) = +6 + (+2), by "changing the sign and adding." Note the minus sign as usual is used in two senses: Applied to a single vector it says take the opposite vector; applied to two vectors it says subtract.

Addition is the "check" on subtraction:

THEOREM 6. **x** + **b** = **a** *if and only if* **x** = **a** − **b**.

Proof. Let **x** = **a** − **b**. By Definition 3, the associative and commutative laws, Definition 2, and the "identity" property of **0**,

$$\mathbf{x} + \mathbf{b} = (\mathbf{a} - \mathbf{b}) + \mathbf{b} = (\mathbf{a} + (-\mathbf{b})) + \mathbf{b} = \mathbf{a} + ((-\mathbf{b}) + \mathbf{b})$$

$$= \mathbf{a} + (\mathbf{b} + (-\mathbf{b})) = \mathbf{a} + \mathbf{0} = \mathbf{a}.$$

Thus **x** + **b** = **a**. Conversely, suppose **x** + **b** = **a**. Then **a** = **x** + **b**. Hence

$$\mathbf{a} - \mathbf{b} = \mathbf{a} + (-\mathbf{b}) = (\mathbf{x} + \mathbf{b}) + (-\mathbf{b})$$

$$= \mathbf{x} + (\mathbf{b} + (-\mathbf{b})) = \mathbf{x} + \mathbf{0} = \mathbf{x}.$$

Thus **x** = **a** − **b** and the proof is complete.

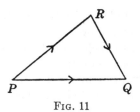

FIG. 11

Theorem 6 says we can "transpose" **b** from one side of the equation to the other as in ordinary algebra.

COROLLARY. $\vec{PQ} - \vec{PR} = \vec{RQ}$.

Proof. This is a geometrical law for subtraction and is derived from relation (1), the geometrical law for addition: $\vec{PR} + \vec{RQ} = \vec{PQ}$ (Fig. 11). Clearly $\vec{RQ} + \vec{PR} = \vec{PQ}$ and $\vec{RQ} = \vec{PQ} - \vec{PR}$ by Theorem 6. Intuitively the result says: To subtract two vectors localized at a point, take the vector "running" from the "terminus" of the second vector to that of the first.

Further Algebraic Properties. It must seem fairly evident by now that the whole apparatus of school algebra, in so far as it concerns addition and subtraction, is applicable to vectors with the null vector **0** playing the role of the number 0. We illustrate this by deriving purely algebraically two familiar principles of school algebra.

THEOREM 7. (Cancellation Law.) *If* **x** + **a** = **y** + **a** *then* **x** = **y**.

Proof. In order to "neutralize" the **a**'s, we add −**a** to both sides in the hypothesis:

$$(\mathbf{x} + \mathbf{a}) + (-\mathbf{a}) = (\mathbf{y} + \mathbf{a}) + (-\mathbf{a}).$$

Rebracketing and applying the properties of −**a** and **0**,

$$\mathbf{x} + (\mathbf{a} + (-\mathbf{a})) = \mathbf{y} + (\mathbf{a} + (-\mathbf{a}))$$

$$\mathbf{x} + \mathbf{0} = \mathbf{y} + \mathbf{0}$$

$$\mathbf{x} = \mathbf{y}.$$

THEOREM 8. −(**a** + **b**) = (−**a**) + (−**b**).

Proof. By definition −(**a** + **b**) is the unique solution **x** of (**a** + **b**) + **x** = **0**. Thus we merely have to solve this equation for **x**. We have **a** + (**b** + **x**) = **0**. Hence **b** + **x** = −**a** by Theorem 5 Corollary 1. Thus **x** + **b** = −**a**. By Theorem 6

$$\mathbf{x} = (-\mathbf{a}) - \mathbf{b} = (-\mathbf{a}) + (-\mathbf{b})$$

and our proof is complete.

Additional algebraic properties are given in the following exercises.

EXERCISES

1. Prove: $-0 = 0; a - a = 0; 0 - a = -a$.
2. Prove: $a - (-b) = a + b$.
3. Prove: $-(a - b) = (-a) + b$.
4. Prove: $a - (b - c) = (a + c) - b$.
5. Prove: $(a + b) + (c + d) = (a + c) + (b + d)$.
6. Prove: $(a + b) - (c + d) = (a - c) + (b - d)$.
7. Prove: $(a - b) - (c - d) = (a + d) - (b + c)$.
8. Prove: $a - b = c - d$ if and only if $a + d = b + c$.

Vectors with the Same Direction. In many geometrical problems we add vectors a, b of the same direction. The laws describing $a + b$ geometrically are exactly analogous to those for adding the (signed) numbers of school algebra. To derive them let $a = \vec{PQ}$, $b = \vec{QR}$ where P, Q, R lie on a line. Then $a + b = \vec{PR}$ and the results can be "read off" from a diagram:

If a, b have a common direction, then $a + b$ has that direction (assuming $a + b \neq 0$). Moreover:

(a) *if a, b have the same sense* (Fig. 12a), *then $a + b$ has that sense, and its magnitude is the sum of their magnitudes;*

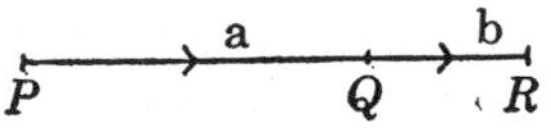

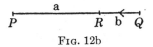

FIG. 12a

(b) *if a, b have different senses and different magnitudes* (Fig. 12b), *then $a + b$ has the sense of the one of greater magnitude, and its magnitude is the difference of their magnitudes;*

$$\overset{a}{\underset{P \qquad\qquad R \; b \; Q}{\rule{5cm}{0.4pt}}}$$

FIG. 12b

(c) *if a, b have different senses and the same magnitude, then $a + b = 0$.*

Application to School Geometry. We show by example how the elements of vector analysis which we have developed so far can be applied to prove theorems or solve exercises in school geometry. First we apply the Conversion Principle (p. 152) to statements (i), (ii) of page 150, obtaining corresponding *vectorial* principles:

(i) *M is the midpoint of segment $\overline{AB}$ if and only if $\vec{AM} = \vec{MB}$.*

(ii) *Quadrilateral $ABCD$ is a parallelogram if and only if $\vec{AB} = \vec{DC}$.*

Now we prove: *The diagonals of parallelogram $ABCD$ bisect each other.*

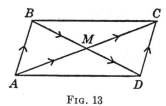

FIG. 13

Let M be the midpoint of $\overline{BD}$ (Fig. 13). We show M is also the midpoint of $\overline{AC}$. By (ii), (i), $\overrightarrow{AB} = \overrightarrow{DC}$, $\overrightarrow{BM} = \overrightarrow{MD}$. Hence

$$\overrightarrow{AM} = \overrightarrow{AB} + \overrightarrow{BM} = \overrightarrow{DC} + \overrightarrow{MD} = \overrightarrow{MD} + \overrightarrow{DC} = \overrightarrow{MC}.$$

By (i) M is the midpoint of $\overline{AC}$. Thus $\overline{AC}$, $\overline{BD}$ have the same midpoint and bisect each other.

We prove a converse proposition: *If the diagonals of quadrilateral ABCD bisect each other, ABCD is a parallelogram.*

Let M be the common midpoint of diagonals $\overline{AC}, \overline{BD}$. Certainly M is the midpoint of $\overline{DB}$. By (i), $\overrightarrow{AM} = \overrightarrow{MC}, \overrightarrow{DM} = \overrightarrow{MB}$ or $\overrightarrow{MB} = \overrightarrow{DM}$. Hence

$$\overrightarrow{AB} = \overrightarrow{AM} + \overrightarrow{MB} = \overrightarrow{MC} + \overrightarrow{DM} = \overrightarrow{DM} + \overrightarrow{MC} = \overrightarrow{DC}$$

and the result follows by (ii).

Finally we derive the familiar theorem that *the segment joining the midpoints of two sides of a triangle equals in length half the third side and is parallel to the third side.*

Let M, N be the respective midpoints of sides $\overline{AB}$, $\overline{AC}$ of triangle ABC (Fig. 14). We seek the relation between $\overrightarrow{BC}$, $\overrightarrow{MN}$. We have $\overrightarrow{BM} = \overrightarrow{MA}$, $\overrightarrow{AN} = \overrightarrow{NC}$.

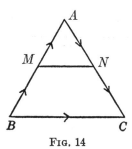

FIG. 14

Using familiar principles and page 159, Exercise 5

$$\overrightarrow{BC} = \overrightarrow{BA} + \overrightarrow{AC} = (\overrightarrow{BM} + \overrightarrow{MA}) + (\overrightarrow{AN} + \overrightarrow{NC})$$

$$= (\overrightarrow{MA} + \overrightarrow{MA}) + (\overrightarrow{AN} + \overrightarrow{AN})$$

$$= (\overrightarrow{MA} + \overrightarrow{AN}) + (\overrightarrow{MA} + \overrightarrow{AN}) = \overrightarrow{MN} + \overrightarrow{MN}.$$

Thus $\overrightarrow{BC} = \overrightarrow{MN} + \overrightarrow{MN}$ and we might be tempted to say $\overrightarrow{BC}$ is *twice* $\overrightarrow{MN}$. We must refrain from doing so, however, at this stage since "doubling a vector" has not yet been defined. To interpret the equation geometrically, we observe $\overrightarrow{MN} + \overrightarrow{MN}$ has the same direction and sense as $\overrightarrow{MN}$ but double the length. Thus $\overrightarrow{BC}$ has twice the length of $\overrightarrow{MN}$. Furthermore lines MN and BC must be parallel or coincide. We see in the diagram (Fig. 14) that MN and BC do not coincide. We can justify this without relying on a diagram. For suppose MN, BC coincide. Then AB and BC have M and B in common and must coincide. Thus MN and BC cannot coincide and $MN\|BC$.

EXERCISES

In solving the following exercises it is good to make diagrams in order to grasp their intuitive geometric significance, even though your solution is purely algebraic. As usual the results hold for all cases including degenerate ones.

1. Prove: $\overrightarrow{AB} + \overrightarrow{BA} = \mathbf{0}; \overrightarrow{AB} + \overrightarrow{BC} + \overrightarrow{CA} = \mathbf{0}$. Generalize.

2. Prove: $\overrightarrow{AB} + \overrightarrow{BC} + \overrightarrow{CD} = \overrightarrow{AD}$. Generalize.

3. Prove: $\overrightarrow{AP} + \overrightarrow{PC} = \overrightarrow{AQ} + \overrightarrow{QC}$.

4. Prove: If M is the midpoint of $\overline{AB}$ then $\overrightarrow{AM} + \overrightarrow{BM} = \mathbf{0}$.

5. Prove: If M is the midpoint of $\overline{BC}$ and A is any point then $\overrightarrow{AB} + \overrightarrow{AC} = \overrightarrow{AM} + \overrightarrow{AM}$.

6. Prove: If M, N are the midpoints of $\overline{AC}$, $\overline{BD}$ respectively then $\overrightarrow{AB} + \overrightarrow{CD} = \overrightarrow{MN} + \overrightarrow{MN}$.

7. Prove: If M, N are the midpoints of $\overline{AC}$, $\overline{BD}$ respectively then

$$\overrightarrow{AB} + \overrightarrow{AD} + \overrightarrow{CB} + \overrightarrow{CD} = \overrightarrow{MN} + \overrightarrow{MN} + \overrightarrow{MN} + \overrightarrow{MN}.$$

MULTIPLICATION OF GEOMETRIC VECTORS BY REAL NUMBERS

In this part we study an operation of multiplying or stretching a vector by a numerical factor. This enables us to treat vectorially the relation between vectors with the same direction. The operation when combined with vector addition enables us to study planar and spatial relations of vectors and greatly increases the scope of vector analysis as a method of solving geometric problems.

Product of Vector and Scalar. The theory we have developed thus far enables us to deal easily with segments on the same or parallel lines which are equal in length. It does not enable us to treat conveniently the case when the segments are unequal in length. In other words, we have no *vectorial* way to express the relation between two vectors **a**, **b** of the same direction and different magnitude. To make our ideas specific, suppose these vectors **a**, **b** have the same sense, and the magnitude of **b**

is three times that of **a**. We might describe the situation *operationally* by thinking of **b** as the result of stretching or multiplying **a** by a factor of 3. Similarly if **b** had *opposite* sense to **a** and three times its magnitude, we might describe **b** as the result of multiplying **a** by -3. Thus we introduce

DEFINITION 4. Let m be a real number and **a** a vector. Then the *product* of **a** by m, denoted m**a**, is defined as follows: (a) if $m > 0$, m**a** is the vector which has the same direction and sense as **a** and magnitude m times that of **a**; (b) if $m < 0$, m**a** has the same direction as **a** but *opposite* sense, and magnitude $|m|$ (the absolute value of m) times that of **a**; (c) we define 0**a** to be **0**.

If in (a) or (b) of Definition 4, **a** happens to be **0**, its magnitude will be 0. Hence m**a** will have magnitude 0, and is determined by this property to be **0**. Thus for any m we have $m\mathbf{0} = \mathbf{0}$.

Multiplication of vectors by numbers is analogous to multiplication of numbers. For example, to multiply a number by -3 we change its "sign" and multiply its absolute value by 3; while to multiply a vector by -3 we change its "sense" and multiply its magnitude by 3. It is very easy to picture the operation geometrically, since if **a** is "localized" on a given line, m**a** can be "localized" on the same line.

As is customary in vector analysis we often refer to real numbers as *scalars* (that is, quantities representable on a linear scale) to distinguish them from vectors. Thus m**a** is called the *product* of *vector* **a** by *scalar* m.

We proceed to develop the algebraic properties of this geometrical kind of multiplication. Clearly a vector is uniquely determined if we know its magnitude, direction, sense. Hence Definition 4 implies

THEOREM 9. m**a** *is a uniquely determined vector for each scalar m and each vector* **a**.

The special elements 0, 1, -1, **0** behave just as you might expect.

THEOREM 10. (a) $0\mathbf{a} = \mathbf{0}$; (b) $m\mathbf{0} = \mathbf{0}$; (c) $1\mathbf{a} = \mathbf{a}$; (d) $(-1)\mathbf{a} = -\mathbf{a}$.

Proof. (a) is part of Definition 4. We have already noted (b). (c) holds since 1**a** and **a** agree in magnitude, direction, sense. For (d), let $\mathbf{a} = \overrightarrow{PQ}$. By Theorem 5, Corollary 3, $-\mathbf{a} = \overrightarrow{QP}$. By definition, $(-1)\mathbf{a}$ has the magnitude and direction of **a**, but opposite sense. Thus $(-1)\mathbf{a}$ and $\overrightarrow{QP}$ have the same magnitude, direction, sense. Hence $(-1)\mathbf{a} = \overrightarrow{QP} = -\mathbf{a}$.

To aid in proving later theorems we introduce the

LEMMA. $(-p)\mathbf{a} = -(p\mathbf{a}) = p(-\mathbf{a})$.

Proof. p here denotes any real number, not necessarily positive. We observe that $(-p)\mathbf{a}$ and $-(p\mathbf{a})$ have the direction of **a** and magnitude equal to $|p|$ times that of **a**. Further we note $(-p)\mathbf{a}$ and $-(p\mathbf{a})$ have sense opposite to that of $p\mathbf{a}$. Thus $(-p)\mathbf{a} = -(p\mathbf{a})$ since they agree in magnitude, direction, sense. The proof is completed by a similar argument.

If we multiply a vector by a scalar the result is a vector. Hence we can multiply again by a scalar. This suggests

THEOREM 11. (Associative Law.) $m(n\mathbf{a}) = (mn)\mathbf{a}$.

Proof. If m or n is 0 the equation holds, since each member is $\mathbf{0}$. Suppose m, n have the same sign, that is both are positive or both are negative. Then both members have the direction and sense of $\mathbf{a}$ and magnitude equal to mn times that of $\mathbf{a}$. Thus the equation holds.

Next suppose m, n have opposite signs. First suppose $m > 0$, $n < 0$. Let $n = -p$; then $p > 0$. Using the Lemma and the preceding cases we have[3]

$$m(n\mathbf{a}) = m((-p)\mathbf{a}) = m(p(-\mathbf{a})) = (mp)(-\mathbf{a}),$$

$$(mn)\mathbf{a} = (m(-p))\mathbf{a} = (-mp)\mathbf{a} = (mp)(-\mathbf{a}),$$

and $m(n\mathbf{a}) = (mn)\mathbf{a}$. The final case $m < 0$, $n > 0$ is treated similarly.

We have been studying properties of multiplication in itself. Now we combine it with addition of scalars.

THEOREM 12. (Distributive Law.) $(m + n)\mathbf{a} = m\mathbf{a} + n\mathbf{a}$.[4]

Proof. There are many cases here because the rules for adding scalars involve many cases. Since a complete proof is not difficult but tedious, we will be content with considering a single case.

Suppose $m > 0$, $n > 0$. To make our ideas specific let $m = 3$, $n = 2$. Then $(m + n)\mathbf{a} = (3 + 2)\mathbf{a} = 5\mathbf{a}$ has the direction and sense of $\mathbf{a}$ and magnitude 5 times that of $\mathbf{a}$. On the other hand, $m\mathbf{a} + n\mathbf{a} = 3\mathbf{a} + 2\mathbf{a}$ is the sum of two vectors of the same direction and sense, namely that of $\mathbf{a}$. Hence (by (a) p. 159) $3\mathbf{a} + 2\mathbf{a}$ has the direction and sense of $\mathbf{a}$ and magnitude equal to the sum of those of $3\mathbf{a}$ and $2\mathbf{a}$, which is 5 times the magnitude of $\mathbf{a}$. Thus the principle holds in the given case.

Observe how this simple algebraic law covers so many different geometrical cases. It is precisely because of this that vector analytic methods are a boon to the study of geometry. Hereafter we need never consider the cases individually—we merely apply our general principle.

COROLLARY. $(m - n)\mathbf{a} = m\mathbf{a} - n\mathbf{a}$.

Proof. By the last theorem

$$(m - n)\mathbf{a} + n\mathbf{a} = ((m - n) + n)\mathbf{a} = m\mathbf{a}.$$

Transposing (Theorem 6) we have $(m - n)\mathbf{a} = m\mathbf{a} - n\mathbf{a}$.

---

[3] The following equality property is tacitly assumed: If equal vectors are multiplied by equal scalars, the products are equal.

[4] The right member stands for $(m\mathbf{a}) + (n\mathbf{a})$. As in school algebra, products of single terms are considered as if enclosed in parentheses.

Now we combine multiplication with vector addition.

THEOREM 13. (Distributive Law.) $m(\mathbf{a} + \mathbf{b}) = m\mathbf{a} + m\mathbf{b}$.

Proof. If $m = 0$, $\mathbf{a} = \mathbf{0}$ or $\mathbf{b} = \mathbf{0}$ the result clearly holds. Suppose $m \neq 0$ and $\mathbf{a}, \mathbf{b} \neq \mathbf{0}$. We consider first the case $m > 0$. Let $\mathbf{a} = \overrightarrow{PQ}$, $\mathbf{b} = \overrightarrow{QR}$, so that $\mathbf{a} + \mathbf{b} = \overrightarrow{PR}$. Suppose P, Q, R do not lie on a line (Fig. 15). Choose Q' on line PQ such that $\overline{PQ'} = m\overline{PQ}$ [5] and Q, Q' are on the

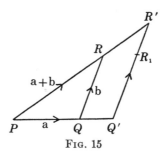

FIG. 15

same side of P. Similarly choose R' on line PR such that $\overline{PR'} = m\overline{PR}$ and R, R' are on same side of P. Then $\overline{PQ'} = m\mathbf{a}$ and $\overrightarrow{PR'} = m(\mathbf{a} + \mathbf{b})$. We show $\overrightarrow{Q'R'} = m\mathbf{b}$. Consider triangles $PQ'R'$, PQR. We have $\angle R'PQ' = \angle RPQ$ and $\overline{PQ'}/\overline{PQ} = \overline{PR'}/\overline{PR} = m$. Thus $\triangle PQ'R' \sim \triangle PQR$ so that $\angle PQR = \angle PQ'R'$ and $\overline{Q'R'}/\overline{QR} = \overline{PQ'}/\overline{PQ} = m$. Hence $Q'R' \| QR$ and $\overline{Q'R'} = m\overline{QR}$. Thus vectors $\overrightarrow{Q'R'}$, $\overrightarrow{QR}$ have the same direction, and the former has magnitude m times that of the latter. Further they have the same sense, since if we localize $\overrightarrow{QR}$ at Q' as $\overrightarrow{Q'R_1}$, we see R' and R_1 are on the same side of Q'. Thus by definition $\overrightarrow{Q'R'} = m\overrightarrow{QR} = m\mathbf{b}$. Since $\overrightarrow{PQ'} + \overrightarrow{Q'R'} = \overrightarrow{PR'}$ we have $m\mathbf{a} + m\mathbf{b} = m(\mathbf{a} + \mathbf{b})$.

The case in which P, Q, R lie on a line is easily disposed of. Finally when $m < 0$ the result is obtained by using the Lemma and Theorem 8.

COROLLARY. $m(\mathbf{a} - \mathbf{b}) = m\mathbf{a} - m\mathbf{b}$.

We prove two "cancellation laws" for products of vectors by scalars. First we consider products equal to "zero":

THEOREM 14. $m\mathbf{a} = \mathbf{0}$ *implies* $m = 0$ *or* $\mathbf{a} = \mathbf{0}$.

Proof. Suppose $m \neq 0$. Then "multiplying" equation $m\mathbf{a} = \mathbf{0}$ by $1/m$ we have $(1/m)(m\mathbf{a}) = (1/m)\mathbf{0}$. Hence $((1/m)m)\mathbf{a} = \mathbf{0}$, $1\mathbf{a} = \mathbf{0}$ and $\mathbf{a} = \mathbf{0}$.

COROLLARY 1. (Cancellation Law.) *If* $m\mathbf{a} = m\mathbf{b}$, $m \neq 0$, *then* $\mathbf{a} = \mathbf{b}$.

Proof. $m\mathbf{a} - m\mathbf{b} = \mathbf{0}$ so that $m(\mathbf{a} - \mathbf{b}) = \mathbf{0}$. Hence the theorem implies $\mathbf{a} - \mathbf{b} = \mathbf{0}$ and $\mathbf{a} = \mathbf{b}$.

---

[5] Heretofore $\overline{PQ'}$ denoted *segment PQ'*. Now it also denotes the distance between P and Q'. The context will indicate the meaning intended.

Similarly we prove

COROLLARY 2. (Cancellation Law.) *If* $m\mathbf{a} = n\mathbf{a}$, $\mathbf{a} \neq \mathbf{0}$, *then* $m = n$.

It must be clear now that the portion of elementary algebra which deals with linear relations can be carried over to "vector algebra." Thus we can define "linear equations" in several vectors with scalar coefficients, for example $m\mathbf{a} + n\mathbf{b} = \mathbf{0}$ or $m\mathbf{a} + n\mathbf{b} = p\mathbf{c} - q\mathbf{d}$. And we can treat these equations as we treat linear equations in school algebra; thus we can "solve" for a term, "multiply" by a scalar, "collect" like terms, and so on.

Application to Geometry. The introduction of multiplication by scalars into our original additive algebra of vectors, makes it a more powerful and convenient instrument for studying geometry. For example we can characterize M as the midpoint of $\overline{AB}$ by $\overrightarrow{AM} = \tfrac{1}{2}\overrightarrow{AB}$. Similarly the trisection points P, Q of $\overline{AB}$ are given by: $\overrightarrow{AP} = \tfrac{1}{3}\overrightarrow{AB}$, $\overrightarrow{AQ} = \tfrac{2}{3}\overrightarrow{AB}$. Further the awkward discussion of the sum of a vector and itself (p. 161) is replaced by a simple algebraic argument:

$$\mathbf{a} + \mathbf{a} = 1\mathbf{a} + 1\mathbf{a} = (1 + 1)\mathbf{a} = 2\mathbf{a}.$$

First we prove the familiar proposition: *The medians of a triangle meet in a point whose distance from each vertex is two thirds the length of the median from that vertex.*

Consider $\triangle ABC$ with medians AA', BB', CC' (Fig. 16). First we show the conclusion holds for medians AA', BB'. Let P, Q, R be chosen

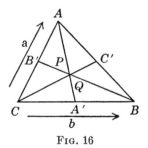

FIG. 16

in $\overline{AA'}$, $\overline{BB'}$, $\overline{CC'}$ respectively such that

$$\overline{AP} = \tfrac{2}{3}\overline{AA'}, \qquad \overline{BQ} = \tfrac{2}{3}\overline{BB'}, \qquad \overline{CR} = \tfrac{2}{3}\overline{CC'}.$$

Then

$$\overrightarrow{AP} = \tfrac{2}{3}\overrightarrow{AA'}, \qquad \overrightarrow{BQ} = \tfrac{2}{3}\overrightarrow{BB'}, \qquad \overrightarrow{CR} = \tfrac{2}{3}\overrightarrow{CC'}.$$

We show $P = Q$. Let $\mathbf{a} = \overrightarrow{CA}$, $\mathbf{b} = \overrightarrow{CB}$. We consider C as a sort of "origin," and express $\overrightarrow{CP}$, $\overrightarrow{CQ}$ in terms of $\mathbf{a}$, $\mathbf{b}$. We have

$$\vec{CP} = \vec{CA} + \vec{AP} = \mathbf{a} + \tfrac{2}{3}\vec{AA'} = \mathbf{a} + \tfrac{2}{3}(\vec{AC} + \vec{CA'})$$

$$= \mathbf{a} + \tfrac{2}{3}(-\mathbf{a} + \tfrac{1}{2}\mathbf{b})$$

$$= \mathbf{a} + \tfrac{2}{3}(-\mathbf{a}) + \tfrac{2}{3}(\tfrac{1}{2}\mathbf{b}) = 1\mathbf{a} + (-\tfrac{2}{3})\mathbf{a} + \tfrac{1}{3}\mathbf{b}$$

$$= (1 + (-\tfrac{2}{3}))\mathbf{a} + \tfrac{1}{3}\mathbf{b}$$

$$= \tfrac{1}{3}\mathbf{a} + \tfrac{1}{3}\mathbf{b}.$$

The same argument is valid for $\vec{CQ}$ but with A, B interchanged, so that $\vec{CQ} = \tfrac{1}{3}\mathbf{a} + \tfrac{1}{3}\mathbf{b}$. Thus $\vec{CP} = \vec{CQ}$ and $P = Q$ by the Representation Principle. Since AA', BB' are arbitrary medians of $\triangle ABC$, similarly we have $Q = R$. Hence P, Q, R coincide and the proposition is justified.

Vector algebraic reasoning is especially effective in spatial situations. We apply it to prove: *The diagonals of a parallelopiped bisect each other.*

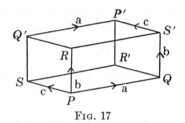

FIG. 17

Consider a parallelopiped with vertices labelled as in Figure 17. We show diagonals $\overline{PP'}$, $\overline{QQ'}$, $\overline{RR'}$ bisect each other. Let P_1, Q_1, R_1 be the respective midpoints of $\overline{PP'}$, $\overline{QQ'}$, $\overline{RR'}$. Let $\mathbf{a} = \vec{PQ}$, $\mathbf{b} = \vec{PR}$, $\mathbf{c} = \vec{PS}$. Note $\mathbf{a} = \vec{RS'} = \vec{Q'P'} = \vec{SR'}$ and the similar relations for $\mathbf{b}$, $\mathbf{c}$. We have

$$\vec{PP_1} = \tfrac{1}{2}\vec{PP'} = \tfrac{1}{2}(\vec{PQ} + \vec{QS'} + \vec{S'P'}) = \tfrac{1}{2}(\mathbf{a} + \mathbf{b} + \mathbf{c}).$$

Likewise we express $\vec{PQ_1}$ in terms of $\mathbf{a}$, $\mathbf{b}$, $\mathbf{c}$.

$$\vec{PQ_1} = \vec{PR} + \vec{RQ'} + \vec{Q'Q_1} = \mathbf{b} + \mathbf{c} + \tfrac{1}{2}\vec{Q'Q}$$

$$= \mathbf{b} + \mathbf{c} + \tfrac{1}{2}(\vec{Q'R} + \vec{RP} + \vec{PQ})$$

$$= \mathbf{b} + \mathbf{c} + \tfrac{1}{2}((-\mathbf{c}) + (-\mathbf{b}) + \mathbf{a})$$

$$= \tfrac{1}{2}(\mathbf{a} + \mathbf{b} + \mathbf{c}).$$

Similarly we show $\vec{PR_1} = \tfrac{1}{2}(\mathbf{a} + \mathbf{b} + \mathbf{c})$. Thus $\vec{PP_1} = \vec{PQ_1} = \vec{PR_1}$, so that $P_1 = Q_1 = R_1$. In the same way we show the midpoint of diagonal $\overline{SS'}$ coincides with P_1, and the result follows.

Observe in these proofs we do not employ geometric ingenuity, as for example clever constructions. Rather we take the given geometric relations and convert them into vector algebraic relations which are amenable to algebraic treatment. It is true that the method used above is not one of discovery—we had to know the geometric conclusion before we verified it vectorially. However this is due only to the incompleteness of our knowledge of vectorial methods; we shall be able to remedy this deficiency after studying linear independence in the next section (p. 170).

EXERCISES

1. Prove the theorem on the segment joining the midpoints of two sides of a triangle (p. 160) using the methods of this section.

2. In parallelogram $ABCD$ let M, N be the respective midpoints of $\overline{AB}$, $\overline{CD}$. Prove $AMCN$ is a parallelogram.

3. Prove: The segments joining the midpoints of the successive sides of a quadrilateral form a parallelogram.

4. Prove: In parallelogram $ABCD$ the lines joining B and D to the respective midpoints of $\overline{CD}$ and $\overline{AB}$ trisect the diagonal $\overline{AC}$. Hint: Let M be the midpoint of $\overline{AB}$; choose P such that $\overrightarrow{AP} = \frac{1}{3}\overrightarrow{AC}$. Show $\overrightarrow{DP}$ is a scalar multiple of $\overrightarrow{DM}$ and infer P on $\overrightarrow{DM}$.

LINEAR INDEPENDENCE AND LINEAR DEPENDENCE OF GEOMETRIC VECTORS

In this part we consider the related concepts, linear independence and linear dependence of vectors. Linear independence signifies geometrically that the vectors are in general or nondegenerate position relative to each other. The idea is applicable to geometrical problems; is related to the idea of *dimensionality*; and leads to the introduction of *cartesian* analytic geometry on a *vector* analytic basis.

Linear Dependence of Vectors. Having developed a theory of linear equations for vectors, you may naturally ask: How are vectors related to each other if they satisfy a linear equation? This suggests

DEFINITION 5. If there exist scalars m_1, $\cdots$, m_r not all 0, such that

$$(4) \qquad\qquad m_1\mathbf{a}_1 + m_2\mathbf{a}_2 + \cdots + m_r\mathbf{a}_r = \mathbf{0}$$

we say vectors $\mathbf{a}_1$, $\cdots$, $\mathbf{a}_r$ are *linearly related* or *linearly dependent* or simply *dependent*. Naturally we do not allow all the m's to be 0, since in such a case any $\mathbf{a}$'s at all would satisfy (4).

In the definition we may have $r = 1$. Then $m_1\mathbf{a}_1 = \mathbf{0}$ but $m_1 \neq 0$. Hence $\mathbf{a}_1 = \mathbf{0}$. Conversely if $\mathbf{a}_1 = \mathbf{0}$, it is dependent, since $1\mathbf{a}_1 = \mathbf{0}$. Thus *a single vector* $\mathbf{a}_1$ *is dependent if and only if* $\mathbf{a}_1 = \mathbf{0}$. That is, $\mathbf{0}$ is the only vector which is dependent.

Dependence of Two Vectors and Collinearity. Observe $\mathbf{a}$, $\mathbf{0}$ are dependent for any choice of $\mathbf{a}$, since $0\mathbf{a} + 1\mathbf{0} = \mathbf{0}$. For a more typical example suppose **a**, **b** $\neq$ **0** and $1\mathbf{a} + 3\mathbf{b} = \mathbf{0}$. Then $\mathbf{a} = (-3)\mathbf{b}$ and **a** and **b** have the same direction, by definition of multiplication. These examples illustrate that two dependent vectors "lie along" a line. Thus if **a**, **b** have different directions they cannot be dependent. For a formal treatment we need a vectorial criterion that a point lie on a line:

THEOREM 15. *Suppose* $A \neq B$. *Then* P *is in line* AB *if and only if*

$$(5) \qquad\qquad \overrightarrow{AP} = t\overrightarrow{AB}$$

holds for some scalar t.

Proof. Suppose P in AB. If $P = A$ then $\overrightarrow{AP} = \mathbf{0} = 0\overrightarrow{AB}$. Suppose $P \neq A$. Let $\overrightarrow{AP}/\overrightarrow{AB} = m$. Then $\overrightarrow{AP}$ has magnitude m times that of $\overrightarrow{AB}$, the same direction as $\overrightarrow{AB}$, and the same or opposite sense as $\overrightarrow{AB}$ according as P and B are on the same or opposite sides of A. Thus by Definition 4, $\overrightarrow{AP} = (\pm m)\overrightarrow{AB}$.

Conversely suppose $\overrightarrow{AP} = t\overrightarrow{AB}$. If $t = 0$, $\overrightarrow{AP} = \mathbf{0}$ and $P = A$ which is in AB. Suppose $t \neq 0$. Then Definition 4 implies $t\overrightarrow{AB}$ or $\overrightarrow{AP}$ has the same direction as $\overrightarrow{AB}$. Thus lines AP, AB must coincide and P is in AB.

Intuitively the theorem says: P is in AB if and only if a "stretch" of $\overrightarrow{AB}$ yields $\overrightarrow{AP}$. Observe (5) is a sort of vector equation for AB, since it holds if and only if P is in AB.

COROLLARY. *If* $A'B'$ *and* AB *have the same direction then* $\overrightarrow{A'B'} = t\overrightarrow{AB}$ *for some scalar* t.

Proof. Localize $\overrightarrow{A'B'}$ at A as $\overrightarrow{AP}$. Then $\overrightarrow{AP}$, $\overrightarrow{AB}$ have the same direction, and AP, AB coincide. Thus P is in AB, and $\overrightarrow{A'B'} = \overrightarrow{AP} = t\overrightarrow{AB}$ by Theorem 15.

Now we easily establish the geometrical significance of dependence of two vectors:

THEOREM 16. *Let* $\mathbf{a} = \overrightarrow{PQ}$, $\mathbf{b} = \overrightarrow{PR}$. *Then* **a**, **b** *are dependent if and only if* P, Q, R *colline, that is, are collinear or lie on a line*.

Proof. Suppose **a**, **b** dependent. Then $m\mathbf{a} + n\mathbf{b} = \mathbf{0}$ where m, n are not both 0. Our plan is to solve for one of **a**, **b** as a multiple of the other, and so infer they "lie along" a line. It is not restrictive to assume $m \neq 0$. Then

$$m\mathbf{a} = -(n\mathbf{b}) = (-n)\mathbf{b}.$$

Multiplying by $1/m$ we have $\mathbf{a} = (-n/m)\mathbf{b} = p\mathbf{b}$, where $p = -n/m$. That is, $\overrightarrow{PQ} = p\overrightarrow{PR}$. If $P = R$ certainly P, Q, R colline. Suppose $P \neq R$. Then by Theorem 15, Q is in PR and P, Q, R colline.

Conversely, suppose P, Q, R colline. If $P = R$ then $\mathbf{b} = \overrightarrow{PP} = \mathbf{0}$, and $m\mathbf{a} + n\mathbf{b} = \mathbf{0}$ holds with $m = 0$, $n = 1$. Suppose $P \neq R$. Then Q is in

PR and by Theorem 15, $\overrightarrow{PQ} = t\overrightarrow{PR}$ or $\mathbf{a} = t\mathbf{b}$. Thus $1\mathbf{a} + (-t)\mathbf{b} = \mathbf{0}$ and the proof is complete.

The intuitive significance of the result is: $\mathbf{a}$, $\mathbf{b}$ are dependent if and only if they have the same direction, or more precisely, if and only if they can be "localized" on a line.

Dependence of Three Vectors and Coplanarity. The coherence established in the last theorem between the ideas "linear dependence of vectors" and "collinearity" is very deep and important. It indicates that (geometrical) linear relations among points can be treated by (algebraic) linear relations among vectors. To establish a similar result for three dependent vectors we use the following criterion, analogous to Theorem 15, that a point be in a plane.

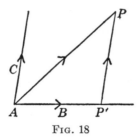

FIG. 18

THEOREM 17. *Suppose A, B, C do not colline. Then P is in plane ABC if and only if $\overrightarrow{AP} = t\overrightarrow{AB} + u\overrightarrow{AC}$ for some scalars t, u.*

Proof. Suppose P in plane ABC. Let line l contain P, have the same direction as line AC, and meet line AB in P' (Fig. 18). Using Theorem 15 and its corollary, we see if $P \neq P'$

$$\overrightarrow{AP'} = t\overrightarrow{AB}, \qquad \overrightarrow{P'P} = u\overrightarrow{AC}.$$

If $P = P'$ these relations hold with $u = 0$. Hence

$$\overrightarrow{AP} = \overrightarrow{AP'} + \overrightarrow{P'P} = t\overrightarrow{AB} + u\overrightarrow{AC}.$$

Conversely, suppose $\overrightarrow{AP} = t\overrightarrow{AB} + u\overrightarrow{AC}$. Let $\overrightarrow{AP'} = t\overrightarrow{AB}$. By Theorem 15, P' is in AB. We have

$$\overrightarrow{AP} = \overrightarrow{AP'} + u\overrightarrow{AC}, \qquad \overrightarrow{AP} = \overrightarrow{AP'} + \overrightarrow{P'P}.$$

Thus $\overrightarrow{AP'} + u\overrightarrow{AC} = \overrightarrow{AP'} + \overrightarrow{P'P}$ and $u\overrightarrow{AC} = \overrightarrow{P'P}$ by the Cancellation Law. If $u = 0$ then $\overrightarrow{P'P} = \mathbf{0}$ and $P = P'$ is in plane ABC. Suppose $u \neq 0$. Then $u\overrightarrow{AC}$ or $\overrightarrow{P'P}$ has the direction of $\overrightarrow{AC}$. Thus $P'P$ and AC have the same direction, that is are parallel or coincident. In either case it follows easily that P is in plane ABC.

Speaking roughly: A vector "lies" in a plane with vectors $\mathbf{a}$, $\mathbf{b}$ if and only if it is expressible in the form $t\mathbf{a} + u\mathbf{b}$.

And now the geometrical significance of dependence of three vectors:

THEOREM 18. *Let* $\mathbf{a} = \overrightarrow{PQ}$, $\mathbf{b} = \overrightarrow{PR}$, $\mathbf{c} = \overrightarrow{PS}$. *Then* $\mathbf{a}$, $\mathbf{b}$, $\mathbf{c}$ *are dependent if and only if* P, Q, R, S *coplane, that is, are coplanar or lie in a plane.*

Proof. Suppose $m\mathbf{a} + n\mathbf{b} + p\mathbf{c} = \mathbf{0}$, where m, n, p are not all 0. It is not restrictive to assume $m \neq 0$. Then we can "solve" the dependency relation for $\mathbf{a}$ getting $\mathbf{a} = q\mathbf{b} + r\mathbf{c}$ where $q = -n/m$, $r = -p/m$. Thus $\overrightarrow{PQ} = q\overrightarrow{PR} + r\overrightarrow{PS}$. If P, R, S colline certainly P, Q, R, S coplane. Suppose P, R, S do not colline. Then Theorem 17 implies Q is in plane PRS and P, Q, R, S coplane.

Conversely, suppose P, Q, R, S coplane. First suppose P, R, S colline. Then Theorem 16 implies $\mathbf{b}$, $\mathbf{c}$ dependent. Thus $n\mathbf{b} + p\mathbf{c} = \mathbf{0}$ where n, p are not both 0. Hence $0\mathbf{a} + n\mathbf{b} + p\mathbf{c} = \mathbf{0}$ and $\mathbf{a}$, $\mathbf{b}$, $\mathbf{c}$ are dependent. Finally, suppose P, R, S do not colline. Then Q is in plane PRS. By Theorem 17 $\overrightarrow{PQ} = q\overrightarrow{PR} + r\overrightarrow{PS}$ or $\mathbf{a} = q\mathbf{b} + r\mathbf{c}$. Thus the dependence of $\mathbf{a}$, $\mathbf{b}$, $\mathbf{c}$ follows.

Thus we may say intuitively: $\mathbf{a}$, $\mathbf{b}$, $\mathbf{c}$ are dependent if and only if they can be "localized" on a plane.

Linear Independence. We say $\mathbf{a}_1, \cdots, \mathbf{a}_r$ are *linearly independent* or *independent* if they are *not* dependent. Suppose $\mathbf{a}_1, \cdots, \mathbf{a}_r$ are independent and satisfy a relation

(6) $$m_1\mathbf{a}_1 + \cdots + m_r\mathbf{a}_r = \mathbf{0}.$$

Then the relation must be *trivial*, that is $m_1 = m_2 = \cdots = m_r = 0$. In other words (6) imposes no restriction on the $\mathbf{a}$'s.

A single vector $\mathbf{a}_1$ is independent if and only if $\mathbf{a}_1 \neq \mathbf{0}$, since $\mathbf{0}$ is the only "dependent" vector. Geometrical significance for independence is obtained merely by restating Theorems 16 and 18. Thus we have precisely: *Let* $\mathbf{a} = \overrightarrow{PQ}$, $\mathbf{b} = \overrightarrow{PR}$; *then* $\mathbf{a}$, $\mathbf{b}$ *are independent if and only if* P, Q, R *do not colline.* Somewhat roughly: $\mathbf{a}$, $\mathbf{b}$ are independent if they have different directions, or cannot be "localized" on a line.

Analysis of Geometric Problems. We now have sufficient knowledge to be able to employ vector algebra not merely as a method of verification of elementary geometric relations but also as a systematic method of discovering them. A good example is the question of the intersection of the medians of a triangle which we considered above. We wish to determine whether medians AA', BB' of $\triangle ABC$ intersect in a single point, and the location of this point in each median if they do so intersect. We use vector algebra literally as a *vector analysis.*

Suppose medians AA', BB' have point S in common (Fig. 19). Then

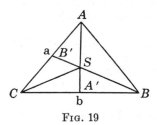

FIG. 19

there exist scalars m, n such that

(7) $\overrightarrow{AS} = m\overrightarrow{AA'}, \qquad \overrightarrow{BS} = n\overrightarrow{BB'}.$

As in the previous discussion of the problem, let $\mathbf{a} = \overrightarrow{CA}$, $\mathbf{b} = \overrightarrow{CB}$ and express the vectors considered in terms of vectors "emanating" from C. We have

$$\overrightarrow{AS} = \overrightarrow{AC} + \overrightarrow{CS} = (-\mathbf{a}) + \overrightarrow{CS}, \quad \overrightarrow{AA'} = \overrightarrow{AC} + \overrightarrow{CA'} = (-\mathbf{a}) + \tfrac{1}{2}\mathbf{b},$$

$$\overrightarrow{BS} = \overrightarrow{BC} + \overrightarrow{CS} = (-\mathbf{b}) + \overrightarrow{CS}, \quad \overrightarrow{BB'} = \overrightarrow{BC} + \overrightarrow{CB'} = (-\mathbf{b}) + \tfrac{1}{2}\mathbf{a}.$$

Substituting in (7) for $\overrightarrow{AS}$, $\overrightarrow{BS}$, $\overrightarrow{AA'}$, $\overrightarrow{BB'}$ we get

$$(-\mathbf{a}) + \overrightarrow{CS} = m((-\mathbf{a}) + \tfrac{1}{2}\mathbf{b}), \qquad (-\mathbf{b}) + \overrightarrow{CS} = n((-\mathbf{b}) + \tfrac{1}{2}\mathbf{a}).$$

We naturally eliminate $\overrightarrow{CS}$, which is "unknown," between these relations, in the hope of obtaining information about m, n. Solving for $\overrightarrow{CS}$ we have

$$\overrightarrow{CS} = (1 - m)\mathbf{a} + (m/2)\mathbf{b} = (1 - n)\mathbf{b} + (n/2)\mathbf{a}.$$

Collecting terms we get

(8) $(1 - m - n/2)\mathbf{a} + (m/2 - 1 + n)\mathbf{b} = \mathbf{0}.$

For a moment it may not be clear how to proceed. But we must remember that an important part of our hypothesis has not been expressed in vector algebraic terms. Namely, that A, B, C are the vertices of a triangle! This means A, B, C do not colline. Hence by Theorem 16 $\overrightarrow{CA}$, $\overrightarrow{CB}$ are independent. That is $\mathbf{a}$, $\mathbf{b}$ are independent. Hence the coefficients in (8) vanish. Thus

$$1 - m - n/2 = 0, \qquad m/2 - 1 + n = 0.$$

Solving these equations we obtain $m = n = \tfrac{2}{3}$. Thus we have discovered: If lines AA', BB' do meet in a point S, then $\overline{AS} = \frac{2}{3}\overline{AA'}$ and $\overline{BS} = \frac{2}{3}\overline{BB'}$. To verify that the medians do meet and complete the proof, we have merely to repeat the discussion of pages 165–66.

Solve the following exercises using the method of analysis illustrated above.
Exercise 1. Prove: The diagonals of a parallelogram bisect each other.
Exercise 2. Prove: In parallelogram $ABCD$ the lines joining B and D to the respective midpoints of $\overline{CD}$ and $\overline{AB}$ trisect the diagonal $\overline{AC}$.

Location or Position Vector. The apparently innocuous idea (pp. 147–48) of representing the location of a point relative to a point by a directed segment, becomes quite important when converted into vectorial terms:

DEFINITION 6. Let O be a given point and P be any point. Then vector $\overrightarrow{OP}$ is called the *position vector* (or *location vector*) of P *relative to origin O*. If O is fixed throughout a discussion we often refer to $\overrightarrow{OP}$ simply as the *position vector* of P. We use the notation $P(\mathbf{r})$ for the point P to indicate that $\mathbf{r}$ is the position vector of P, that is $\mathbf{r} = \overrightarrow{OP}$. (Compare the notation $P(x, y)$ in plane analytic geometry.)

If we associate to each point P its position vector $\mathbf{r}$, we have a one-to-one correspondence between the set of all points of space and the set of all vectors. For (1) each point has a uniquely determined position vector; and (2) each vector $\mathbf{a}$ is the position vector of a unique point, since by the Representation Principle $\mathbf{a}$ can be localized at O in the form $\mathbf{a} = \overrightarrow{OA}$, in a unique way.

Vector Equations of Line, Plane. This simple method of associating vectors to points enables us to systematize the procedures for expressing geometrical relations algebraically, by reducing them to algebraic relations connecting the position vectors of the points involved. As an illustration we derive vector equations for lines and planes.

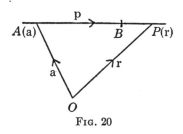

FIG. 20

THEOREM 19. *Let $A \neq B$, $\mathbf{a} = \overrightarrow{OA}$, $\mathbf{p} = \overrightarrow{AB}$, (Fig. 20). Then $P(\mathbf{r})$ is in line AB if and only if $\mathbf{r}$ satisfies*

(9) $$\mathbf{r} = \mathbf{a} + t\mathbf{p}$$

for some scalar t.

Proof. We merely have to restate Theorem 15 in terms of position vectors. By Theorem 15, $P(\mathbf{r})$ is in AB if and only if

(10) $$\overrightarrow{AP} = t\overrightarrow{AB} = t\mathbf{p}.$$

By Theorem 6, Corollary

$$\overrightarrow{AP} = \overrightarrow{OP} - \overrightarrow{OA} = \mathbf{r} - \mathbf{a}.$$

Hence (10) holds if and only if $\mathbf{r} - \mathbf{a} = t\mathbf{p}$. Since this is equivalent to (9) our proof is complete.

Observe (9) represents AB in terms of $\mathbf{a}$, the position vector of A, and $\mathbf{p}$, a vector directed along AB. That is we get (9) by knowing a *fixed point* of the line and its *direction*. Note if A coincides with O the line goes through the origin, and (9) takes on the simple form $\mathbf{r} = t\mathbf{p}$.

Let us see how line AB and equation (9) are related. If $P(\mathbf{r})$ is in AB, then $\mathbf{r}$ satisfies (9) for some scalar t. Conversely if in (9) we assign a real value to t, we obtain a vector $\mathbf{r}$ which is the position vector of a uniquely determined point P, and P is in AB. Thus as the values of t range over the real number system, the positions of P range over line AB. Adopting language from cartesian analytic geometry we call (9) a *vector equation* of line AB, and say equation (9) *represents* line AB. Since P is determined, when $\mathbf{a}$, $\mathbf{p}$ are given, by the assignment of real values to the arbitrary variable or so-called *parameter t*, (9) is often called a *parametric* equation of line AB. An analogous situation in cartesian analytic geometry is the representation of the line $y = 2x$ by *parametric equations* $x = t, y = 2t$.

To derive a vector representation of a *plane* we apply the argument of the last theorem to Theorem 17 and easily get

THEOREM 20. *Let A, B, C be noncollinear, $\mathbf{a} = \overrightarrow{OA}, \mathbf{p} = \overrightarrow{AB}, \mathbf{q} = \overrightarrow{AC}$. Then $P(\mathbf{r})$ is in plane ABC if and only if $\mathbf{r}$ satisfies*

(11) $$\mathbf{r} = \mathbf{a} + t\mathbf{p} + u\mathbf{q}$$

for some scalars t, u.

As above we call (11) a *vector equation* of plane ABC. Observe that now P is determined by *two* arbitrary variables or *parameters t, u*.

Geometric Application. Using position vectors and vector equations we can solve many geometric problems; as an illustration we find the point of intersection of the diagonals of a parallelogram. Choosing one vertex of the parallelogram as origin O, we may denote it as $OACB$

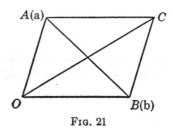

FIG. 21

(Fig. 21). Let the position vectors of A, B be **a**, **b** respectively. Then

$$\vec{OC} = \mathbf{a} + \mathbf{b}; \qquad \vec{AB} = \vec{OB} - \vec{OA} = \mathbf{b} - \mathbf{a}.$$

Applying Theorem 19, we obtain equations

(12) $\mathbf{r} = t(\mathbf{a} + \mathbf{b})$ for OC,

(13) $\mathbf{r} = \mathbf{a} + u(\mathbf{b} - \mathbf{a})$ for AB.

OC and AB meet if and only if there exist scalars t, u which yield the *same* vector **r** when substituted in (12), (13) respectively. Thus so to speak we solve (12), (13) *simultaneously* by equating the values of **r** and determining t, u. We get

$$t(\mathbf{a} + \mathbf{b}) = \mathbf{a} + u(\mathbf{b} - \mathbf{a}).$$

Multiplying out, transposing and collecting terms we have

(14) $(t + u - 1)\mathbf{a} + (t - u)\mathbf{b} = \mathbf{0}.$

Clearly **a**, **b** are independent. Hence (14) implies

$$t + u - 1 = 0, \qquad t - u = 0.$$

Solving we have $t = u = \frac{1}{2}$. Substituting these values in (12), (13) we get the same result for **r**, namely $\mathbf{r} = \frac{1}{2}(\mathbf{a} + \mathbf{b})$. Thus OC, AB meet in the point P whose position vector is $\frac{1}{2}(\mathbf{a} + \mathbf{b})$. It is easy to show P is the midpoint of $\overline{OC}$ and of $\overline{AB}$.

Dependence of Four Vectors. Let us return to the theme of linear dependence. We seek geometrical significance for dependence of four vectors comparable to Theorems 16, 18 for two and three vectors. This suggests a preparatory theorem (comparable to Theorems 15, 17) giving a vectorial criterion that a point be in a 3-space. But all points are in a 3-space, since our basic geometry is Euclidean solid geometry. Thus the desired theorem takes on the following form:

THEOREM 21. *Suppose A, B, C, D do not coplane. Then for any point P, $\vec{AP} = t\vec{AB} + u\vec{AC} + v\vec{AD}$ for some scalars t, u, v.*

We dispense with the proof which can be based on Theorem 15 and its Corollary, using the method of Theorem 17.

Now consider four vectors $\mathbf{a} = \overrightarrow{PQ}$, $\mathbf{b} = \overrightarrow{PR}$, $\mathbf{c} = \overrightarrow{PS}$, $\mathbf{d} = \overrightarrow{PT}$. By analogy with Theorems 16, 18 we would be tempted to say $\mathbf{a}$, $\mathbf{b}$, $\mathbf{c}$, $\mathbf{d}$ are dependent if and only if P, Q, R, S, T lie in a 3-space. This suggests any four vectors are dependent which we now prove.

THEOREM 22. *Any vectors* $\mathbf{a}$, $\mathbf{b}$, $\mathbf{c}$, $\mathbf{d}$ *are dependent.*

Proof. If three of $\mathbf{a}$, $\mathbf{b}$, $\mathbf{c}$, $\mathbf{d}$ are dependent then all are dependent, since the fourth vector can be introduced, with a zero coefficient, into the dependency relation which binds the other three. Thus we may assume $\mathbf{b}$, $\mathbf{c}$, $\mathbf{d}$ are not dependent. We show $\mathbf{a}$ is "linearly expressible" in terms of $\mathbf{b}$, $\mathbf{c}$, $\mathbf{d}$. Let $\mathbf{a} = \overrightarrow{PQ}$, $\mathbf{b} = \overrightarrow{PR}$, $\mathbf{c} = \overrightarrow{PS}$, $\mathbf{d} = \overrightarrow{PT}$. Then Theorem 18 implies P, R, S, T do not coplane. Hence by Theorem 21

$$\overrightarrow{PQ} = t\overrightarrow{PR} + u\overrightarrow{PS} + v\overrightarrow{PT}.$$

That is $\mathbf{a} = t\mathbf{b} + u\mathbf{c} + v\mathbf{d}$ and the conclusion follows.

COROLLARY. *Any vectors* $\mathbf{a}_1$, $\mathbf{a}_2$, $\cdots$, $\mathbf{a}_r$, $r \geq 4$, *are dependent.*

We have often had occasion in the preceding discussion to solve a linear relation in several vectors for a certain vector in terms of the others. We now derive a general result of this nature.

THEOREM 23. *If* $\mathbf{a}_1$, $\cdots$, $\mathbf{a}_{r+1}$ *are dependent and* $\mathbf{a}_1$, $\cdots$, $\mathbf{a}_r$ *are independent then* $\mathbf{a}_{r+1}$ *is expressible in the form* $p_1\mathbf{a}_1 + \cdots + p_r\mathbf{a}_r$.

Proof. We have by hypothesis

$$(15) \qquad m_1\mathbf{a}_1 + \cdots + m_r\mathbf{a}_r + m_{r+1}\mathbf{a}_{r+1} = \mathbf{0}$$

where one of the m's is not 0. We see $m_{r+1} \neq 0$; since otherwise $m_1\mathbf{a}_1 + \cdots + m_r\mathbf{a}_r = \mathbf{0}$ where one of the coefficients is not 0, which contradicts the independence of $\mathbf{a}_1$, $\cdots$, $\mathbf{a}_r$. Thus we may multiply (15) by $1/m_{r+1}$ and solve for $\mathbf{a}_{r+1}$ to yield the desired result.

The last two theorems enable us to derive an important representation of an arbitrary vector in terms of three basic ones.

THEOREM 24. *Let* $\mathbf{u}$, $\mathbf{v}$, $\mathbf{w}$ *be independent vectors. Then any vector* $\mathbf{r}$ *is expressible in the form* $x\mathbf{u} + y\mathbf{v} + z\mathbf{w}$ *in one and only one way.*

Proof. By Theorem 22, $\mathbf{u}$, $\mathbf{v}$, $\mathbf{w}$, $\mathbf{r}$ are dependent. By hypothesis $\mathbf{u}$, $\mathbf{v}$, $\mathbf{w}$ are independent. Hence Theorem 23 implies $\mathbf{r} = x\mathbf{u} + y\mathbf{v} + z\mathbf{w}$. To prove the uniqueness condition we suppose $\mathbf{r} = x'\mathbf{u} + y'\mathbf{v} + z'\mathbf{w}$, and show $x = x'$, $y = y'$, $z = z'$. We have

$$x\mathbf{u} + y\mathbf{v} + z\mathbf{w} = x'\mathbf{u} + y'\mathbf{v} + z'\mathbf{w}.$$

Transposing and collecting terms

$$(x\mathbf{u} - x'\mathbf{u}) + (y\mathbf{v} - y'\mathbf{v}) + (z\mathbf{w} - z'\mathbf{w}) = 0,$$

$$(x - x')\mathbf{u} + (y - y')\mathbf{v} + (z - z')\mathbf{w} = 0.$$

Since $\mathbf{u}$, $\mathbf{v}$, $\mathbf{w}$ are independent $x - x' = y - y' = z - z' = 0$. Hence $x = x'$, $y = y'$, $z = z'$ and our proof is complete.

Coordinates by Means of Vectors. We give the last result direct geometric meaning in terms of position vectors. The methods of cartesian analytic geometry appear as a by-product.

Let O, U, V, W be four noncoplanar points; P be any point of space (Fig. 22). We obtain an expression for the position vector (relative to O)

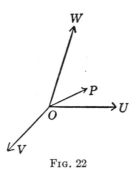

Fɪɢ. 22

of P in terms of those of U, V, W. Let $\mathbf{u} = \overrightarrow{OU}$, $\mathbf{v} = \overrightarrow{OV}$, $\mathbf{w} = \overrightarrow{OW}$, $\mathbf{r} = \overrightarrow{OP}$. By Theorem 18, $\mathbf{u}$, $\mathbf{v}$, $\mathbf{w}$ are independent. Hence $\mathbf{r} = \overrightarrow{OP} = x\mathbf{u} + y\mathbf{v} + z\mathbf{w}$ where x, y, z are uniquely determined. Thus we may assert

THEOREM 25. *If points O, U, V, W do not coplane, then any vector $\overrightarrow{OP}$ is uniquely representable in the form*

(16) $$\overrightarrow{OP} = x\overrightarrow{OU} + y\overrightarrow{OV} + z\overrightarrow{OW}.$$

We naturally think of x, y, z as "coordinates" which serve to locate P by means of O, U, V, W. Thus we introduce

DEFINITION 7. Let O, U, V, W be given noncoplanar points and P be an arbitrary point. Then the numbers x, y, z of (16) are called the *coordinates of P with respect to O, U, V, W* or in the *frame* (O, U, V, W). We refer to OU, OV, OW as the x, y, z *axes* respectively. If O, U, V, W are fixed throughout a discussion we call x, y, z simply the *coordinates* of P and use the familiar notation $P(x, y, z)$. Clearly the association of x, y, z to P, effects a one-to-one correspondence between the set of all points of space and the set of all number triples (x, y, z).

It is not hard to see that the coordinates of P as defined vectorially, are actually the familiar cartesian coordinates of P relative to the (oblique) axes OU, OV, OW in which U, V, W are the "unit points" on these axes respectively. Thus the vector analytic treatment of geometry not merely has a deeper and more truly geometric basis than the cartesian analytic treatment, but actually includes the latter. As an illustration we derive representations of lines of solid analytic geometry.

Let O, U, V, W be fixed noncoplanar points and consider the equation of line AB as given in Theorem 19:

$$(17) \qquad \mathbf{r} = \mathbf{a} + t\mathbf{p},$$

where $\mathbf{r} = \overrightarrow{OP}, \mathbf{a} = \overrightarrow{OA}, \mathbf{p} = \overrightarrow{AB}$. Let $\mathbf{u} = \overrightarrow{OU}, \mathbf{v} = \overrightarrow{OV}, \mathbf{w} = \overrightarrow{OW}$ and express $\mathbf{r}, \mathbf{a}, \mathbf{p}$ in terms of $\mathbf{u}, \mathbf{v}, \mathbf{w}$. We have

$$\mathbf{r} = x\mathbf{u} + y\mathbf{v} + z\mathbf{w}, \qquad \mathbf{a} = x_1\mathbf{u} + y_1\mathbf{v} + z_1\mathbf{w}, \qquad \mathbf{p} = p_1\mathbf{u} + p_2\mathbf{v} + p_3\mathbf{w}.$$

Substituting in (17) we get

$$x\mathbf{u} + y\mathbf{v} + z\mathbf{w} = x_1\mathbf{u} + y_1\mathbf{v} + z_1\mathbf{w} + t(p_1\mathbf{u} + p_2\mathbf{v} + p_3\mathbf{w}),$$

and

$$(18) \quad x\mathbf{u} + y\mathbf{v} + z\mathbf{w} = (x_1 + tp_1)\mathbf{u} + (y_1 + tp_2)\mathbf{v} + (z_1 + tp_3)\mathbf{w}.$$

By the uniqueness condition in Theorem 25 we may equate "corresponding" coefficients in (18), getting

$$(19) \qquad x = x_1 + tp_1, \qquad y = y_1 + tp_2, \qquad z = z_1 + tp_3.$$

Assuming p_1, p_2, p_3 are not 0, we have

$$(20) \qquad (x - x_1)/p_1 = (y - y_1)/p_2 = (z - z_1)/p_3 = t.$$

Relations (19), (20) are standard cartesian representations of a line. We can easily retrace our steps from (20) to (19) to (17) so that (20) and (19) truly represent line AB.

Linear Independence and Dimensionality. Returning to the geometrical significance of independence, note that 2 independent vectors can be localized in a plane (or 2-space) but never in a line (or 1-space). Likewise 3 independent vectors are in 3-space but never in a 2-space. This indicates a relation between independence and dimensionality which we wish to amplify.

In order to focus attention on *vector* properties, let us call the set of all geometric vectors a *spatial set of vectors* and denote it S_3. Thus S_3 is the set of all vectors $\overrightarrow{XY}$ where X, Y are points of our underlying 3-space. Then S_3 contains 3 independent vectors $\mathbf{a}_1, \mathbf{a}_2, \mathbf{a}_3$. By Theorem 24 S_3 consists of all vectors expressible in the form $t_1\mathbf{a}_1 + t_2\mathbf{a}_2 + t_3\mathbf{a}_3$. Moreover 3 is the maximum number of independent vectors in S_3, since by Theorem 22 any 4 vectors of S_3 are dependent. To summarize: S_3 *contains* 3 *independent vectors* $\mathbf{a}_1$, $\mathbf{a}_2$, $\mathbf{a}_3$ *and consists of all linear combinations of* $\mathbf{a}_1$, $\mathbf{a}_2$, $\mathbf{a}_3$. *The maximum number of independent vectors in S_3 is* 3.

Similar results hold for a line. Let us call the set of all vectors $\overrightarrow{XY}$, where X, Y lie in line AB, a *lineal set of vectors*, denoted S_1. Any vector of S_1 can be localized as $\overrightarrow{AP}$, where P is in line AB. Let $\mathbf{a}_1 = \overrightarrow{AB}$. Then $\mathbf{a}_1 \neq \mathbf{0}$ and so is independent. By Theorem 15, S_1 consists of all vectors $t_1 \mathbf{a}_1$. Finally, any 2 vectors of S_1 are dependent by Theorem 16. In summary: S_1 *contains* 1 *independent vector* $\mathbf{a}_1$ *and consists of all linear combinations of* $\mathbf{a}_1$. *The maximum number of independent vectors in S_1 is* 1.

Similarly we can define a *planar set of vectors* S_2, and get the analogous result. Thus the Euclidean *dimension number of a manifold* (line, plane or the whole space) is precisely the maximum number of independent vectors "lying" in the manifold.

THE SCALAR PRODUCT OF TWO GEOMETRIC VECTORS[6]

The vector algebra we have developed enables us to study purely linear relations like concurrence and parallelism of lines, and to compare and relate distances on lines which have the same direction. It does not provide a general theory of metrical questions, for example, it does not enable us to compare distances in general, or to compare or measure angles, or even to test for perpendicularity of lines. We introduce now a powerful vectorial operation which is applicable to these situations and is based on the notion perpendicularity.

Scalar Product. We base the definition of the operation on the idea "directed distance" on a line, which we now recall. Let there be assigned to a given line l a fixed sense of description, which we call the *positive* or *standard sense* on l. If A, B are any points of l, not necessarily distinct, the *directed distance* from A to B (dd. AB) is a number equal to the distance $\overline{AB}$ taken positively or negatively according as the sense from A to B is the same as or opposite to the standard sense on l. Thus dd. $AB = -$dd. BA. We recall the following principle which is easily proved:

For any points A, B, C of l, dd. $AB +$ dd. $BC =$ dd. AC.

Next we define a notion called the *scalar projection* or simply the *projection* of $\mathbf{a}$ on $\mathbf{b}$. Suppose $\mathbf{b} \neq \mathbf{0}$. Let $\mathbf{a} = \overrightarrow{PQ}$, $\mathbf{b} = \overrightarrow{PR}$ (Fig. 23). Let Q' be the foot of the perpendicular from Q to PR. Then the *projection* of $\mathbf{a}$ on $\mathbf{b}$ is dd. PQ', where the positive sense on PR is that of $\mathbf{b}$,

---

[6] This section is important for geometric vector analysis, but not essential to an understanding of the remainder of this chapter.

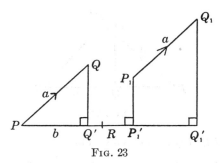

FIG. 23

and so is from P to R. If $\mathbf{b} = \mathbf{0}$ we define the *projection* of $\mathbf{a}$ on $\mathbf{b}$ to be the number 0.

Finally we have

DEFINITION 8. Denote the magnitude of any vector $\mathbf{r}$ by $|\,\mathbf{r}\,|$. Then the *scalar product* of $\mathbf{a}$ and $\mathbf{b}$, denoted $\mathbf{a} \cdot \mathbf{b}$, is the product of the projection of $\mathbf{a}$ on $\mathbf{b}$ multiplied by $|\,\mathbf{b}\,|$. Thus in the notation above when $\mathbf{b} \neq \mathbf{0}$, $\mathbf{a} \cdot \mathbf{b} = (\text{dd. } PQ')(\overline{PR})$. We write $\mathbf{a}^2$ for $\mathbf{a} \cdot \mathbf{a}$. Observe $\mathbf{a}^2 = |\,\mathbf{a}\,|^2$.

In forming the projection of $\mathbf{a}$ on $\mathbf{b}$, according to our definition $\mathbf{a}$ and $\mathbf{b}$ must be localized at the same point P. Often, however, it is very convenient to localize $\mathbf{a}$ and $\mathbf{b}$ at different points. Suppose we localize $\mathbf{a}$ (see Fig. 23) as $\overrightarrow{P_1Q_1}$ where P_1, Q_1 and PR coplane. Then if we project P_1, Q_1 perpendicularly on PR to get P_1', Q_1', we see that dd. $P_1'Q_1' =$ dd. PQ' and so equals the projection of $\mathbf{a}$ on $\mathbf{b}$. This result holds if P_1, Q_1 and PR do not coplane, but in this case it is not so obvious since $P_1Q_1Q_1'P_1'$ is not a plane figure and we shall not assume the result.

We see easily that the projection of $\mathbf{a}$ on $\mathbf{b}$ and the scalar product $\mathbf{a} \cdot \mathbf{b}$ are uniquely determined scalars. Concerning products involving "zero," we have $\mathbf{a} \cdot \mathbf{0} = \mathbf{0} \cdot \mathbf{a} = 0$; that is, any vector "times" the null vector is the number zero. If $\mathbf{a}$, $\mathbf{b} \neq \mathbf{0}$ the value of $\mathbf{a} \cdot \mathbf{b}$ is determined by the size of $\angle QPR$ and the distances $\overline{PQ}$, $\overline{PR}$ (Fig. 23), so that it seems well suited to the study of problems involving measurement of angles and distances. Observe $\mathbf{a} \cdot \mathbf{b}$ is positive, zero or negative according as $\angle QPR$ is acute, right or obtuse. From this we infer the

LEMMA. *Let* $\mathbf{a} = \overrightarrow{PQ}$, $\mathbf{b} = \overrightarrow{PR}$ *be vectors not equal to* $\mathbf{0}$. *Then* $\angle QPR$ *is a right angle if and only if* $\mathbf{a} \cdot \mathbf{b} = 0$.

Algebraic Properties. We consider the algebraic properties of the scalar product.

THEOREM 26. $\mathbf{a} \cdot \mathbf{b} = \mathbf{b} \cdot \mathbf{a}$.

Proof. Let $\mathbf{a} = \overrightarrow{PQ}$, $\mathbf{b} = \overrightarrow{PR}$. Suppose P, Q, R do not colline (Fig. 24). If $\angle QPR$ is a right angle then $\mathbf{a} \cdot \mathbf{b} = 0 = \mathbf{b} \cdot \mathbf{a}$ by the Lemma. Suppose then $\angle QPR$ is not a right angle. Let Q', R' be the perpendicular projec-

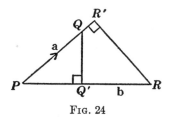

tions of Q, R on PR, PQ respectively. Then

$$\mathbf{a \cdot b} = (\pm\overline{PQ'})\overline{PR}, \qquad \mathbf{b \cdot a} = (\pm\overline{PR'})\overline{PQ}$$

where the sign is $+$ or $-$ according as $\angle QPR$ is acute or obtuse. Clearly $\triangle PQQ' \sim \triangle PRR'$ so that

$$\overline{PQ'}/\overline{PQ} = \overline{PR'}/\overline{PR}, \qquad (\overline{PQ'})(\overline{PR}) = (\overline{PR'})(\overline{PQ})$$

and $\mathbf{a \cdot b} = \mathbf{b \cdot a}$ holds. The degenerate cases where P, Q, R colline are easily disposed of.

There is little more to say about the formal properties of the operation scalar multiplication of vectors in itself. This is because it associates to two vectors a *scalar* rather than a vector. However it has, in conjunction with the other vector operations, important algebraic properties which we now consider.

THEOREM 27. $(m\mathbf{a}) \cdot \mathbf{b} = m(\mathbf{a \cdot b})$.

Proof. This is easily justified by considering separately the cases $m > 0$, $m = 0$, $m < 0$.

COROLLARY. $(-\mathbf{a}) \cdot \mathbf{b} = -(\mathbf{a \cdot b})$.

We derive the distributive law for scalar multiplication with respect to addition in a restricted form and later remove the restriction.

THEOREM 28. (Restricted Distributive Law.) $(\mathbf{a} + \mathbf{b}) \cdot \mathbf{c} = \mathbf{a \cdot c} + \mathbf{b \cdot c}$ *provided* $\mathbf{a}$, $\mathbf{b}$, $\mathbf{c}$ *are dependent.*

Proof. If $\mathbf{a}$, $\mathbf{b}$ or $\mathbf{c} = \mathbf{0}$ the result is immediate. Suppose $\mathbf{a}$, $\mathbf{b}$, $\mathbf{c} \neq \mathbf{0}$. Let $\mathbf{a} = \overrightarrow{PQ}$, $\mathbf{b} = \overrightarrow{PR}$, $\mathbf{c} = \overrightarrow{PS}$. By Theorem 18 P, Q, R, S are in a plane, say π (Fig. 25). Relocalize $\mathbf{b}$ as $\overrightarrow{QT}$, and let Q', T' be the respective pro-

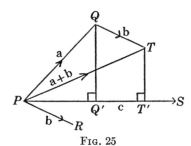

jections of Q, T on PS. Since P, Q, R are in π, T also is in π. Thus Q, T and PS *coplane*. Hence by our remark following Definition 8, the projection of $\mathbf{b}$ on $\mathbf{c}$ is dd. $Q'T'$. Clearly the projection of $\mathbf{a} + \mathbf{b}$ on $\mathbf{c}$ is dd. PT', and that of $\mathbf{a}$ on $\mathbf{c}$ is dd. PQ'. Hence

$$(\mathbf{a} + \mathbf{b})\cdot\mathbf{c} = (\text{projection of } \mathbf{a} + \mathbf{b} \text{ on } \mathbf{c})| \mathbf{c} |$$
$$= (\text{dd. } PT')| \mathbf{c} | = (\text{dd. } PQ' + \text{dd. } Q'T')| \mathbf{c} |$$
$$= (\text{dd. } PQ')| \mathbf{c} | + (\text{dd. } Q'T')| \mathbf{c} | = \mathbf{a}\cdot\mathbf{c} + \mathbf{b}\cdot\mathbf{c}.$$

COROLLARY 1. $(\mathbf{a} - \mathbf{b})\cdot\mathbf{c} = \mathbf{a}\cdot\mathbf{c} - \mathbf{b}\cdot\mathbf{c}$ *provided* $\mathbf{a}$, $\mathbf{b}$, $\mathbf{c}$ *are dependent.*
Proof. Clearly $\mathbf{a}$, $-\mathbf{b}$, $\mathbf{c}$ are dependent. Hence

$$(\mathbf{a} - \mathbf{b})\cdot\mathbf{c} = (\mathbf{a} + (-\mathbf{b}))\cdot\mathbf{c} = \mathbf{a}\cdot\mathbf{c} + (-\mathbf{b})\cdot\mathbf{c}$$
$$= \mathbf{a}\cdot\mathbf{c} + (-(\mathbf{b}\cdot\mathbf{c})) = \mathbf{a}\cdot\mathbf{c} - \mathbf{b}\cdot\mathbf{c}^{\displaystyle.}$$

COROLLARY 2. $(\mathbf{a} \pm \mathbf{b})^2 = \mathbf{a}^2 \pm 2(\mathbf{a}\cdot\mathbf{b}) + \mathbf{b}^2.$
Proof. Applying the distributive law purely formally we have

$$(\mathbf{a} + \mathbf{b})^2 = (\mathbf{a} + \mathbf{b})\cdot(\mathbf{a} + \mathbf{b}) = \mathbf{a}\cdot(\mathbf{a} + \mathbf{b}) + \mathbf{b}\cdot(\mathbf{a} + \mathbf{b})$$
$$= \mathbf{a}\cdot\mathbf{a} + \mathbf{a}\cdot\mathbf{b} + \mathbf{b}\cdot\mathbf{a} + \mathbf{b}\cdot\mathbf{b} = \mathbf{a}^2 + 2\mathbf{a}\cdot\mathbf{b} + \mathbf{b}^2.$$

Clearly $\mathbf{a}$, $\mathbf{b}$, $\mathbf{a} + \mathbf{b}$ are dependent since $1\mathbf{a} + 1\mathbf{b} + (-1)(\mathbf{a} + \mathbf{b}) = \mathbf{0}$. Likewise the triples $\mathbf{a}$, $\mathbf{b}$, $\mathbf{a}$ and $\mathbf{a}$, $\mathbf{b}$, $\mathbf{b}$ are dependent. Thus the applications of the distributive law are covered by the theorem. The reasoning is identical for $(\mathbf{a} - \mathbf{b})^2$.

COROLLARY 3. $\mathbf{a}\cdot\mathbf{b} = \tfrac{1}{4}((\mathbf{a} + \mathbf{b})^2 - (\mathbf{a} - \mathbf{b})^2).$
Proof. Form the difference of the equations in Corollary 2 and divide by 4.

Now we remove the restriction in Theorem 28. This could be done by following the proof given and showing, as an exercise in solid geometry, that the projection of $\mathbf{b}$ on $\mathbf{c}$ is dd. $Q'T'$ even if Q, T and PS do not coplane. We prefer however to derive the result by vector algebraic reasoning to illustrate its elegance and power.

THEOREM 29. (Distributive Law.) $(\mathbf{a} + \mathbf{b})\cdot\mathbf{c} = \mathbf{a}\cdot\mathbf{c} + \mathbf{b}\cdot\mathbf{c}.$
Proof. Leaning heavily on the last two corollaries, we have

$$(\mathbf{a} + \mathbf{b})\cdot\mathbf{c} = \tfrac{1}{4}((\mathbf{a} + \mathbf{b} + \mathbf{c})^2 - (\mathbf{a} + \mathbf{b} - \mathbf{c})^2)$$
$$= \tfrac{1}{4}(\mathbf{a}^2 + 2\mathbf{a}\cdot(\mathbf{b} + \mathbf{c}) + (\mathbf{b} + \mathbf{c})^2$$
$$- (\mathbf{a}^2 + 2\mathbf{a}\cdot(\mathbf{b} - \mathbf{c}) + (\mathbf{b} - \mathbf{c})^2))$$
$$= \tfrac{1}{4}(2\mathbf{a}\cdot(\mathbf{b} + \mathbf{c}) - 2\mathbf{a}\cdot(\mathbf{b} - \mathbf{c}) + (\mathbf{b} + \mathbf{c})^2 - (\mathbf{b} - \mathbf{c})^2)$$
$$= \tfrac{1}{2}(\mathbf{a}\cdot(\mathbf{b} + \mathbf{c}) - \mathbf{a}\cdot(\mathbf{b} - \mathbf{c})) + \tfrac{1}{4}((\mathbf{b} + \mathbf{c})^2 - (\mathbf{b} - \mathbf{c})^2)$$

$$= \tfrac{1}{8}((\mathbf{a} + \mathbf{b} + \mathbf{c})^2 - (\mathbf{a} - \mathbf{b} - \mathbf{c})^2$$
$$- ((\mathbf{a} + \mathbf{b} - \mathbf{c})^2 - (\mathbf{a} - \mathbf{b} + \mathbf{c})^2)) + \mathbf{b} \cdot \mathbf{c}$$
$$= \tfrac{1}{8}((\mathbf{a} + \mathbf{c} + \mathbf{b})^2 + (\mathbf{a} + \mathbf{c} - \mathbf{b})^2$$
$$- (\mathbf{a} - \mathbf{c} + \mathbf{b})^2 - (\mathbf{a} - \mathbf{c} - \mathbf{b})^2) + \mathbf{b} \cdot \mathbf{c}$$
$$= \tfrac{1}{8}(2(\mathbf{a} + \mathbf{c})^2 + 2\mathbf{b}^2 - 2(\mathbf{a} - \mathbf{c})^2 - 2\mathbf{b}^2) + \mathbf{b} \cdot \mathbf{c}$$
$$= \tfrac{1}{4}((\mathbf{a} + \mathbf{c})^2 - (\mathbf{a} - \mathbf{c})^2) + \mathbf{b} \cdot \mathbf{c}$$
$$= \mathbf{a} \cdot \mathbf{c} + \mathbf{b} \cdot \mathbf{c}.$$

COROLLARY. $\mathbf{a} \cdot (\mathbf{b}_1 + \cdots + \mathbf{b}_n) = \mathbf{a} \cdot \mathbf{b}_1 + \cdots + \mathbf{a} \cdot \mathbf{b}_n$.

Geometric Properties. Many important geometric consequences flow from the distributive law. They are easily derived by vectorializing the treatment of angular magnitude. For this purpose we introduce

DEFINITION 9. Suppose $\mathbf{a}, \mathbf{b} \neq \mathbf{0}$. Let $\mathbf{a} = \vec{PQ}$, $\mathbf{b} = \vec{PR}$, then the *angle between vectors* $\mathbf{a}, \mathbf{b}$, denoted $(\mathbf{a}, \mathbf{b})$, is defined to be the magnitude of $\measuredangle QPR$. If $\mathbf{a}$ or $\mathbf{b}$ is $\mathbf{0}$ we do not find it convenient to define $(\mathbf{a}, \mathbf{b})$. Note that $(\mathbf{a}, \mathbf{b})$ can be the magnitude of any angle from a zero angle to a straight angle, inclusive.

We use this idea to reformulate the definition of $\mathbf{a} \cdot \mathbf{b}$.

THEOREM 30. *Suppose* $\mathbf{a}, \mathbf{b} \neq \mathbf{0}$. *Then* $\mathbf{a} \cdot \mathbf{b} = |\mathbf{a}| |\mathbf{b}| \cos (\mathbf{a}, \mathbf{b})$.

Proof. Let $\mathbf{a} = \vec{PQ}$, $\mathbf{b} = \vec{QR}$, Q' be the projection of Q on PR (Fig. 26). Then

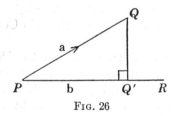

FIG. 26

$$\mathbf{a} \cdot \mathbf{b} = (\mathrm{dd}.\ PQ') |\mathbf{b}| = \frac{\mathrm{dd}.\ PQ'}{PQ} \overline{PQ} |\mathbf{b}| =$$

$$(\cos \measuredangle QPR) |\mathbf{a}| |\mathbf{b}| = |\mathbf{a}| |\mathbf{b}| \cos (\mathbf{a}, \mathbf{b}).$$

Solving for $\cos (\mathbf{a}, \mathbf{b})$ we obtain an important formula for determining the angle between two lines in terms of vectors directed along them.

COROLLARY. *Suppose* $\mathbf{a}, \mathbf{b} \neq \mathbf{0}$. *Then* $\cos (\mathbf{a}, \mathbf{b}) = \dfrac{\mathbf{a} \cdot \mathbf{b}}{|\mathbf{a}| |\mathbf{b}|}$.

To obtain a corresponding formula for the distance between two points

$A(\mathbf{a})$, $B(\mathbf{b})$ in terms of their position vectors, we need merely express geometrically the formula for the square of the difference of two vectors (Theorem 28, Corollary 2). Let $\mathbf{c} = \overrightarrow{AB}$ (Fig. 27). Then $\mathbf{c} = \mathbf{b} - \mathbf{a}$ and

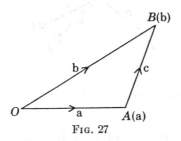

FIG. 27

$|\,\mathbf{a}\,|$, $|\,\mathbf{b}\,|$, $|\,\mathbf{c}\,|$ are the respective distances $\overline{OA}$, $\overline{OB}$, $\overline{AB}$. We have

$$\overline{AB}^2 = |\,\mathbf{c}\,|^2 = \mathbf{c}^2 = (\mathbf{b} - \mathbf{a})^2 = \mathbf{a}^2 - 2\mathbf{a}\cdot\mathbf{b} + \mathbf{b}^2.$$

Now suppose $\mathbf{a}, \mathbf{b} \neq \mathbf{0}$. Then by Theorem 30 and Definition 9

$$|\,\mathbf{c}\,|^2 = \mathbf{a}^2 + \mathbf{b}^2 - 2\,|\,\mathbf{a}\,|\,|\,\mathbf{b}\,|\cos(\mathbf{a},\mathbf{b})$$

$$= |\,\mathbf{a}\,|^2 + |\,\mathbf{b}\,|^2 - 2\,|\,\mathbf{a}\,|\,|\,\mathbf{b}\,|\cos \measuredangle\ AOB.$$

This is, if O, A, B do not colline, the Law of Cosines of elementary trigonometry.

Moreover if $\measuredangle AOB$ is a right angle, $\mathbf{a}\cdot\mathbf{b} = 0$, and we have

$$|\,\mathbf{c}\,|^2 = |\,\mathbf{a}\,|^2 + |\,\mathbf{b}\,|^2,$$

which justifies the Pythagorean Theorem.

The fact that these two important geometric theorems are merely corollaries of the distributive law is an indication of the depth and power of vector analysis. Observe the contrast with cartesian analytic geometry where these principles are first proved by the methods of classical geometry, and then are used to derive important formulas like those for the distance between two points and the angle between two lines.

Geometric Application. Our "extended" vector algebra is applicable to school geometry. First consider the following basic theorem of solid geometry which we have not assumed in the development of our theory: *If a line is perpendicular to each of two distinct intersecting lines at their intersection, it is perpendicular to each line in their plane passing through the intersection.*

Suppose $OP \perp OA, OB$; OA, OB are distinct lines; and OC is in plane OAB (Fig. 28). We show $OP \perp OC$. Let $\mathbf{r} = \overrightarrow{OP}$, $\mathbf{a} = \overrightarrow{OA}$, $\mathbf{b} = \overrightarrow{OB}$,

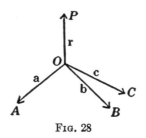

$\mathbf{c} = \overrightarrow{OC}$. By the Lemma $\mathbf{r}\cdot\mathbf{a} = \mathbf{r}\cdot\mathbf{b} = 0$. Since C is in plane OAB, Theorem 17 implies

$$\mathbf{c} = \overrightarrow{OC} = t\overrightarrow{OA} + u\overrightarrow{OB} = t\mathbf{a} + u\mathbf{b}.$$

Hence

$$\mathbf{r}\cdot\mathbf{c} = \mathbf{r}\cdot(t\mathbf{a} + u\mathbf{b}) = \mathbf{r}\cdot(t\mathbf{a}) + \mathbf{r}\cdot(u\mathbf{b})$$

$$= t(\mathbf{r}\cdot\mathbf{a}) + u(\mathbf{r}\cdot\mathbf{b}) = 0$$

and $OP \perp OC$ by the Lemma (p. 179).

Another example interesting in the interplay of linear dependence and vector operations is a sort of converse. Assuming as above $OP \perp OA$, OB and OA, OB are distinct, we suppose $OP \perp OC$ and show OA, OB, OC coplane. Using the notation above, the hypothesis implies by the Lemma,

(21) $\mathbf{r}\cdot\mathbf{a} = \mathbf{r}\cdot\mathbf{b} = \mathbf{r}\cdot\mathbf{c} = 0.$

By Theorem 22 $\mathbf{a}$, $\mathbf{b}$, $\mathbf{c}$, $\mathbf{r}$ are dependent. Thus we have

(22) $x\mathbf{a} + y\mathbf{b} + z\mathbf{c} + w\mathbf{r} = \mathbf{0},$

where x, y, z, w are not all zero. Taking the scalar product of both members of (22) with $\mathbf{r}$, we have

$$x(\mathbf{r}\cdot\mathbf{a}) + y(\mathbf{r}\cdot\mathbf{b}) + z(\mathbf{r}\cdot\mathbf{c}) + w(\mathbf{r}\cdot\mathbf{r}) = 0.$$

Hence (21) implies $w(\mathbf{r}\cdot\mathbf{r}) = w\,|\,\mathbf{r}\,|^2 = 0$. Since $\mathbf{r} = \overrightarrow{OP} \neq \mathbf{0}$, we see $w = 0$. Substituting in (22) we see $\mathbf{a}$, $\mathbf{b}$, $\mathbf{c}$ are dependent. Thus (Theorem 18) O, A, B, C coplane and so do OA, OB, OC.

Next as an illustration of the vectorial study of distances we show: *If a triangle has two equal medians it is isosceles.* Consider $\triangle OAB$ with $\overline{AB'} = \overline{BA'}$ (Fig. 29) where A', B' are the respective midpoints of OA, OB. Suppose $A(\mathbf{a})$, $B(\mathbf{b})$.[7] Then $A'(\tfrac{1}{2}\mathbf{a})$, $B'(\tfrac{1}{2}\mathbf{b})$. Thus $\overrightarrow{AB'} =$

---

[7] That is $\mathbf{a}$ and $\mathbf{b}$ are the respective position vectors of A and B.

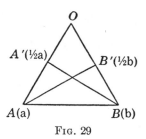

FIG. 29

$\frac{1}{2}\mathbf{b} - \mathbf{a}$, $\overrightarrow{BA'} = \frac{1}{2}\mathbf{a} - \mathbf{b}$. The equality $\overline{AB'} = \overline{BA'}$ expressed vectorially is $|\overrightarrow{AB'}| = |\overrightarrow{BA'}|$ or $|\frac{1}{2}\mathbf{b} - \mathbf{a}| = |\frac{1}{2}\mathbf{a} - \mathbf{b}|$. Squaring we get

$$|\tfrac{1}{2}\mathbf{b} - \mathbf{a}|^2 = |\tfrac{1}{2}\mathbf{a} - \mathbf{b}|^2.$$

Hence

$$(\tfrac{1}{2}\mathbf{b} - \mathbf{a})^2 = (\tfrac{1}{2}\mathbf{a} - \mathbf{b})^2$$

$$\tfrac{1}{4}\mathbf{b}^2 - \mathbf{b}\cdot\mathbf{a} + \mathbf{a}^2 = \tfrac{1}{4}\mathbf{a}^2 - \mathbf{a}\cdot\mathbf{b} + \mathbf{b}^2.$$

Collecting terms $\frac{3}{4}\mathbf{a}^2 = \frac{3}{4}\mathbf{b}^2$ and $\mathbf{a}^2 = \mathbf{b}^2$. Thus $|\mathbf{a}| = |\mathbf{b}|$ and $\overline{OA} = \overline{OB}$.

Exercise. Prove the formula $(\mathbf{a} + \mathbf{b})\cdot(\mathbf{a} - \mathbf{b}) = \mathbf{a}^2 - \mathbf{b}^2$ and interpret it geometrically.

Rectangular Coordinates. We conclude this part by deriving the familiar formulas of solid analytic geometry for the distance between two points and the angle between two lines in rectangular coordinates. In the present context rectangular are preferable to oblique coordinates since they enable us to conveniently exploit the "rectangular" properties of the scalar product.

Thus we choose a so-called *rectangular coordinate frame* (O, U, V, W) which has the properties that OU, OV, OW are mutually perpendicular and $\overrightarrow{OU}, \overrightarrow{OV}, \overrightarrow{OW}$ are equal to a given unit of length. Let $\mathbf{i} = \overrightarrow{OU}$, $\mathbf{j} = \overrightarrow{OV}$, $\mathbf{k} = \overrightarrow{OW}$. Let (x, y, z) be the coordinates of P in this given frame (O, U, V, W); then by Definition 7, $\overrightarrow{OP} = x\mathbf{i} + y\mathbf{j} + z\mathbf{k}$. Applying the operation "scalar product" to the "basic" vectors $\mathbf{i}, \mathbf{j}, \mathbf{k}$ we have

$$\mathbf{i}^2 = \mathbf{j}^2 = \mathbf{k}^2 = 1, \qquad \mathbf{i}\cdot\mathbf{j} = \mathbf{j}\cdot\mathbf{k} = \mathbf{i}\cdot\mathbf{k} = 0.$$

Thus we get an especially simple formula for the scalar product:

$$(23) \quad (x_1\mathbf{i} + y_1\mathbf{j} + z_1\mathbf{k})\cdot(x_2\mathbf{i} + y_2\mathbf{j} + z_2\mathbf{k}) = x_1x_2 + y_1y_2 + z_1z_2.$$

This easily yields a formula for the magnitude of a vector:

$$(24) \quad | x\mathbf{i} + y\mathbf{j} + z\mathbf{k} | = \sqrt{(x\mathbf{i} + y\mathbf{j} + z\mathbf{k})^2} = \sqrt{x^2 + y^2 + z^2}.$$

Now let P_1, P_2 be arbitrary points with respective coordinates (x_1, y_1, z_1), (x_2, y_2, z_2). Then $\overrightarrow{OP_1} = x_1\mathbf{i} + y_1\mathbf{j} + z_1\mathbf{k}$, $\overrightarrow{OP_2} = x_2\mathbf{i} + y_2\mathbf{j} + z_2\mathbf{k}$ and employing (24)

$$\overrightarrow{P_1P_2} = | \overrightarrow{P_1P_2} | = | \overrightarrow{OP_2} - \overrightarrow{OP_1} |$$
$$= | (x_2 - x_1)\mathbf{i} + (y_2 - y_1)\mathbf{j} + (z_2 - z_1)\mathbf{k} |$$
$$= \sqrt{(x_2 - x_1)^2 + (y_2 - y_1)^2 + (z_2 - z_1)^2}.$$

Suppose $P(\mathbf{r})$ and $\mathbf{r} = x\mathbf{i} + y\mathbf{j} + z\mathbf{k} \neq \mathbf{0}$. Then

$$\mathbf{r}\cdot\mathbf{i} = (x\mathbf{i} + y\mathbf{j} + z\mathbf{k})\cdot\mathbf{i} = x.$$

Applying Theorem 30, Corollary, we have

$$(25) \qquad\qquad \cos(\mathbf{r}, \mathbf{i}) = \frac{\mathbf{r}\cdot\mathbf{i}}{|\mathbf{r}||\mathbf{i}|} = \frac{x}{|\mathbf{r}|}.$$

Similarly we show

$$(26) \qquad\qquad \cos(\mathbf{r}, \mathbf{j}) = \frac{y}{|\mathbf{r}|}, \qquad \cos(\mathbf{r}, \mathbf{k}) = \frac{z}{|\mathbf{r}|}.$$

Observe that $(\mathbf{r}, \mathbf{i})$ is the angle formed by $\mathbf{r}$ and $\overrightarrow{OU}$ or, equivalently, by the directed segment (O, P) and the x-axis positively sensed. Similar statements hold for $(\mathbf{r}, \mathbf{j})$ and $(\mathbf{r}, \mathbf{k})$ with respect to the y and z-axes. Thus $(\mathbf{r}, \mathbf{i})$, $(\mathbf{r}, \mathbf{j})$, $(\mathbf{r}, \mathbf{k})$ are merely the familiar *direction angles* of $\mathbf{r}$ or of directed segment (O, P). Thus (25), (26) are formulas for the cosines of these angles—the so-called *direction cosines* of $\mathbf{r}$ or of (O, P).

Finally let $\mathbf{r}_1 = x_1\mathbf{i} + y_1\mathbf{j} + z_1\mathbf{k}$ and $\mathbf{r}_2 = x_2\mathbf{i} + y_2\mathbf{j} + z_2\mathbf{k}$ be any vectors not $\mathbf{0}$. Let l_1, m_1, n_1 and l_2, m_2, n_2 be their respective direction cosines. By Theorem 30, Corollary, and (23), (25), (26)

$$\cos(\mathbf{r}_1, \mathbf{r}_2) = \frac{\mathbf{r}_1\cdot\mathbf{r}_2}{|\mathbf{r}_1||\mathbf{r}_2|} = \frac{x_1x_2 + y_1y_2 + z_1z_2}{|\mathbf{r}_1||\mathbf{r}_2|}$$

$$= \frac{x_1}{|\mathbf{r}_1|}\frac{x_2}{|\mathbf{r}_2|} + \frac{y_1}{|\mathbf{r}_1|}\frac{y_2}{|\mathbf{r}_2|} + \frac{z_1}{|\mathbf{r}_1|}\frac{z_2}{|\mathbf{r}_2|}$$

$$= l_1l_2 + m_1m_2 + n_1n_2.$$

INTRODUCTION TO ABSTRACT VECTOR ALGEBRA

In this part we introduce the concept of vector space. This is an autonomous mathematical idea abstracted from the *algebraic* properties of geometric vectors. It is applicable to many nongeometric systems which behave *algebraically* like geometric vectors.

An Algebra of Points. As we have seen (p. 172) if we associate to point P its position vector $\overrightarrow{OP}$, we have a one-to-one correspondence between the set of all points and the set of all geometric vectors. We used this correspondence to reduce the "geometry of points" to the "algebra of vectors." Now we "reverse" the correspondence and associate to each vector $\mathbf{r} = \overrightarrow{OP}$, the point P whose position vector is $\mathbf{r}$. Then any property of *geometric vectors* $\overrightarrow{OA}, \overrightarrow{OB}, \overrightarrow{OC}, \cdots$, will be *reflected* as some property of the corresponding *points* $A, B, C, \cdots$. Just as any motion a child makes before a mirror yields a corresponding motion of his image. For example if $\overrightarrow{OC} = \overrightarrow{OA} + \overrightarrow{OB}$, how is C related to A and B? We now consider this question.

DEFINITION 10. If $\overrightarrow{OA} + \overrightarrow{OB} = \overrightarrow{OC}$ we call C the (vectorial) *sum* of points A and B relative to origin O, or when the choice of O is fixed in a discussion, simply the (vectorial) *sum* of A and B and write $A + B = C$.

This is not really unfamiliar. This is how we operate on points when we add complex numbers graphically. What are the basic properties of this operation of adding points? We shall see that they are formally the same as those of addition of geometric vectors as established in Theorems $1, \cdots, 5$. Clearly we have

1. (*Closure Law*): $A + B$ *is a uniquely determined point.*

Next we prove

2. (*Associative Law*): $(A + B) + C = A + (B + C)$.

By the associative law for adding geometric vectors we have

$$(27) \qquad (\overrightarrow{OA} + \overrightarrow{OB}) + \overrightarrow{OC} = \overrightarrow{OA} + (\overrightarrow{OB} + \overrightarrow{OC}) = \overrightarrow{OD}.$$

Let $\overrightarrow{OA} + \overrightarrow{OB} = \overrightarrow{OP}$. Substituting in (27) we get $\overrightarrow{OP} + \overrightarrow{OC} = \overrightarrow{OD}$. The last two relations yield for addition of points

$$A + B = P, \qquad P + C = D.$$

Substituting for P in the second equation its "value" as given in the first, we get $(A + B) + C = D$.

Similarly letting $\overrightarrow{OB} + \overrightarrow{OC} = \overrightarrow{OQ}$ we have by (27) $\overrightarrow{OA} + \overrightarrow{OQ} = \overrightarrow{OD}$. Thus

$$B + C = Q, \qquad A + Q = D.$$

Eliminating Q we have $A + (B + C) = D$ so that $(A + B) + C = A + (B + C)$.

Similarly using the commutative law for addition of vectors we can derive

3. (*Commutative Law*): $A + B = B + A$.

Furthermore the vector relation $\overrightarrow{OA} + \overrightarrow{OO} = \overrightarrow{OA}$ implies the "point" relation

4. (*Identity Law*): $A + O = A$ *for each point* A.

Finally we show

5. (*Inverse Law*): *For each point* A *the equation* $A + X = O$ *has one and only one solution* X.

To prove the existence of such a point X, consider the relation $\overrightarrow{OA} + (-\overrightarrow{OA}) = \mathbf{0} = \overrightarrow{OO}$. Localize $-\overrightarrow{OA}$ at O as $\overrightarrow{OA'}$. Then $\overrightarrow{OA} + \overrightarrow{OA'} = \overrightarrow{OO}$. Hence $A + A' = O$, and A' satisfies the given equation. To prove uniqueness suppose $A + X = O$. By Definition 10 this means $\overrightarrow{OA} + \overrightarrow{OX} = \overrightarrow{OO}$. By Theorem 5 this equation has a unique solution. Thus $\overrightarrow{OX} = \overrightarrow{OA'}$ and $X = A'$.

Observe A', the point "inverse" to A, is gotten graphically by "reflecting" A in origin O.

Similarly we study multiplication of points by scalars:

DEFINITION 11. If $m\overrightarrow{OA} = \overrightarrow{OB}$ we call B the *product* of point A by scalar m (relative to origin O) and write $mA = B$.

By "translating" vector properties into "point" properties, we can easily show the basic laws for multiplication of vectors by scalars hold for multiplication of points by scalars:

6. mA *is a uniquely determined point*.

7. $m(nA) = (mn)A$.

8. $(m + n)A = mA + nA$.

9. $m(A + B) = mA + mB$.

10. $1A = A$.

A few words on the graphical significance of our operations: Consider the effect on an arbitrary point X of adding to it a fixed point A. That is consider the "geometric" function, $X' = X + A$. We have (Fig. 30)

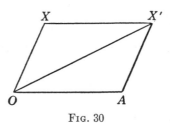

FIG. 30

$$\overrightarrow{OX'} = \overrightarrow{OX} + \overrightarrow{OA}, \qquad \overrightarrow{OX'} = \overrightarrow{OX} + \overrightarrow{XX'}.$$

Thus $\overrightarrow{XX'} = \overrightarrow{OA}$ and $(X, X') \equiv (O, A)$. Hence X' arises from X by "sliding" X a distance OA along the direction of OA, in the sense from O to A. That is $X' = X + A$ represents a *translation* of space.

Similarly $X' = mX$ represents a *dilatation* or "stretching transformation" of space in which the distance from X to O is "stretched" by a factor of m.

An Algebra of Number Triples. We have seen that the basic *algebraic* properties of geometric vectors hold for points of space when suitable operations are defined. They have no essential relation to the fact that geometric vectors are an abstraction from directed segments, or are "quantities with magnitude, direction and sense." This is seen in an even more striking way in the following construction of a "vectorial system" whose elements are purely algebraic in character. We employ the correspondence between points and their coordinates (see p. 176) to convert our "algebra of points" into an algebra of number triples.

Choose a frame (O, U, V, W) arbitrarily and associate to each point P its coordinate triple (x, y, z). Suppose $A + B = C$ where A (x_1, y_1, z_1), $B(x_2, y_2, z_2)$, $C(x_3, y_3, z_3)$. Then we call (x_3, y_3, z_3) the *sum* of (x_1, y_1, z_1) and (x_2, y_2, z_2) and write

$$(x_1, y_1, z_1) + (x_2, y_2, z_2) = (x_3, y_3, z_3).$$

Likewise if $mD = E$ and $D(x, y, z)$, $E(x', y', z')$ we call (x', y', z') the *product* of (x, y, z) by m, and write $m(x, y, z) = (x', y', z')$.

Although we have defined these operations geometrically, they are expressible purely in algebraic terms. For, recalling the definition of coordinates of a point, we have

$$\overrightarrow{OA} = x_1\overrightarrow{OU} + y_1\overrightarrow{OV} + z_1\overrightarrow{OW},$$
$$\overrightarrow{OB} = x_2\overrightarrow{OU} + y_2\overrightarrow{OV} + z_2\overrightarrow{OW},$$
$$\overrightarrow{OC} = x_3\overrightarrow{OU} + y_3\overrightarrow{OV} + z_3\overrightarrow{OW}.$$

$A + B = C$ implies $\overrightarrow{OA} + \overrightarrow{OB} = \overrightarrow{OC}$. Substituting in this for $\overrightarrow{OA}$, $\overrightarrow{OB}$, $\overrightarrow{OC}$ and collecting terms

$$(x_1 + x_2)\overrightarrow{OU} + (y_1 + y_2)\overrightarrow{OV} + (z_1 + z_2)\overrightarrow{OW} = x_3\overrightarrow{OU} + y_3\overrightarrow{OV} + z_3\overrightarrow{OW}.$$

Equating coefficients we get

$$x_1 + x_2 = x_3, \qquad y_1 + y_2 = y_3, \qquad z_1 + z_2 = z_3.$$

Thus we have an algebraic characterization of the sum of two triples:

$$(28) \quad (x_1, y_1, z_1) + (x_2, y_2, z_2) = (x_1 + x_2, y_1 + y_2, z_1 + z_2).$$

Similarly we can show

$$(29) \qquad m(x_1, x_2, x_3) = (mx_1, mx_2, mx_3).$$

It is easily seen from the *geometric* definitions of these operations in terms of corresponding operations on points that Properties 1, $\cdots$, 10 (pp. 187–88) when suitably rephrased hold for number triples. Thus Property 1 is rephrased as: The sum of two number triples is a uniquely determined number triple. Property 2 becomes the associative law for addition of number triples. The "zero" element is the triple (0, 0, 0) and the "inverse" of $(a_1, a_2, a_3) = (-a_1, -a_2, -a_3)$. Actually these ten properties can most easily be derived algebraically by reasoning directly from (28), (29) and should be so verified by the reader.

The Vector Space Concept. We have thus constructed three vectorial algebraic systems denoted V_G, V_P, V_3, whose elements are respectively *geometric vectors, points, number triples*. Each system has its own operations of addition and multiplication by a scalar and they share the ten basic properties we have enumerated. This suggests the introduction of a concept of abstract vector system, of which the three given systems will be concrete examples. Thus we adopt

DEFINITION 12. A *vector space*[8] is a system V, consisting of a set S of elements $\alpha, \beta, \gamma, \cdots$ and two operations which respectively associate to any two elements α, β of S a *sum* $\alpha + \beta$ and to any element α of S and any real number m a *product* $m\alpha$, which has the following properties:

V1. *$\alpha + \beta$ is a uniquely determined element of S;*

V2. $(\alpha + \beta) + \gamma = \alpha + (\beta + \gamma);$

V3. $\alpha + \beta = \beta + \alpha;$

V4. *There exists an element Θ in S such that $\alpha + \Theta = \alpha$ for each element α of S;*

V5. *For each element α of S the equation $\alpha + \xi = \Theta$ has a unique solution ξ in S;*

V6. *$m\alpha$ is a uniquely determined element of S;*

V7. $m(n\alpha) = (mn)\alpha;$

V8. $(m + n)\alpha = m\alpha + n\alpha;$

V9. $m(\alpha + \beta) = m\alpha + m\beta;$

V10. $1\alpha = \alpha.$

The elements of S are called *vectors* or *points*, the real numbers, *scalars*. A vector space is determined when its element set S *and* its operations are specified. Sometimes we refer to the elements of S as elements of V. If in a discussion the operations are given, we often refer to the vector space simply as S.

The systems V_G, V_P, V_3 discussed above clearly are examples of

---

[8] A better name might be "linear vector system," since, as defined, it is an algebraic concept. The term "vector space" indicates the historical origin of the idea.

vector spaces. Further examples abound. We may take S to be the set of complex numbers $a + bi$ and define addition and multiplication by real numbers in the natural way:

$$(a + bi) + (c + di) = (a + c) + (b + d)i; \qquad m(a + bi) = ma + mbi.$$

Properties $V1, \cdots , V10$ then become familiar principles of the algebra of complex numbers. Likewise the set of all polynomials $a_n x^n + a_{n-1} x^{n-1} + \cdots + a_0$ ($n \geqq 0$) with real coefficients becomes a vector space, if we take the operations to be addition of polynomials and multiplication of polynomials by real numbers.

There is a simple generalization of the vector space V_3 of number triples to a vector space V_n whose elements are n-tuples $(a_1 , \cdots , a_n)$ of real numbers, using the definitions:

$$(a_1 , \cdots , a_n) + (b_1 , \cdots , b_n) = (a_1 + b_1 , \cdots , a_n + b_n),$$

$$m(a_1 , \cdots , a_n) = (ma_1 , \cdots , ma_n).$$

Finally an example from calculus: Let S be the set of functions $f(x)$ defined for all real numbers x, and the operations addition of functions and multiplication of functions by real numbers.

ELEMENTS OF THE THEORY OF VECTOR SPACES

In this final part we introduce some of the important ideas and results of the theory of vector spaces.

Algebraic Properties. First we indicate that the algebraic formalism for geometric vectors, which we discussed earlier, holds for any vector space V, that is, it is derivable from properties $V1, \cdots , V10$.

We begin by considering the uniqueness of the "zero" element Θ in vector space V. Suppose Θ, Θ' satisfy $V4$, the Identity Law, so that $\alpha + \Theta = \alpha$, $\alpha + \Theta' = \alpha$ for each α in V. Then taking $\alpha = \Theta'$ in the first equation and $\alpha = \Theta$ in the second equation we have $\Theta' + \Theta = \Theta'$, $\Theta + \Theta' = \Theta$. Then $\Theta' + \Theta = \Theta + \Theta'$ implies $\Theta' = \Theta$.[9] Thus V has a uniquely determined "zero" element.

In our treatment of addition of geometric vectors we verified properties $V1 , \cdots , V5$ of a vector space geometrically as Theorems 1, $\cdots , 5$ and then derived the remaining algebraic properties by algebraic reasoning from these theorems. Thus we can carry over to the case of an arbitrary vector space V the purely algebraic portion of the theory of geometric vectors.

We define $-\alpha$ the *inverse* or *negative* of α to be the unique solution ξ

---

[9] We are tacitly assuming the properties of equality. (See page 155.)

of $\alpha + \xi = \Theta$. As usual $\alpha - \beta$ denotes $\alpha + (-\beta)$. We obtain, as on page 157, $-(-\alpha) = \alpha$ and the relation between addition and subtraction: $\xi + \beta = \alpha$ if and only if $\xi = \alpha - \beta$. The Cancellation Law, $\xi + \alpha = \eta + \alpha$ implies $\xi = \eta$, and the law for the negative of a sum, $-(\alpha + \beta) = (-\alpha) + (-\beta)$, follow as in Theorems 7, 8 and it follows that the familiar properties of addition and subtraction hold in V.

We proceed to develop the properties of multiplication of elements of V by scalars. We have

$$1\alpha + \Theta = 1\alpha = (1 + 0)\alpha = 1\alpha + 0\alpha$$

so that $0\alpha = \Theta$ by the Cancellation Law. A similar argument shows $m\Theta = \Theta$. Hence

$$\Theta = 0\alpha = (1 + (-1))\alpha = 1\alpha + (-1)\alpha = \alpha + (-1)\alpha$$

so that $(-1)\alpha = -\alpha$ by definition of inverse. Using this it is not hard to derive the distributive laws for subtraction:

$$(m - n)\alpha = m\alpha - n\alpha; \qquad m(\alpha - \beta) = m\alpha - m\beta.$$

Now the formal properties of multiplication of geometric vectors by scalars, (see pp.161–65) are easily seen to hold for any vector space V, and so the formal theory of linear equations holds as well.

Subspaces. Since the algebraic formalism is disposed of, let us consider some of the basic concepts of vector space theory. We begin with the idea "subsystem." Consider the notion "lineal" set of geometric vectors (p. 178). Let S_1 be the "lineal" set composed of all geometric vectors determined by pairs of points of a given line. Then S_1 consists of all geometric vectors $t\mathbf{a}_1$, where $\mathbf{a}_1$ is fixed, and is "part" of V_G. Suppose we apply to S_1, the operations of V_G. That is, we form a new system with element set S_1 and operations those of V_G restricted to the elements of S_1.

This system is a vector space. For the formal laws $V2$, $V3$, $V7$, $V8$, $V9$, $V10$ automatically hold in the new system, since they hold in V_G. $V1$, the Closure Law, holds since $t\mathbf{a}_1 + u\mathbf{a}_1 = (t + u)\mathbf{a}_1$ is an element of S_1. Similarly for $V6$. The Identity and Inverse Laws $V4$, $V5$ are valid since $V6$ is. For if S_1 contains $\mathbf{r}$, $V6$ implies that S_1 contains $0\mathbf{r} = \mathbf{0}$ and $(-1)\mathbf{r} = -\mathbf{r}$.

Similarly S_2, the "planar" set composed of all geometric vectors determined by pairs of points of a given plane, consists of all geometric vectors $t_1\mathbf{a}_1 + t_2\mathbf{a}_2$ for fixed $\mathbf{a}_1$, $\mathbf{a}_2$ and yields a vector space when the operations of V_G are applied to it. These examples suggest

DEFINITION 13. Vector space V' is a (vector) *subspace* of vector space V if S' the set of elements of V' is a subset of that of V, and the operations of V' are those of V applied to the elements of V'. Since V' is determined when S' is given, we often refer to S' (instead of V') as a *subspace* of V.

The discussion above suggests and can be used to justify the following criterion for a subspace:

Let S be the element set of vector space V; then S' a (nonvoid) subset of S is a subspace of V if and only if S' contains with any elements α, β the sum $\alpha + \beta$ and the product $m\alpha$ for any scalar m.

Using the criterion one easily shows that if $\alpha_1, \cdots, \alpha_n$ are elements of vector space V the elements of the form $t_1\alpha_1 + \cdots + t_n\alpha_n$ constitute a subspace of V. Referring to pages 190–91, the criterion implies that the polynomials of the form $ax + b$ and the continuous functions are subspaces respectively of the vector spaces of polynomials and of functions. In V_P the lines and planes through O (considered as sets of points) are subspaces.

Linear Independence and Dimensionality. The notions linear independence and dependence are carried over from the previous discussion (pp. 167–78) without change, since the definitions employed there are purely algebraic. The discussion of the relation between dimensionality and independence properties (pages 177–178) suggests

DEFINITION 14. If there is a maximum number n of independent elements in vector space V, we say *vector space V is finite dimensional* and has *dimension* or *rank* n. If there exists no such maximum number, that is, if V contains m independent elements for $m = 1, 2, 3, \cdots$ we say *vector space V is infinite dimensional.*

Since the maximum number of independent geometric vectors is 3, V_G has dimension 3. It can be shown that the dimension of V_P is 3 and that of V_n is n. The vector spaces of polynomials and of functions are infinite dimensional.

Basis of a Vector Space. Suppose vector space V has dimension n. Then there exists a set of n independent elements of V. Let $\alpha_1, \cdots, \alpha_n$ be such a set. Let ρ be any element of V. Then $\alpha_1, \cdots, \alpha_n, \rho$ cannot be independent. For, if $\rho = \alpha_i$ for some i, $1 \leq i \leq n$, then $1\alpha_i + (-1)\rho = 0$ and it follows that $\alpha_1, \cdots, \alpha_n, \rho$ are dependent. If $\rho \neq \alpha_i$ for all i, $1 \leq i \leq n$, then $\alpha_1, \cdots, \alpha_n, \rho$ consists of $n + 1$ elements, which cannot be independent since n is the maximum number of independent elements of V. The argument of Theorem 23 now applies to show

(30) $\rho = x_1\alpha_1 + \cdots + x_n\alpha_n$.

Thus V is completely determined by $\alpha_1, \cdots, \alpha_n$. This suggests

DEFINITION 15. Let $\alpha_1, \cdots, \alpha_n$ be independent elements of vector

194   INSIGHTS INTO MODERN MATHEMATICS

space V, such that each element ρ of V is expressible in the form (30). Then we say $\alpha_1, \cdots, \alpha_n$ form a *basis* of V.

Thus we have shown: *If vector space V has dimension n it has a basis of n elements*. The converse also is true but is harder to prove: *If vector space V has a basis of n elements it has dimension n.* This implies an important uniqueness condition: *Any two bases of a vector space contain the same number of elements.*

Coordinates. Let $\alpha_1, \cdots, \alpha_n$ be a basis of vector space V. By definition $\alpha_1, \cdots, \alpha_n$ are independent and (30) holds for any element ρ of V. Hence we can apply the proof of uniqueness in Theorem 24 to show ρ is *uniquely* representable in the form (30); that is $x_1, \cdots, x_n$ in (30) are uniquely determined. Thus we call $x_1, \cdots, x_n$ the *coordinates* of ρ *relative to the basis* $\alpha_1, \cdots, \alpha_n$ or if $\alpha_1, \cdots, \alpha_n$ are fixed in a discussion, simply the *coordinates* of ρ, and write $\rho\,(x_1, \cdots, x_n)$. Clearly the association of $x_1, \cdots, x_n$ to ρ effects a one-to-one correspondence between the elements of V and the n-tuples of real numbers $(x_1, \cdots, x_n)$, that is between the element sets of V and of V_n.

Isomorphism. The correspondence just introduced is of an important type. For simplicity in studying it let $n = 3$. Then V has a basis $\alpha_1, \alpha_2, \alpha_3$ of 3 elements and we have a one-to-one correspondence "between" V and V_3 which associates to each element of V its coordinate triple. Suppose $\rho\,(x_1, x_2, x_3)$, $\sigma\,(y_1, y_2, y_3)$. Then

$$\rho = x_1\alpha_1 + x_2\alpha_2 + x_3\alpha_3, \qquad \sigma = y_1\alpha_1 + y_2\alpha_2 + y_3\alpha_3.$$

Forming $\rho + \sigma$ and $m\rho$ we have

$$\rho + \sigma = (x_1 + y_1)\alpha_1 + (x_2 + y_2)\alpha_2 + (x_3 + y_3)\alpha_3,$$
$$m\rho = (mx_1)\alpha_1 + (mx_2)\alpha_2 + (mx_3)\alpha_3.$$

Hence the coordinate triple for $\rho + \sigma$ is

$$(x_1 + y_1, x_2 + y_2, x_3 + y_3) = (x_1, x_2, x_3) + (y_1, y_2, y_3)$$

and that for $m\rho$ is

$$(mx_1, mx_2, mx_3) = m(x_1, x_2, x_3).$$

Thus the one-to-one correspondence has the following properties: (a) *It associates to the sum of two elements of V the sum of their associates in V_3;* (b) *it associates to the product of an element of V by a scalar, the product of its associate in V_3 by the same scalar.* This suggests

DEFINITION 16. If vector spaces V', V'' are related by a one-to-one correspondence which satisfies (a), (b) above, we say V' is *isomorphic* to

V'', and we call the correspondence an *isomorphism*. (See pages 112–13 for the concept of isomorphism of groups.)

V_G is easily seen to be isomorphic to V_P. In fact we constructed V_P with this in mind. Likewise V_P is isomorphic to V_3. Isomorphic vector spaces have in a sense the same internal structure, since in effect the application of an operation to elements in one can be performed by applying the corresponding operation to the associated elements of the other.

Isomorphism of vector spaces has the basic properties of equality: (a) *V' is isomorphic to V'; (b) if V' is isomorphic to V'' then V'' is isomorphic to V'; (c) if V' is isomorphic to V'' and V'' is isomorphic to V''' then V' is isomorphic to V'''.*

Our discussion at the beginning of this section showed if V has a basis of 3 elements it is isomorphic to V_3. The reasoning is perfectly general and justifies: *If V has a basis of n elements then V is isomorphic to V_n.* This in combination with the existence of a basis for a vector space of dimension n yields the important result: *If V has dimension n then V is isomorphic to V_n.* Furthermore applying the "equality" properties of isomorphism we obtain: *Any two vector spaces of dimension n are isomorphic.*

Geometry of a Vector Space. In defining vector space V_P we constructed an algebraic system whose elements are the points of Euclidean 3-space, by defining geometrically addition of points and scalar multiples of points. Now we reverse the process. Given an arbitrary vector space V, we convert it into a kind of geometry, by defining *line* and *plane* in terms of the algebraic operations of V. To see how to do this we characterize the lines and planes which are already present in V_P in terms of its operations.

Theorem 19 on the vector equation of a line may be stated: *If $A \neq B$, point P is in line AB if and only if*

$$(31) \qquad\qquad \overrightarrow{OP} = \overrightarrow{OA} + t\overrightarrow{AB}.$$

Localize $\overrightarrow{AB}$ at O as $\overrightarrow{OM}$. Then $\overrightarrow{AB} \neq \mathbf{0}$ implies $M \neq O$; and (31) is equivalent to $\overrightarrow{OP} = \overrightarrow{OA} + t\overrightarrow{OM}$. In terms of the operations of V_P this is equivalent to

$$(32) \qquad\qquad P = A + tM.$$

Thus P is in AB if and only if (32) holds as an equation in V_P. Thus in V_P the set of points of a line is precisely the set of points $A + tM$ where A, M are fixed and $M \neq O$.

Similarly Theorem 20 implies that in V_P the set of points of a plane is the set $A + tM + uN$ where A, M, N are fixed and M, N are linearly independent. Thus we adopt

DEFINITION 17. Let V be any vector space (not necessarily finite dimensional). We refer to the elements of V as *points*. The set of elements of V of the form $\alpha + t\mu$ where α, μ are fixed and $\mu \neq \mathcal{O}$ is called a *line*. The set of elements of V of the form $\alpha + t\mu + u\nu$ where α, μ, ν are fixed and μ, ν are linearly independent is a *plane*.

Observe if $\alpha = \mathcal{O}$ in these expressions the lines and planes contain $\mathcal{O}$. They are clearly subspaces of V (p. 192) and can be shown to have dimensions 1 and 2 respectively. We now have the basis for a rudimentary kind of geometry using points, lines, and planes. The familiar alignment properties of these entities are derivable from the algebraic properties of V. Thus we can show two points belong to a unique line, three non-collinear points to a unique plane. Parallel lines are defined in the usual manner as two lines contained in a plane which have no common point. The geometry thus established is called *affine geometry* and may be described roughly as Euclidean geometry with general metrical properties omitted, or the geometry of *linearity* and *parallelism*.

Linear Transformations. Now we consider the notion linear transformation of a vector space which is related to the concept isomorphism. First we point out the formal use of the term *transformation* to denote a rule or law which associates to each element of a given set of elements S a uniquely determined element of a set S'. The translations and dilatations of space mentioned on pages 188–89 are types of transformations. Denoting a given transformation by a letter, say T, we employ the functional notation $T(x)$ to denote the element which T associates to x. Essentially the notion *transformation* is the same as that of *correspondence* or *function*.

Let V be a vector space and T a transformation which associates to each ξ of V a corresponding element $T(\xi)$ of V such that for all vectors ξ, η, *of* V and all scalars m

$$T(\xi + \eta) = T(\xi) + T(\eta), \qquad T(m\xi) = mT(\xi).$$

Then we call T a *linear transformation* of V *into* V or simply of V.

Clearly an isomorphism between V and V itself is a linear transformation of V which is one-to-one. To construct a "geometrical" linear transformation let R be a rotation on plane π, about point O of π, through an angle of measure θ. That is, each point P of π is "transformed" into a new position $P' = R(P)$ by rotating P about O through an angle of measure θ; we are of course making the usual agreements associating to θ a counterclockwise or clockwise sense of rotation depending on the sign of θ. One can show, and it is intuitively evident, that R transforms collinear points into collinear points, and equivalent directed segments into equivalent directed segments; that is if $(A, B) \equiv (C, D)$ then $(A', B') \equiv (C', D')$. Thus R induces a transformation of the set of geometric vectors of π—that is if we define $R(\overrightarrow{PQ}) = \overrightarrow{P'Q'}$ we have associated to each geometric vector $\overrightarrow{PQ}$ of π a unique geometric vector $\overrightarrow{P'Q'}$ of π. Thus R effects a transformation of the vector space V determined by the geometric vectors of π. We have

$$R(\overrightarrow{AB} + \overrightarrow{BC}) = R(\overrightarrow{AC}) = \overrightarrow{A'C'} = \overrightarrow{A'B'} + \overrightarrow{B'C'} = R(\overrightarrow{AB}) + R(\overrightarrow{BC}).$$

Furthermore we can show if $\overrightarrow{OD} = m\overrightarrow{OA}$ then $R(\overrightarrow{OD}) = mR(\overrightarrow{OA})$. Thus R effects a linear transformation of V.

An algebraic example: In V_2, the vector space of number couples (x, y), it is easy to verify that T defined by $T((x, y)) = (x', y')$ where $x' = ax + by$, $y' = cx + dy$ is a linear transformation.

For the remainder of our discussion of linear transformations we assume the dimension of V is 2. The results are typical of those for vector spaces of dimension n. Let T be a linear transformation of V and $T(\rho) = \rho'$. V has a basis of 2 elements, say ϵ_1, ϵ_2. Then

$$(33) \qquad \rho = x_1\epsilon_1 + x_2\epsilon_2, \qquad \rho' = x_1'\epsilon_1 + x_2'\epsilon_2,$$

holds. In terms of coordinates relative to ϵ_1, ϵ_2 we have $\rho(x_1, x_2)$ and $\rho'(x_1', x_2')$. We wish to determine the algebraic relations between the coordinates of ρ and of ρ', its corresponding element. We have

$$\rho' = T(\rho) = T(x_1\epsilon_1 + x_2\epsilon_2) = T(x_1\epsilon_1) + T(x_2\epsilon_2)$$
$$= x_1T(\epsilon_1) + x_2T(\epsilon_2).$$

$T(\epsilon_1)$ and $T(\epsilon_2)$ being elements of V are representable in terms of ϵ_1, ϵ_2. Let

$$T(\epsilon_1) = a_{11}\epsilon_1 + a_{21}\epsilon_2, \qquad T(\epsilon_2) = a_{12}\epsilon_1 + a_{22}\epsilon_2.$$

Substituting these values in the preceding relation, multiplying out and collecting terms we have

$$(34) \qquad \rho' = (a_{11}x_1 + a_{12}x_2)\epsilon_1 + (a_{21}x_1 + a_{22}x_2)\epsilon_2.$$

Comparing the expressions for ρ' in (33), (34) we have, by the principle of uniqueness of representation of an element in terms of a basis,

$$(35) \qquad x_1' = a_{11}x_1 + a_{12}x_2, \qquad x_2' = a_{21}x_1 + a_{22}x_2.$$

Thus having fixed the basis ϵ_1, ϵ_2, T is represented by, and can be studied by means of, the simple linear equations (35). Actually equations

(35) determine a linear transformation T^* of V_2 by the rule

$$T^*((x_1, x_2)) = (x_1', x_2').$$

Detaching the coefficients in (35) from the variables we form the *matrix*, or rectangular array of numbers,

$$\begin{pmatrix} a_{11} & a_{12} \\ a_{21} & a_{22} \end{pmatrix}$$

which we associate to T. This association is a one-to-one correspondence between the set of linear transformations of V and the set of all such matrices, and enables us to reduce the theory of linear transformations to that of matrices. We close our discussion of linear transformations at the threshold of the classical treatment of linear algebra, which studies matrices, determinants, simultaneous linear equations, etc.

Vector Spaces Over a Field. Looking back over our treatment of the theory of vector spaces, it becomes evident that it does not depend on the requirement that the scalars be real numbers but rather on their algebraic properties. More specifically it rests on the fact that the "rational" operations addition, subtraction, multiplication and division can be performed in the real number system and satisfy familiar formal properties. Thus our theory would be valid if we replaced the real numbers as scalars by the rational numbers or the complex numbers; or even the system of integers modulo 2 whose elements are 0, 1 and which are combined according to the rules: $0 + 0 = 1 + 1 = 0, 0 + 1 = 1 + 0 = 1,$ $0 \times 0 = 0 \times 1 = 1 \times 0 = 0, 1 \times 1 = 1.$ In fact the modern theory of *vector spaces over a field* permits the scalars to be the elements of an arbitrary "field," which is a number system in which the rational operations behave in the familiar manner. (See pages 125–27.)

The construction of an affine geometry in a vector space is valid in the more general theory and gives rise to geometries in which the coordinates of a point need not be real numbers. For example if we take the scalars to be the system of integers modulo 2 we get a geometry in which each line contains exactly 2 points.

This completes our outline of the evolution of the vector concept. We have traced the development from its origin in the physical notion of "quantity with magnitude, direction and sense" through the precise idea of geometric vector to its culmination in the powerful algebraic abstraction of vector space. References for further study are given below.

BIBLIOGRAPHY

Geometrical Vector Analysis:
1. WEATHERBURN, C. E. *Elementary Vector Analysis*. London: George Bell and Sons, 1935.

Theory of Vector Spaces:

2. BIRKHOFF, GARRETT, and MacLANE, SAUNDERS. *A Survey of Modern Algebra.*
   New York: The Macmillan Co., 1953, Chapter 7.
3. HALMOS, P. R. *Finite Dimensional Vector Spaces.* Princeton, N. J.; Princeton
   University Press, 1948.
4. STOLL, R. R. *Linear Algebra and Matrix Theory.* New York: McGraw-Hill
   Book Co., 1952, Chapter 2.

VII

Limits

JOHN F. RANDOLPH

ENO in the 5th century B. C. was probably first to express the idea of approaching a limit in his Achilles paradox:

Achilles cannot overtake the tortoise for Achilles must first reach the place from which the tortoise started. By that time the tortoise will have moved a little way. Achilles must then transverse that, and still the tortoise will be ahead. He is always nearer yet never catches up to it.

Early attempts at finding areas of circles and segments of curved surfaces were made by using ideas of sequences and limits. It was not until some great mathematicians of the 19th century pointed out the inconsistencies and lack of rigor in the earlier intuitive proofs that rigorous definitions were given and the basic concepts of limits were completely understood. Basic concepts of the theory of limits are sequences, infinite series and the theory of convergence.

SEQUENCES

Basic Concepts. The array

$$1, \frac{1}{2}, \frac{1}{3}, \cdots, \frac{1}{n}, \cdots .$$

means an endless succession of numbers where, for each positive integer n, the nth number is obtained by taking the reciprocal of n. By a *sequence* of numbers is meant a law such that for each positive integer n a number a_n is determined. A sequence is usually displayed in the form

$$a_1, a_2, a_3, \cdots, a_n, \cdots .$$

For example the sequence with defining relation (or law) given as

(1) $$a_n = 1 + \frac{(-1)^n}{n}$$

is visualized by the array

$$0, \frac{3}{2}, \frac{2}{3}, \frac{5}{4}, \frac{4}{5}, \cdots, 1 + \frac{(-1)^n}{n}, \cdots$$

On some intelligence tests a question such as the following is asked: What is the next term of the sequence

$$1, \tfrac{1}{2}, \tfrac{1}{3}, \cdots ?$$

The answer expected is $\tfrac{1}{4}$ since it seems natural that the law should be $a_n = 1/n$ for each positive integer n. The question is, in a sense, not a fair one since

$$(2) \qquad\qquad a_n = \frac{1}{n^3 - 6n^2 + 12n - 6}$$

is also a possible law. Note that:

for $n = 1$, $\quad \dfrac{1}{1^3 - 6 \cdot 1^2 + 12 \cdot 1 - 6} = \dfrac{1}{1 - 6 + 12 - 6} = 1,$

for $n = 2$, $\quad \dfrac{1}{2^3 - 6 \cdot 2^2 + 12 \cdot 2 - 6} = \dfrac{1}{8 - 24 + 24 - 6} = \dfrac{1}{2},$

for $n = 3$, $\quad \dfrac{1}{3^3 - 6 \cdot 3^2 + 12 \cdot 3 - 6} = \dfrac{1}{27 - 54 + 36 - 6} = \dfrac{1}{3}.$

Thus from (2) the first three terms are $1, \tfrac{1}{2}, \tfrac{1}{3}$, but the fourth term is not $\tfrac{1}{4}$ since

for $n = 4$, $\quad \dfrac{1}{4^3 - 6 \cdot 4^2 + 12 \cdot 4 - 6} = \dfrac{1}{64 - 96 + 48 - 6} = \dfrac{1}{10}.$

This example is given to show that a sequence is not determined by giving any number of its terms; *in order that a sequence be determined its law must be known.*

It may be helpful to consider the terms of a sequence plotted on an axis (or number line) in order to see relations between the first few terms and to help visualize the behavior of later terms. Figure 1 illustrates this graphical representation of the sequence whose law is (1). The points representing terms of this sequence cluster around the point 1 in such

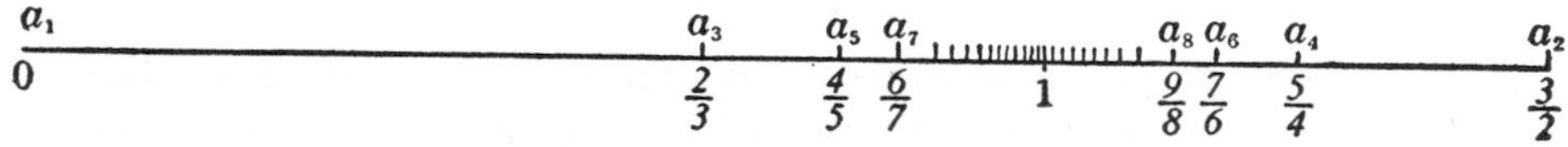

FIG. 1

a way that any open interval, no matter how short, with its midpoint at 1 will contain all except possibly "a few" of these points. In particular

$$1 - 0.1 < 1 + \frac{(-1)^n}{n} < 1 + 0.1$$

is not satisfied if n is any of the ten integers, 1, 2, 3, $\cdots$, 10, but is satisfied if n is an integer greater than 10. The inequality

$$1 - 0.01 < a_n < 1 + 0.01$$

is not satisfied by the first hundred terms, but all remaining terms do satisfy these inequalities. In fact if e is any fixed positive number, then

$$1 - e < 1 + \frac{(-1)^n}{n} < 1 + e$$

is not satisfied by the integers n such that $n \leq 1/e$, but is satisfied by any integer n such that $n > 1/e$.

Other examples of sequences which have properties we wish to point out are:

$$\sqrt{1}, \sqrt{2}, \sqrt{3}, \cdots, a_n = \sqrt{n}, \cdots,$$

$$2, \quad 2, \quad 2, \cdots, a_n = \quad 2, \cdots,$$

$$-1, \quad \frac{1}{2}, -\frac{1}{3}, \cdots, a_n = \frac{(-1)^n}{n}, \cdots,$$

The first of these sequences is such that no matter what large fixed positive number G is given, there are at most a finite number of terms at the beginning whose value is less than or equal to G while all remaining terms exceed G. All terms of the second sequence have the same value; this is an example of a *constant sequence*. The third sequence, having terms alternating in sign, is said to be an *alternating sequence*. The first and second sequences are such that for each positive integer n, $a_n \leq a_{n+1}$. A sequence a_1 , a_2 , a_3 , $\cdots$, a_n , ..., which is such that

$$a_1 \leq a_2 \leq \cdots \leq a_n \leq a_{n+1} \leq \cdots,$$

is said to be *monotonically increasing,* but if

$$a_1 \geq a_2 \geq \cdots \geq a_n \geq a_{n+1} \geq \cdots$$

the sequence is said to be *monotonically decreasing.* The third sequence above is neither monotonically increasing nor monotonically decreasing.

The law of a sequence need not be given by the same rule for each integer. For example

(3)
$$b_n = \begin{cases} \dfrac{1}{n} & \text{if } n \text{ is odd} \\[2mm] 1 - \dfrac{1}{n} & \text{if } n \text{ is even} \end{cases}$$

states definitely how to compute each term.[1] Here again the first three terms are 1, $\frac{1}{2}$, $\frac{1}{3}$, since

$$b_1 = \frac{1}{1}, \qquad b_2 = 1 - \frac{1}{2} = \frac{1}{2}, \quad \text{and} \quad b_3 = \frac{1}{3}.$$

What is the fourth term? For this sequence there are odd terms as close to 0 as desired and even terms as close to 1 as desired.

In some cases the first term of a sequence is given and thereafter each term depends by a given law on the immediately preceding term. For example let $a_1 = 2$ and for each integer $n > 1$ let

$$a_n = \frac{1}{2}\left(a_{n-1} + \frac{3}{a_{n-1}} \right).$$

The first three terms of this sequence are

$$a_1 = 2, \quad a_2 = \frac{1}{2}\left(2 + \frac{3}{2} \right) = \frac{1}{2}\left(\frac{4+3}{2} \right) = \frac{7}{4}.$$

$$a_3 = \frac{1}{2}\left(\frac{7}{4} + \frac{3}{7/4} \right) = \frac{1}{2}\left(\frac{7}{4} + \frac{12}{7} \right) + \frac{1}{2}\left(\frac{49+48}{28} \right) = \frac{97}{56}.$$

By division we find $\dfrac{97}{56} = 1.732$ to three decimal places and note that to three decimal places $\sqrt{3} = 1.732$.

Later on we shall show that with A any positive number, a_1 any number greater than $\sqrt{A}$, and

$$a_n = \frac{1}{2}\left(a_{n-1} + \frac{A}{a_{n-1}} \right), \qquad n = 2, 3, 4, \cdots$$

then a_2, a_3, $\cdots$, a_n, $\cdots$ are better and better approximations of $\sqrt{A}$. Moreover given any tolerance from $\sqrt{A}$ whatever, there is a term in the sequence which missed $\sqrt{A}$ by less than this tolerance and also any later term will be within this tolerance.

The law of a sequence need not even be given by means of formulas,

---

[1] We could write $b_n = \dfrac{1 + (-1)^n}{2} - \dfrac{(-1)^n}{n}$ but the form (3) seems preferable.

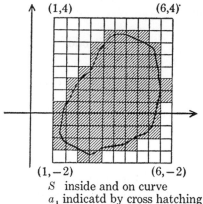

S inside and on curve
a_1 indicatd by cross hatching
FIG. 2

but may be given as a verbal description. The following two examples illustrate such verbal descriptions and also are used later on.

Example 1. In a plane let S be a point set (see page 36) and R a rectangle which includes S, where R has its sides parallel to the axes and has integers for both coordinates of its four corners. The rectangle R is made up of an integral number of squares each having sides one unit long. Divide each of these unit squares into four squares each with side $\frac{1}{2}$ and let $\bar{a}_1$ be $(\frac{1}{2})^2$ times the number of such squares having a point in common with S. (See Fig. 2.) Next divide each of these smaller squares into four equal squares and let $\bar{a}_2$ be $(\frac{1}{2^2})^2$ times the number of such squares having a point in common with S. At the nth stage, the squares which were selected at the $(n-1)$th stage are divided into four equal squares each of side $\frac{1}{2}^n$ and $\bar{a}_n$ is $(\frac{1}{2}^n)^2$ times the number of these squares which have at least one point in common with S.

The sequence $\bar{a}_1, \bar{a}_2, \cdots, \bar{a}_n, \cdots$ is well-defined in the sense that (with S given) any two persons working independently will agree on a value of $\bar{a}_1$, a value for $\bar{a}_2$, etc. It should be seen that

$$\bar{a}_1 \geqq \bar{a}_2 \geqq \cdots \geqq \bar{a}_{n-1} \geqq \bar{a}_n \geqq \cdots$$

Example 2. Let S and R be the same as in Example 1, and let T denote the set of all points of R which are not points of S. In set notation (see Chapter III)

$$T = R - S.$$

Now apply the same process to T that was applied to S in Example 1, and obtain another sequence $\bar{b}_1, \bar{b}_2, \cdots, \bar{b}_n, \cdots$ where again the b-sequence is monotonically decreasing.

THE LIMIT PROCESS

Limit of a Sequence. In some of the work which follows, statements involving the words "finite number" will be made. Any number is in fact finite so the adjective "finite" is superfluous, but we use it anyway as a bow to convention and the hope that anyone who is tempted to think of ∞ and $-\infty$ as numbers will know that these are definitely excluded. Thus any number, no matter how large, is finite and the sum of two numbers is finite (even the national debt added to the number of atoms in the universe is still finite).

A property we shall use repeatedly without saying so is:

Any finite set of numbers has a smallest and a largest element. Thus if a property holds for all except a finite number of integers and a second property holds except for a finite number of integers, then both properties hold except for a finite number of integers.

Given a sequence whose nth term is denoted by a_n, we shall use the single letter a to denote the whole sequence.

For the sequence a defined by $a_n = 1 + (-1)^n/n$ it was seen on page 202 that whatever positive number e is given, then for

$$n > \frac{1}{e} \text{ it follows that } 1 - e < a_n < 1 + e$$

that is, the inequalities $1 - e < a_n < 1 + e$ hold except possibly for a finite set of integers n. In the terminology defined below, it will be said that the sequence a (which may be displayed as $0, \frac{3}{2}, \frac{2}{3}, \frac{5}{4}, \cdots$, $a_n = 1 + ((-1)^n/n) \cdots$) *converges* to 1.

By a *symmetric neighborhood* of a number L is meant the set of all numbers x such that

$$L - e < x < L + e$$

for some positive number e. Thus each positive number e determines a symmetric neighborhood of L. A symmetric neighborhood of L may be visualized on a number line as an open interval with midpoint having coordinate L.

DEFINITION 1. A sequence a is said to *converge* to L if each symmetric neighborhood of L contains all except possibly a finite number of terms of the sequence a.

Note that a logical consequence of this definition is:

A sequence a does not converge to L if there is at least one symmetric neighborhood of L outside of which are infinitely many terms of the sequence.

There are sequences which do not converge to any number. For example the sequence defined by

$$b_n = \begin{cases} \dfrac{1}{n} \text{ if } n \text{ is odd} \\[2ex] 1 - \dfrac{1}{n} \text{ if } n \text{ is even} \end{cases}$$

is such that no matter where an open interval of length less than $\frac{1}{2}$ is placed there will be infinitely many points representing elements of the sequence outside this interval since infinitely many odd terms differ from 0 by less than $\frac{1}{4}$ while infinitely many even terms differ from 1 by less than $\frac{1}{4}$.

Another direct consequence of the above definition is:

If a sequence converges to some number, then this sequence does not converge to any other number.

For let a sequence converge to a number L and take any number M where $M \neq L$. Then $| L - M | > 0$ and $| L - M |/2 > 0$. With the points having coordinates L and M as centers take open intervals each of length $| L - M |/2$. These intervals have no point in common and represent symmetric neighborhoods of L and M. All except a finite number of terms of the sequence a are within the symmetric neighborhood of L so that any given number of terms can be found outside the symmetric neighborhood of M; that is, the sequence a does not converge to M. Since M was any number different from L we see that the sequence a does not converge to any number other than L.

If a sequence a converges to L, then we say that the limit of the sequence a exists and is L and we write

$$\lim a = L.$$

Arithmetic and Limit Processes. In ordinary arithmetic, calculations are made by adding, subtracting, multiplying, or dividing two numbers (no division by zero allowed) or any finite combination of these operations. When we consider any of these operations repeated without end we are no longer in the domain of ordinary arithmetic. For example, the symbolism

$$(4) \qquad\qquad 1 + \frac{1}{2} + \frac{1}{4} + \cdots + \frac{1}{2^{n-1}} + \cdots$$

has no meaning in arithmetic. True we may add $1 + \frac{1}{2}$ then add $\frac{1}{4}$, then $\frac{1}{8}$ etc., but no matter at what term we stop there will be another plus sign and another term to add. Of course we can stop, say with the term $\frac{1}{2}^{n-1}$, write

$$s_n = 1 + \frac{1}{2} + \frac{1}{2^2} + \cdots + \frac{1}{2^{n-1}}$$

and then multiply both sides by $\frac{1}{2}$ to obtain

$$\frac{1}{2} s_n = \frac{1}{2} + \frac{1}{2^2} + \cdots + \frac{1}{2^{n-1}} + \frac{1}{2^n}.$$

Now by subtraction $s_n - \frac{1}{2}s_n = \frac{1}{2}s_n = 1 - \frac{1}{2}^n$ and hence upon multiplying by 2

$$s_n = 2 - \frac{1}{2^{n-1}}.$$

We thus see that no matter at what term we stop, the sum of the terms taken will be less than 2. On the other hand no matter what number less than 2 is chosen (such as 1.99 or 1.99999) we can continue adding terms and eventually stop with a sum greater than that chosen number less than 2.

Man, being a reasonable and ingenious animal, is then willing to admit defeat but capitalize on it by saying:

a. No one can add all of the terms in (4).

b. The number 2 will be called the *sum* of the series in (4).

These are not contradictory statements when (as should be) the word "add" in the first statement has its ordinary arithmetic sense while the second statement extends the use of the word "sum."

This extension was first precisely formulated by Augustin Louis Cauchy (1789–1859). Paraphrased, he said that given

(5) $\qquad\qquad a_1 + a_2 + a_3 + \cdots + a_n + \cdots$

(where each a_n is a number), form the sequence s defined by

$$s_1 = a_1$$

$$s_2 = s_1 + a_2$$

$$s_3 = s_2 + a_3$$

$$\vdots$$

$$s_n = s_{n-1} + a_n$$

$$\vdots$$

and if this sequence converges, let $L = \lim s_n$ and call L the *sum* of (5).

Note the phrase "if this sequence converges." There are in fact series for which the corresponding "sum sequence" does not converge. As an example consider the formally written series

$$(6) \qquad\qquad 1 + \frac{1}{2} + \frac{1}{3} + \cdots + \frac{1}{n} + \cdots.$$

Group the first few terms as follows:

$$(1) + (\tfrac{1}{2}) + (\tfrac{1}{3} + \tfrac{1}{4}) + (\tfrac{1}{5} + \tfrac{1}{6} + \tfrac{1}{7} + \tfrac{1}{8})$$
$$\qquad + (\tfrac{1}{9} + \tfrac{1}{10} + \tfrac{1}{11} + \tfrac{1}{12} + \tfrac{1}{13} + \tfrac{1}{14} + \tfrac{1}{15} + \tfrac{1}{16}) + (\cdots)$$

It should be seen that the terms in each group add up to $\tfrac{1}{2}$ or more and it should further be seen that as many more groups as desired may be obtained each adding up to $\tfrac{1}{2}$ or more. Thus given any large number G whatever, we may pick a term in (6) and add to it the preceding terms to obtain a number greater than G. Thus there is no number we could call the "sum" of (6).

Limiting processes have long played an important and indispensable role in mathematics. Without a limiting process how could the area of a circle be defined? Even the area of a rectangle in the so-called incommensurable case is obtained by a limiting process, although considerable effort is made to hide this fact. Trigonometric and logarithmic tables are computed from approximations which depend directly on limiting processes.

Early notions of limits were intuitive in the sense that "beliefs" or "feelings" were expressed. For example it is not hard to look at $\sqrt{x}$ and visualize what "it gets close to" as x "changes with time and snuggles up to 4." These early notions of limits were dynamic in the sense that a time element was present, although kept in the background. These intuitive notions served very well in the hands of some unusually gifted men until situations arose too complicated for even their mastery and certainly too involved to be transmitted to others in the language and terminology then available. It was Cauchy who freed the notion of a limit from any time element and gave a "static" definition. To reiterate the definition given above, we now say:

$$\lim a = L$$

means that whatever positive number e is given, there is an integer N such that for each integer $n > N$ it follows that $|a_n - L| < e$.

Note that this definition contains no word with a time connotation and the only rudiment of time is in the abbreviation "lim." Mathematics is of course independent of time, but we hope that intuitive thinking based on time considerations will not be stifled by our comments. The main point is that after intuitive reasoning one should clinch the arguments to himself and be able to convey his ideas to others by precise means.

Sets Bounded Above. Let S be a set of numbers. If there is a number B such that for each number s in S it follows that $s \leq B$, then S is said to be *bounded above* and B is said to be an *upper bound* of S.

There are sets of numbers which are not bounded above. For example, the set of integers is not bounded above.

If a set S is bounded above and B is an upper bound of S, then any number greater than B is also an upper bound of S, but there may possibly be some number less than B which is also an upper bound of S. If S consists of all positive numbers less than 1, then 2 is certainly an upper bound of S and also $3/2$, $4/3$, and even 1 are upper bounds of S, but there is no upper bound of S less than 1.

In addition to the properties of numbers upon which ordinary arithmetic is based, we also postulate the following property of numbers.

AXIOM. *For a set of numbers which is non-empty and bounded above there is a least upper bound.*

Sequence Limit Theorems. The following two theorems are examples of "existence theorems"; they give conditions under which it is known that a sequence has a limit without giving a readily practical method for determining the limit.

THEOREM 1. *Any monotonically increasing sequence which is bounded above has a limit.*

Proof. Let a be a monotonically increasing sequence bounded above and let B be an upper bound so that

$$a_1 \leq a_2 \leq \cdots \leq a_n \leq \cdots$$

and $a_n \leq B$ for every positive integer n. The set of numbers representing terms of the sequence is therefore bounded above and consequently, by the axiom, this set has a least upper bound. We denote this least upper bound by L and will show that $\lim a = L$. Take an arbitrary symmetric neighborhood of L. Since L is an upper bound, no element of the sequence is in the upper half of this neighborhood. Since L is the least upper bound, there is at least one element of the sequence in the lower half of this neighborhood: let a_{n_0} be such an element. Since the sequence is monotonically increasing it follows that if n is a positive integer such that $n_0 \leq n$, then $a_{n_0} \leq a_n$ and hence a_n also lies in the lower half of the neighborhood of L. Since there are only a finite number of n with $n < n_0$, the neighborhood of L contains all except possibly a finite number of elements of the sequence. Thus the sequence converges (as we wished to prove) and in fact converges to L; that is, lim $a = L$.

The following theorem is a companion to Theorem 1 and the proof is similar so will not be given.

THEOREM 2. *A monotonically decreasing sequence which is bounded below has a limit.*

For an illustration of the use of this theorem we refer to the sequences $\bar{a}$ and $\bar{b}$ of Examples 1 and 2, respectively (p. 204). Both sequences are monotonically decreasing and bounded below (by zero) and hence each has a limit. Consequently numbers $\bar{A}$ and $\bar{B}$ exist such that

$$\lim \bar{a} = \bar{A} \quad \text{and} \quad \lim \bar{b} = \bar{B}.$$

The set S might be defined ever so precisely (such as all points in and on a circle) and still it could be physically impossible to draw accurate enough figures to obtain $\bar{a}_n$ and $\bar{b}_n$ for more than a dozen or so values of n. We know, nevertheless, that a limit exists for each of the well defined sequences $\bar{a}$ and $\bar{b}$. Later on methods will be given whereby these limits may be obtained for a large variety of sets.

The following theorem will be used later on.

THEOREM 3. *If a is a convergent sequence, then there are numbers $\underline{c}$ and $\bar{c}$ such that $\underline{c} < a_n < \bar{c}$ for all positive integers n.*

Proof. Let a be a convergent sequence and let L be its limit. In particular for all except a finite number of integers n

$$L - 1 < a_n < L + 1.$$

If there are any terms of a such that $a_n \leqq L - 1$ there will be only a finite number of them and hence a smallest one which we denote by a_{n_1}. In this case let $\underline{c}$ be the smaller of $a_{n_1} - 1$ and $L - 1$. In case $L - 1 < a_n$ for all terms of a we let $\underline{c} = L - 1$. In either case $\underline{c} < a_n$ for each term of the sequence a. It should now be seen how to show the existence of a number $\bar{c}$ such that $a_n < \bar{c}$ for all terms of a and thus to complete the proof.

Another statement of this last theorem is: *If a sequence converges, then the sequence is bounded.* As a logical consequence we thus state: *If a sequence is unbounded, then the sequence does not converge.* For example the sequence

$$\sqrt{1}, \sqrt{2}, \sqrt{3}, \cdots, \sqrt{n}, \cdots$$

is bounded below (in particular each term is positive), but is unbounded above and thus does not have a limit. It is customary, nevertheless, to write (according to the following definition) lim $a = \infty$.

DEFINITION 2. If a sequence a fails to have a limit but is such that for each positive number G the inequality $G < a_n$ holds for all except a finite number of terms of the sequence, then the sequence is said to become *positively infinite* and we write

$$\lim a = \infty.$$

If, on the other hand, for each positive number G the inequality $a_n < -G$ holds for all except a finite number of terms, then the sequence is said to become *negatively infinite* and we write

$$\lim a = -\infty.$$

As a caution, we point out that the sequence

$$-1,\, 2,\, -3,\, 4,\, \cdots,\, (-1)^n n,$$

neither becomes positively infinite nor does it become negatively infinite.

THEOREM 4. *If a and b are convergent sequences with* $\lim a = A$ *and* $\lim b = B$ *and if for each positive integer n the inequalities $a_n \leqq b_n$ hold, then also $A \leqq B$.*

Proof. In this proof we shall use the so-called "contrapositive" method; that is, we shall assume the only alternative to the stated conclusion and will reach a contradiction. Assume that $A > B$. Consequently $A - B > 0$ and $(A - B)/2 > 0$. Now, for all except a finite number of n, both sets of inequalities

$$A - \frac{A - B}{2} < a_n < A + \frac{A - B}{2} \qquad \text{and}$$

$$B - \frac{A - B}{2} < b_n < B + \frac{A - B}{2}$$

hold. Consequently, there are values of n for which both sets of inequalities hold and we let n_0 be one such integer. Hence, using only half of what is known in each case, we have

$$A - \frac{A - B}{2} < a_{n_0} \qquad \text{and} \qquad b_{n_0} < B + \frac{A - B}{2}.$$

Since $A - (A - B)/2 = (2A - A + B)/2 = (A + B)/2$ and also $B + (A - B)/2 = (A + B)/2$ we have

$$\frac{A + B}{2} < a_{n_0} \qquad \text{and} \qquad b_{n_0} < \frac{A + B}{2}.$$

Hence $b_{n_0} < \frac{1}{2}(A + B) < a_{n_0}$ which says that $b_{n_0} < a_{n_0}$. This is a contradiction since part of the hypothesis is that $a_n \leqq b_n$ for each positive integer n, so in particular $a_{n_0} \leqq b_{n_0}$. The assumption that $A > B$ is therefore wrong and the only alternative; namely $A \leqq B$, is therefore correct.

As an illustration of the use of Theorem 4, consider again the set S of Example 1 of page 204. Let S^* be another set such that

$$S \subset S^*;$$

i.e. each point of S is also a point of S^*. As $\bar{a}_n$ was formed for S, now form $\bar{a}_n^*$ for S^*. If a square contributed to $\bar{a}_n$, then it has a point in common with S so certainly has a point in common with S^*. Consequently

(7) $\bar{a}_n \leqq \bar{a}_n^*$ for $n = 1, 2, \cdots$.

By the same reasoning which led to a number $\bar{A}$ such that $\lim \bar{a} = \bar{A}$, it follows that there is a number $\bar{A}^*$ such that $\lim \bar{a}^* = \bar{A}^*$. From Theorem 4 and (7), it then follows that $\bar{A} \leqq \bar{A}^*$.

Given a sequence a and a sequence b, we form three sequences s, p, and q defined respectively by giving the nth term as

$$s_n = a_n + b_n , \qquad p_n = a_n \cdot b_n , \qquad\qquad \text{and}$$

$$q_n = \begin{cases} 1 & \text{if} \quad b_n = 0, \\ a_n/b_n & \text{if} \quad b_n \neq 0. \end{cases}$$

Extending the use of algebraic signs we also write

$$s = a + b, \qquad p = a \cdot b, \quad \text{and} \quad q = a \div b.$$

The double definition for the sequence q may seem strange, but the only use we shall make of this sequence is when it is known that $b_n \neq 0$ for all except a finite number of n.

Also for a sequence a and a number c, we use ca to denote the sequence whose nth term is ca_n and we use c/a to denote the sequence whose nth term is c/a_n if $a_n \neq 0$ and 1 if $a_n = 0$.

We now prove the following theorem, which in a sense gives the simple algebra of limits.

THEOREM 5. *Let a and b be convergent sequences and let c be a number. Then the sequences ca, $a + b$, and $a \cdot b$ are also convergent and*

(i) $\lim ca = c \lim a,$

(ii) $\lim (a + b) = \lim a + \lim b,$

(iii) $\lim (a \cdot b) = (\lim a)(\lim b).$

Let, for an additional result, $\lim b \neq 0$. Then $b_n \neq 0$ for all except a finite number of n and

(iv) $\lim (a \div b) = \dfrac{\lim a}{\lim b}.$

Proof. For convenience let $\lim a = A$ and $\lim b = B$. Also let e be any positive number. We now prove each part of the theorem separately.

Proof of (i). If $c = 0$, then $ca_n = 0$ for each n so the sequence ca is the constant zero sequence and hence converges to $0 = 0A = cA$.

We thus consider the case $c \neq 0$. In this case $|c| > 0$. Hence for all except a finite number of n we have

$$|a_n - A| < \frac{e}{|c|}, |c||a_n - A| < e, |ca_n - cA| < e$$

which establishes the desired result.

Proof of (ii). Note that $e/2$ is positive. Since the sequences a and b converge to A and B, respectively, we know that

$$A - e/2 < a_n < A + e/2 \quad \text{and} \quad B - e/2 < b_n < B + e/2$$

each (and hence both) hold for all except a finite number of n. If n is such that all four inequalities hold we know for this n that

$$(A - e/2) + (B - e/2) < a_n + b_n < (A + e/2) + (B + e/2).$$

Thus for all except a finite number of n

$$(A + B) - e < a_n + b_n < (A + B) + e$$

which, by definition, says that the sequence $a + b$ converges to $A + B$; that is, (ii) holds.

Proof of (iii). Note that $1, e/(2|A| + 2)$, and $e/(2|B| + 2)$ are all positive. Each of the following then holds for all except a finite number of n:

$$-(|A| + 1) < A - 1 < a_n < A + 1 \leq |A| + 1,$$

$$A - e/(2|B| + 2) < a_n < A + e/(2|B| + 2), \qquad \text{and}$$

$$B - e/(2|A| + 2) < b_n < B + e/(2|A| + 2).$$

Hence for all except a finite number of n we know that all of these inequalities hold and that

$$|a_n b_n - AB| = |a_n b_n - a_n B + a_n B - AB|$$

$$\leq |a_n b_n - a_n B| + |a_n B - AB|$$

$$= |a_n||b_n - B| + |B||a_n - A|$$

$$< (|A| + 1)\frac{e}{2|A| + 2} + |B|\frac{e}{2|B| + 2}$$

$$< \frac{e}{2} + \frac{e}{2} = e.$$

Consequently the sequence $a \cdot b$ converges to AB:

$$\lim (a \cdot b) = AB = (\lim a)(\lim b).$$

Proof of (iv). Since for this part $B \neq 0$, we have $|B|/2 > 0$. Hence for all except a finite number of n

(8) $|b_n - B| < |B|/2.$

With n an integer for which (8) holds we have

$$|B| = |B - b_n + b_n| \leq |B - b_n| + |b_n|$$

$$< |B|/2 + |b_n| \quad \text{so} \quad |b_n| > |B|/2.$$

Therefore for all except a finite number of n

(9) $|b_n| > |B|/2.$

Hence $b_n \neq 0$ for all except a finite number of n.

Toward finishing the proof of (iv) we prove two auxiliary results the first of which is that the sequence B/b is convergent and

(10) $\lim (B/b) = 1.$

To prove this, note that $e |B|/2 > 0$. Hence for all except a finite number of n

(11) $|B - b_n| < e |B|/2.$

Let n be an integer for which both (9) and (11) hold so that $b_n \neq 0$ and

$$\left| \frac{B - b_n}{b_n} \right| = \frac{|B - b_n|}{|b_n|} < \frac{e |B|}{2 |b_n|} = \frac{e |B|}{2} \frac{1}{|b_n|} < e |b_n| \frac{1}{|b_n|} = e.$$

Hence for all except a finite number of n

(12) $\left| \frac{B}{b_n} - 1 \right| < e.$

which proves the first auxiliary result so that (10) holds.

Since B^{-1} is a number and B/b is a convergent sequence, we know from Part (i) that $B^{-1}(B/b)$ is a convergent sequence and that

$$\lim B^{-1}(B/b) = B^{-1} \lim (B/b) = B^{-1}(1) = 1/B.$$

Hence (this is the second auxiliary result) the sequence $1/b$ is convergent and

$$\lim (1/b) = 1/B.$$

Now since the sequence a and the sequence $1/b$ are convergent, we know from Part (iii) that the sequence $a \cdot (1/b)$, which is the sequence $a \div b$, is convergent and

$$\lim (a \div b) = \lim a(1/b) = (\lim a)(\lim (1/b)) = A\,\frac{1}{B} = \frac{A}{B} = \frac{\lim a}{\lim b}.$$

This finishes the proof of the theorem.

We are now in a position to prove:

For A a positive number, and for $a_1 > \sqrt{A}$ and for

$$a_n = \frac{1}{2}\left(a_{n-1} + \frac{A}{a_{n-1}}\right), \qquad n = 2, 3, 4, \cdots$$

it follows that the sequence a so defined is convergent and

$$\lim a = \sqrt{A}.$$

First for $a_1 > \sqrt{A}$ we have a_1 is positive. Hence $a_2 = \frac{1}{2}(a_1 + A/a_1)$ is positive, $a_3 = \frac{1}{2}(a_2 + A/a_2)$ is positive etc. Also for $n \geq 2$

$$a_n - \sqrt{A} = \frac{1}{2}\left(a_{n-1} + \frac{A}{a_{n-1}}\right) - \sqrt{A}$$

$$= \frac{1}{2}\left(\frac{a_{n-1}^2 + A - 2a_{n-1}\sqrt{A}}{a_{n-1}}\right) = \frac{(a_{n-1} - \sqrt{A})^2}{2a_{n-1}} \geq 0$$

since $a_{n-1} > 0$ and certainly $(a_{n-1} - \sqrt{A})^2 \geq 0$. Thus $a_n - \sqrt{A} \geq 0$ so that $a_n \geq \sqrt{A} > 0$ and $a_n^2 \geq A$. Hence $A \leq a_{n-1}^2$ so that

$$a_n = \frac{1}{2}\left(a_{n-1} + \frac{A}{a_{n-1}}\right) \leq \frac{1}{2}\left(a_{n-1} + \frac{a_{n-1}^2}{a_{n-1}}\right) = \frac{1}{2}(a_{n-1} + a_{n-1}) = a_{n-1}.$$

Consequently $a_n \leq a_{n-1}$. Thus

$$a_1 \geq a_2 \geq \cdots \geq a_{n-1} \geq a_n \geq \cdots \geq \sqrt{A}.$$

The sequence a is therefore monotonically decreasing and bounded below by $\sqrt{A} > 0$ and thus has a limit $L \geq \sqrt{A}$; that is

$$\lim a = L \geq \sqrt{A} > 0.$$

Now let b be the sequence defined by $b_n = a_{n-1}$, for $n = 2, 3, 4, \cdots$. Note that the sequence b also converges to L:

$$\lim b = L.$$

Since

$$a_n = \frac{1}{2}\left(a_{n-1} + \frac{A}{a_{n-1}}\right) = \frac{1}{2}\left(b_n + \frac{A}{b_n}\right)$$

we have by using the various parts of Theorem 5 (in particular $L > 0$ so (iv) may be used)

$$L = \lim a = \lim \frac{1}{2}\left(b + \frac{A}{b}\right) = \frac{1}{2}\lim\left(b + \frac{A}{b}\right)$$

$$= \frac{1}{2}\left[\lim b + \lim \frac{A}{b}\right] = \frac{1}{2}\left[L + \frac{A}{\lim b}\right] = \frac{1}{2}\left[L + \frac{A}{L}\right]$$

But the positive solution of the equation

$$L = \frac{1}{2}\left[L + \frac{A}{L}\right]$$

is $L = \sqrt{A}$ so that, as we wished to prove

$$\lim a = \sqrt{A}.$$

The "squeeze" process of the next theorem is sometimes useful in finding the limit of a sequence.

THEOREM 6. *Let a be a sequence. If two convergent sequences b and c with the same limit can be found such that $b_n \leqq a_n \leqq c_n$ for each n, then the sequence a also converges and has the same limit.*

Proof. Assume sequences b and c have been found and set

$$A = \lim b = \lim c.$$

In an arbitrary neighborhood of A will be found all except a finite number of terms of the sequence b and the sequence c and hence (since $b_n \leqq a_n \leqq c_n$ for all n) in this neighborhood of A will be found all except a finite number of terms of the sequence a. Consequently the sequence a also converges and

$$\lim a = A.$$

ELEMENTARY FIGURES[2]

Area. A square with each side of unit length has area 1 square unit. A rectangle with an integral number of units on each side has area the number of unit squares into which it can be divided. A square of side $\frac{1}{2}^p$ has area $(\frac{1}{2}^p)^2$ square units.

By an *elementary figure* is meant any plane point set which, for some integer p, can be divided into a finite number of squares each of side $\frac{1}{2}^p$ and the *area* of this elementary figure is defined to be $(\frac{1}{2}^p)^2$ times the number of such squares.

---

[2] It seems appropriate to provide here a glimpse of the modern studies of the so-called elementary figures. This section of the chapter is not essential to the understanding of the remaining sections.

For example the square with opposite corners at the points with coordinates $(0, 0)$ and $(\sqrt{2}, \sqrt{2})$ is not an elementary figure.

In Example 1 (p. 204), for each integer n, $\bar{a}_n$ is the area of an elementary figure which includes the point set S.

DEFINITION 3. The number $\bar{A}$ such that $\lim \bar{a} = \bar{A}$, is defined to be the *outer* area of S. The outer area of S is denoted by $\bar{a}(S)$ so that

$$\bar{a}(S) = \bar{A}.$$

Remember that the rectangle R used in this earlier example was chosen with sides parallel to the axes and with all corners having integers for both coordinates. This rectangle R is therefore an elementary figure and we denote its area by $a(R)$. The above definition being perfectly general, we know how the outer area of the set $T = R - S$ is defined.

DEFINITION 4. For the point set S (and the rectangle R containing S) considered earlier, the *inner area* of S is denoted by $\underline{a}(S)$ and is defined by

$$\underline{a}(S) = a(R) - \bar{a}(R - S).$$

If a point set S is such that $\bar{a}(S) = \underline{a}(S)$, then this common value is called the *area* of S and is denoted by $a(S)$.

As an example of a point set S whose inner and outer areas are different (and thus whose area does not exist), let R be the unit square with opposite corners $(0, 0)$ and $(1, 1)$ and let S be the set of all points in this square having rational numbers as both coordinates. Now for each integer n, divide R into squares each with side $\frac{1}{2}^n$. Each of these little squares contains both points of S and points of $T = R - S$. Thus, in the terminology of the general case, both $\bar{a}_n = 1$ and $\bar{b}_n = 1$. Hence $\bar{A} = 1$ and $\bar{B} = 1$; that is

$$\bar{a}(S) = 1 \quad \text{and} \quad \bar{a}(T) = 1.$$

Since also $a(R) = 1$ in this case we have

$$\underline{a}(S) = a(R) - \bar{a}(T) = 1 - 1 = 0.$$

Consequently this set S has inner area 0 and outer area 1.

THEOREM 7. *For any set S, $\underline{a}(S) \leq \bar{a}(S)$.*

Proof. With the situation as considered above, $\bar{a}_n$ and $\bar{b}_n$ are the areas of elementary figures including S and $T = R - S$, respectively. These elementary figures may have some squares in common, namely those squares containing both points of S and T. The elementary figure whose area is $\bar{a}_n$ together with the elementary figure whose area is $\bar{b}_n$ make up all of R so that $\bar{a}_n + \bar{b}_n \geq a(R)$. Consequently

$$\bar{A} + \bar{B} = \lim \bar{a} + \lim \bar{b} = \lim (\bar{a} + \bar{b}) \geq a(R);$$

that is, $\overline{a}(S) + \overline{a}(T) \geqq a(R)$. Since $T = R - S$, it follows that

$$\overline{a}(S) \geqq a(R) - \overline{a}(R - S) = \underline{a}(S)$$

as we wished to prove.

THEOREM 8. *For S^* a subset of S, that is $S^* \subset S$, it follows that*

$$\overline{a}(S^*) \leqq \overline{a}(S) \quad \text{and} \quad \underline{a}(S^*) \leqq \underline{a}(S).$$

Proof. For each integer n, the rectangle R including S is divided into squares of side $\frac{1}{2}^n$. These small squares which intersect S form an elementary figure of area $\bar{a}_n$ and the ones which intersect S^* form an elementary figure whose area we call $\bar{a}_n{}^*$. If a square contains a point of S^* then it certainly contains a point of S. Thus the elementary figure for S^* contains fewer squares (if there is any difference) than the elementary figure for S so that $\bar{a}_n{}^* \leqq \bar{a}_n$. Hence

$$\overline{a}(S^*) = \lim \bar{a}^* \leqq \lim \bar{a} = \overline{a}(S)$$

thus establishing the first inequality of the theorem.

With $T = R - S$ and $T^* = R - S^*$ it follows that $T \subset T^*$ since $S \supset S^*$. Thus by the first part of this proof $\overline{a}(T) \leqq \overline{a}(T^*)$. Consequently

$$a(R) - \overline{a}(T) \geqq a(R) - \overline{a}(T^*).$$

By definition the left side is $\underline{a}(S)$ while the right side is $\underline{a}(S^*)$ so we have $\underline{a}(S) \geqq \underline{a}(S^*)$; that is $\underline{a}(S^*) \leqq \underline{a}(S)$ which is the second of the inequalities of the theorem.

THEOREM 9. *Let the point set S be divided into two subsets S^* and S' by a line perpendicular to the x-asis. Then*

$$\overline{a}(S) = \overline{a}(S^*) + \overline{a}(S').$$

The same result holds if S is divided into two subsets by a line parallel to the x-axis.

Proof. As usual S is contained in the rectangle R (whose altitude we now denote by H). For each integer n the rectangle R is divided into squares of side $\frac{1}{2}^n$, and $\bar{a}_n$ is the sum of the areas of these squares which intersect S. Let $\bar{a}_n{}^*$ and $\bar{a}_n{}'$ be the sum of the areas of those squares which intersect S^* and S', respectively. Note the squares which intersect both S^* and S' and see that they certainly lie in a rectangle which is an elementary figure of base $2(\frac{1}{2}^n)$ and altitude the integer H. Consequently

$$\bar{a}_n \leqq \bar{a}_n{}^* + \bar{a}_n{}' \leqq \bar{a}_n + 2(\frac{1}{2}^n)H.$$

Since each of these sequences converges and we denote $\overline{a}(S^*) = \lim \bar{a}^*$ and $\overline{a}(S') = \lim \bar{a}'$, it follows from Theorem 4 and the fact that the sequence whose nth term is $H/2^{n-1}$ converges to 0, that

$$\bar{a}(S) \leq \bar{a}(S^*) + \bar{a}(S') \leq \bar{a}(S).$$

Thus the equalities must hold and the theorem is proved.

THEOREM 10. *For a rectangle with sides parallel to the axes the area exists (i.e. the inner and outer areas are the same) and is the product of the lengths of the base and altitude.*

Proof. Let the set S be in particular a rectangle with sides parallel to the axes and let b and h be the lengths of the base and altitude. For each integer n the squares of side $\frac{1}{2}^n$ intersecting S are arranged, now that S is a rectangle, into a rectangle of p columns of squares with q squares in each column where p and q are the integers such that

$$b \leq p \cdot (\tfrac{1}{2}^n) \leq b + \tfrac{2}{2}^n \quad \text{and} \quad h \leq q \cdot (\tfrac{1}{2}^n) \leq h + \tfrac{2}{2}^n.$$

Since $\bar{a}_n = p \cdot q \cdot (\tfrac{1}{2}^n)^2$ we thus have

$$b \cdot h \leq \bar{a}_n \leq (b + \tfrac{1}{2}^{n-1})(h + \tfrac{1}{2}^{n-1}).$$

Considering the left member as the nth term of a constant sequence we have

$$\bar{a}(S) = \lim \bar{a}_n = b \cdot h$$

by Theorem 4, since by Theorem 5 (iii) the sequences whose nth terms are $b + \tfrac{1}{2}^{n-1}$ and $h + \tfrac{1}{2}^{n-1}$ converge to b and h, respectively.

Hence for any rectangle with sides parallel to the axes, the outer area is the length of the base times the length of the altitude.

As usual S is in a rectangle R with an integral number of units in its base and altitude. Since for this theorem, S is a rectangle, the set $T = R - S$ may be divided into eight rectangles by extending all sides of S until they meet sides of R. These extended sides being parallel or perpendicular to the x-axis it follows from Theorem 9 that the outer area of T is the sum of the outer areas of these eight rectangles and moreover

$$a(R) = \bar{a}(S) + \bar{a}(T).$$

Since by definition $\underline{a}(S) = a(R) - \bar{a}(T)$ it follows that $\underline{a}(S) = \bar{a}(S)$ so the area of the rectangle exists. Also since the first part of this proof showed that $\bar{a}(S) = b \cdot h$ the proof of the theorem is complete.

LIMIT PROCESS AND FUNCTIONS

Functions. So far in this chapter we have tacitly assumed, and we now explicitly state for the remainder of the chapter, that all numbers considered are restricted to the real number system. Thus for $x < 0$ no meaning is attached to $\sqrt{x}$. To be perfectly definite, for $x \geqq 0$ we use $\sqrt{x}$ to mean not just any number whose square is x, but the one and only non-negative number whose square is x. Note in particular that if x is negative then $\sqrt{x^2}$ is not equal to x, but regardless of whether x is positive, negative, or zero

$$\sqrt{x^2} = |x|.$$

By an *ordered pair* of numbers is meant a pair of numbers one of which has been designated as the first. An ordered pair is written (x, y).

A single letter (such as S) has been used to designate a set of numbers or a set of points. We shall be considering sets of ordered pairs and will find it convenient to designate a set of ordered pairs by a symbol, or letter, or even a combination of letters. For example $\sqrt{}$ will be used to symbolize the set of all ordered pairs of numbers where in each ordered pair both numbers are non-negative and the square of the second is equal to the first. In this sense of pairing numbers, $\sqrt{}$ is a function according to the following definition.

DEFINITION 5. A set f of ordered pairs of numbers is called a *function* provided f has the property that if (a, b) and (a, c) are in the set f, then $b = c$. The set of all first numbers in f is called the *domain* of f while the set of all second numbers is called the *range* of f. Also for x a definite number in the domain of f, the unique number paired with x is represented by $f(x)$ and is called the *value* of f at x.

For the function $\sqrt{}$ the domain and range coincide in that each is the set of all non-negative numbers.

Given any number x, by sin x is meant the number obtained by constructing an angle of x radians and taking the sine of this angle. Thus sin is a function whose domain is the set of all real numbers while the range is the set of all numbers from -1 to $+1$ inclusive.

For x any number $[x]$ is sometimes used to mean the greatest integer less than or equal to x. In this sense $[]$ is a function whose domain is the set of all numbers and whose range is the set of integers.

Consider the set of ordered pairs defined by saying that if an ordered pair has its first number rational then the second number is 0, but if the first number is irrational then the second number is 1. This set of ordered pairs is therefore a function since no two pairs of the set have the same first element. The domain is the set of all numbers and the range consists of the two numbers 0 and 1.

Note that a sequence is a function whose domain is the set of positive integers.

Sometimes one function is used in the definition of another function.

For example (and this notion will be used later) let f be a given function and let x be a number in the domain of f. We now define a function g by saying that for each number h such that $h \neq 0$ and $x + h$ is in the domain of f, then

$$g(h) = \frac{f(x + h) - f(x)}{h}.$$

Note that 0 is not in the domain of g and that the domain of g may depend upon the number x chosen in the domain of f. Since the function g depends both on f and on x, it is customary to use a notation which displays this fact by using $Qf(x)$ instead of g. Thus

$$Qf(x)(h) = \frac{f(x + h) - f(x)}{h}.$$

In particular for the function $\sqrt{}$ and for $x \geq 0$,

$$Q\sqrt{x}(h) = \frac{\sqrt{x + h} - \sqrt{x}}{h}$$

and the domain of the function $Q\sqrt{x}$ consists of all numbers except zero which are greater than or equal to $-x$.

Since a function f is a set of ordered pairs of numbers and since each point in a coordinate plane has an ordered pair of numbers as its coordinates, it is natural to represent the ordered pairs of the function by points in the plane. The resulting set of points is then called the *rectangular graph* of the function. Note that the condition for a function f "If (a, b) and (a, c) are in the set f, then $b = c$" has the geometric interpretation "If a line perpendicular to the x-axis intersects the graph of f, then this line intersects the graph in one and only one point."

Also for f a function and x a number in the domain of f, there is a geometric interpretation of $Qf(x)(h)$; namely, the slope of the line joining the points represented by $(x, f(x + h))$ and $(x, f(x))$. Also if $|h|$ is "small" this line appears to be nearly tangent to the graph of f at the point represented by $(x, f(x))$.

Functions, Limits, and Continuity. Given a number c and a positive number e, we shall use the notation $N(c, e)$ to mean the set of all numbers each of which is both greater than $c - e$ and less than $c + e$: that is $N(c, e)$ is the symmetric neighborhood of c which visualized as an interval of the number line has length $2e$. Also by $\mathring{N}(c, e)$ we shall mean all of the set $N(c, e)$ except the number c itself and shall refer to $\mathring{N}(c, e)$ as a *punctured neighborhood* of c.

DEFINITION 6. Given a function f, a number c, and a number L, then the function f is said to have *limit* L at c if:

1. Every punctured neighborhood of c contains at least one point of the domain of f, and
2. For each number $e > 0$ there is a number $d > 0$ such that whatever number x is both in the domain of f and in $\overset{\circ}{N}(c, d)$ the number $f(x)$ is in $N(L, e)$.

If f has limit L at c we write

$$L = \lim_c f.$$

Notice that c need not be in the domain of f and even if c is in the domain of f it may be that $f(c) \neq L$.

The above definition is only a definition so do not expect it to produce a limit. The definition is a formalization of the intuitive notion of a limit and is used to prove theorems which in turn may reveal methods for determining limits. Also the definition is used to check whether a suspected value is or is not a limit. For example, we certainly expect for $c > 0$ that the function $\sqrt{}$ has a limit at c and that this limit is $\sqrt{c}$. To establish this let e be a positive number and for convenience take $e < \sqrt{c}$. Then $\sqrt{c} - e > 0$, so that $\sqrt{(\sqrt{c} - e)^2} = \sqrt{c} - e$. Let $(\sqrt{c} - e)^2 = x_1$ so $\sqrt{x_1} = \sqrt{c} - e$. Also let $(\sqrt{c} + e)^2 = x_2$ so $\sqrt{x_2} = \sqrt{c} + e$. Now for x any number such that $x_1 < x < x_2$ we have $\sqrt{x_1} < \sqrt{x} < \sqrt{x_2}$ so that

$$\sqrt{c} - e < \sqrt{x} < \sqrt{c} + e.$$

which states that $\sqrt{x}$ is in $N(\sqrt{c}, e)$. Consequently upon choosing d as the smaller of the positive numbers $\sqrt{c} - x_1$ and $\sqrt{x_2} - c$ we see that for x in $\overset{\circ}{N}(c, d)$, $\sqrt{x}$ is in $N(\sqrt{c}, e)$. Hence from the definition of a limit, $\sqrt{c}$ is the limit at c of the function $\sqrt{}$, *i.e.*

$$\lim_c \sqrt{} = \sqrt{c}.$$

Given functions f and g with a common domain, we define another function with the same domain as the set of ordered pairs each of which has first element in the given domain and second element the sum of the corresponding second elements of f and g. This function is called the *sum* of f and g and is denoted by

$$f + g.$$

By replacing the word sum by *product* in the above definition we obtain the function denoted by

$$f \cdot g.$$

If the domain of g contains a subset such that every ordered pair of g with first element in this subset has second element different from 0, then

$$f \div g$$

is the function with this subset as domain and each second element of an ordered pair the quotient of corresponding second elements of f and g.

Also with f a function and k a number we use

$$kf$$

as the function each of whose second elements is k times the corresponding second element of f.

The value of $f \cdot g$ at x is denoted by $(f \cdot g)(x)$ so that

$$(f \cdot g)(x) = f(x)g(x).$$

Analogous notation for $f + g$, $f \div g$, and kf at x is used.

The following theorem is similar to one proved for sequences.

THEOREM 11. *Let f and g be functions with the same domain, let k be a number, and let c be a number such that the limit of f at c and the limit of g at c both exist. Then the limit of kf, the limit of $f + g$, and the limit of $f \cdot g$ at c exist and*

(i) $$\lim_c kf = k \lim_c f$$

(ii) $$\lim_c (f + g) = \lim_c f + \lim_c g$$

(iii) $$\lim_c (f \cdot g) = (\lim_c f)(\lim_c g).$$

Also if $\lim_c g \neq 0$, then there is a number $e > 0$ such that for x in the intersection of $N(c, e)$ and the domain of g, $g(x) \neq 0$. Moreover the limit of $f \div g$ at c exists and

(iv) $$\lim_c (f \div g) = (\lim_c f)/(\lim_c g).$$

Since the proof of this theorem is similar to the proof of the corresponding theorem for sequences, we prove only (ii) as an illustration of modifications necessary. Let $L_1 = \lim_c f$ and $L_2 = \lim_c g$. Now let e be any positive number. Corresponding to $e/2$ let $d_1 > 0$ and $d_2 > 0$ be such that whatever number x is taken in:

  (a) the domain of f and in $\overset{\circ}{N}(c, d_1)$, then $f(x)$ is in $N(L_1, e/2)$,
  (b) the domain of g and in $\overset{\circ}{N}(c, d_2)$, then $g(x)$ is in $N(L_2, e/2)$.

Let d be the smaller of d_1 and d_2 so that $d > 0$. Since f and g have a common domain, for x in this domain and also in $\overset{\circ}{N}(c, d)$, then both $f(x)$ is in $N(L_1, e/2)$ and $g(x)$ is in $N(L_2, e/2)$; that is,

$$L_1 - e/2 < f(x) < L_1 + e/2 \quad \text{and} \quad L_2 - e/2 < g(x) < L_2 + e/2$$

so that $L_1 + L_2 - e < f(x) + g(x) < L_1 + L_2 + e$ and hence

$$f(x) + g(x) \quad \text{is in} \quad N(L_1 + L_2, e).$$

Consequently by definition $\lim_{c} (f + g)$ exists and is equal to $L_1 + L_2$ so that (ii) is established.

For c a number and e a positive number let $N_+(c, e)$ be the set of all numbers greater than or equal to c but less than $c + e$. Let $\overset{\circ}{N}_+(c, e)$ be the set $N_+(c, e)$ with c removed. $N_+(c, e)$ is called a *right neighborhood* of c and $\overset{\circ}{N}_+(c, e)$ a *right punctured neighborhood* of c. Now in the definition of a limit we replace N by N_+ and $\overset{\circ}{N}$ by $\overset{\circ}{N}_+$ and obtain the definition of the *right limit* of f at c and for this use the notation

$$\lim_{c+} f.$$

It should be seen how Theorem 11 may be modified to obtain a theorem about right limits. Left-handed remarks are too obvious to make.

A function f is said to be *monotonically increasing* if whatever numbers x_1 and x_2 with $x_1 < x_2$ are in the domain of f, then $f(x_1) < f(x_2)$.

A function f is said to be *bounded above* if there is a number G such that $f(x) \leq G$ for each number x in the domain of f.

All theorems about limits of sequences may now be duplicated for functions. For example the theorem "A monotonically increasing sequence bounded above has a limit" has the following counterpart whose proof is so similar that it is not given:

THEOREM 12. *If f is a monotonically increasing function bounded above and if c is a number such that every left punctured neighborhood of c contains a number in the domain of f, then the left limit of f at c exists.*

At some number a function may have a left limit and a right limit, but these limits may be different. For example the function [ ] (defined by $[x]$ is the greatest integer less than or equal to x) is such that

$$\lim_{1-} [\ ] = 0 \quad \text{and} \quad \lim_{1+} [\ ] = 1.$$

THEOREM 13. *Let f be a function and c a number such that the left and right limits of f at c exist and are equal. Then the limit of f at c exists and these three limits are equal.*

Proof. Let $L = \lim_{c-} f = \lim_{c+} f$. Let e be a positive number. There is then a number $d_1 > 0$ such that if x is in $\overset{\circ}{N}_-(c, d_1)$, then $f(x)$ is in $N(L, e)$ and a number $d_2 > 0$ such that if x is in $\overset{\circ}{N}_+(c, d_2)$ then $f(x)$ is in $N(L, e)$. Let d be the smaller of d_1 and d_2 so that $d > 0$. Note that (see Chapter III for the meaning of $\subset$ and $\cup$)

$$\overset{\circ}{N}(c, d) \subset \{\overset{\circ}{N}_-(c, d_1) \cup \overset{\circ}{N}_+(c, d_2)\}$$

so that if x is in $\overset{\circ}{N}(c, d)$ then x is in either $\overset{\circ}{N}_-(c, d_1)$ or else in $\overset{\circ}{N}_+(c, d_2)$ and then certainly $f(x)$ is in $N(L, e)$. Hence the theorem is proved.

If f is a function and c is a number such that $\lim_c f$ exists, it may be (as mentioned earlier) that c is not in the domain of f. For example, let g be the function defined by

$$(13) \quad g(h) = \frac{\sqrt{2 + h} - \sqrt{2}}{h} \quad \text{for} \quad h \neq 0 \quad \text{and} \quad h \geq -2.$$

and let $c = 0$. For $h \neq 0$ and $h \geq -2$ we have by simple algebra

$$\frac{\sqrt{2 + h} - \sqrt{2}}{h} = \frac{\sqrt{2 + h} - \sqrt{2}}{h} \cdot \frac{\sqrt{2 + h} + \sqrt{2}}{\sqrt{2 + h} + \sqrt{2}}$$

$$= \frac{(2 + h) - 2}{h(\sqrt{2 + h} + \sqrt{2})} = \frac{1}{\sqrt{2 + h} + \sqrt{2}} \cdot \frac{h}{h}.$$

We thus see that

$$(14) \quad g(h) = \frac{1}{\sqrt{2 + h} + \sqrt{2}} \quad \text{for} \quad h \neq 0 \quad \text{and} \quad h \geq -2.$$

In this second form the reason for excluding 0 from the domain of g is not apparent, but the expressions on the right of (13) and (14) are equivalent only if both are defined. The second expression itself needs only the restriction $h \geq -2$, but the first expression needs the additional restriction $h \neq 0$. That (13) and (14) are equivalent is because of the simple fact that $h/h = 1$ if and only if $h \neq 0$. Now the result proved earlier (pp. 220–22) about the function $\sqrt{}$ allows us to state that the function defined by the denominator of (14) has limit at 0 equal to $\sqrt{2} + \sqrt{2}$. Since $2\sqrt{2} \neq 0$ we may now use Theorem 11 (iv) to see from (14) that

$$\lim_c g = \frac{1}{2\sqrt{2}}$$

Thus even though 0 is not in the domain of the function g defined by (13), nevertheless the limit of this function at 0 exists.

If f is a function and c is a number in the domain of f (so that $f(c)$ is defined) and if the limit of f at c exists, it may still be, as mentioned earlier, that $\lim_c f \neq f(c)$.

DEFINITION 7. *If f is a function and c is a number in the domain of f such that the limit of f at c exists and if furthermore*

$$\lim_c f = f(c),$$

then f is said to be *continuous* at c. If f is continuous at each number in its domain, then f is said to be a *continuous function*.

Many of the functions arising from physical considerations are continuous, but others are not. An object flying through air at supersonic speed creates shock waves at which functions connected with pressure, density, entropy, etc., are not continuous and for this reason some important problems in supersonic flow await solutions until further mathematical theory is developed.

There are some simple-sounding theorems about continuous functions whose proofs are too complicated for this short chapter. We state the following theorem without proof, but will use the stated facts in the next section.

THEOREM 14. *If $a < b$ are numbers and f is a continuous function whose domain is the set of all numbers x such that $a \leq x \leq b$, then f has a maximum value and a minimum value. Also if $f(a) < 0$ and $f(b) > 0$, then there is at least one number x_0 such that $a < x_0 < b$ and $f(x_0) = 0$. If g is also a continuous function with the same domain as f, then $f + g$, and $f \cdot g$ are also continuous (but it may be that $f \div g$ is not continuous).*

DIFFERENTIATION AND INTEGRATION

Derivatives. Let f be a function and let x be a number in the domain of f. We now define two new functions whose domain consists of all numbers h such that $h \neq 0$ and such that $x + h$ is in the domain of f. These new functions are denoted by $\Delta f(x)$ and $Qf(x)$ and are defined by

$$\Delta f(x)(h) = f(x + h) - f(x) \qquad \text{and}$$

$$Qf(x)(h) = \frac{f(x + h) - f(x)}{h}$$

with the above restrictions on h.

A pictorial concept of the relation between the function f and the function $Qf(x)$ may be obtained as follows:

1. Sketch the graph of f.
2. With x any number in the domain of f, spot the point with coordinates $(x, f(x))$.

3. Next with h a number such that $h \neq 0$ and $x + h$ is in the domain of f, spot the point with coordinates $(x + h, f(x + h))$.

4. Find the slope of the line joining the two points obtained in 2 and 3; this slope is

$$\frac{f(x + h) - f(x)}{(x + h) - x} = \frac{f(x + h) - f(x)}{h} = Qf(x)(h).$$

Thus the value of $Qf(x)$ at h is the same as the slope of this line.

Consider a particle moving along a line on which coordinates are given say in feet and consider time as measured in seconds. Let f be that function which gives the coordinate of the particle at any designated time. Hence for t a number, then t seconds from the beginning of the experiment the particle is $f(t)$ feet from the origin of the line. For h a number such that $h \neq 0$, at time $t + h$ seconds the particle is $f(t + h)$ feet from the origin, $f(t + h) - f(t)$ feet is the change in position from t seconds to $t + h$ seconds and for this time interval the average velocity of the particle is

$$Qf(t)(h) = \frac{f(t + h) - f(t)}{h} \text{ ft/sec.}$$

Thus we have a physical concept of the relation between the function f and the function $Qf(x)$.

DEFINITION 8. For x in the domain of f, if the limit of $Qf(x)$ at 0 exists, we denote this limit by $f'(x)$; that is,

$$f'(x) = \lim_0 Qf(x).$$

Thus f' is a function whose domain (if it contains any numbers at all) is included in the domain of f. The function f' is called the *derived function* of f and for each number x for which $f'(x)$ exists, $f'(x)$ is called the *derivative* of f at x.

For x in the domain of both f and f', then no matter how small h is, so long as $h \neq 0$ and $x + h$ is in the domain of f, the points with coordinates $(x, f(x))$ and $(x + h, f(x + h))$ are distinct and determine a line which seems to "barely" cut the graph of f. It is thus natural to define:

The line through the point P with coordinates $(x, f(x))$ and slope $f'(x)$ is said to be *tangent* to the graph of f at P.

We thus have a geometric interpretation of the derivative.

For f giving the position of a particle on a line as described above, we define $f'(x)$ ft/sec to be the *instantaneous velocity* of the particle at time t. We therefore have a physical interpretation of the derivative.

By following the procedure of page 225, it should be seen that:
For f defined by $f(x) = \sqrt{x}$, $x \geq 0$, then f' is defined by

$$f'(x) = \frac{1}{2\sqrt{x}}, x > 0.$$

Note that for this function, the domain of f' does not include the number 0. Even if the reason for this is not clear, proceed to the next example.

Now let f be the function defined by $f(x) = \sqrt[3]{x}$. Then, with the usual restrictions,

$$Qf(x)(h) = \frac{\sqrt[3]{x+h} - \sqrt[3]{x}}{h}.$$

To handle this situation we rationalize the numerator (remembering the factor formula $a^3 - b^3 = (a - b)(a^2 + ab + b^2)$):

$$\frac{\sqrt[3]{x+h} - \sqrt[3]{x}}{h}$$

$$= \frac{\sqrt[3]{x+h} - \sqrt[3]{x}}{h} \cdot \frac{(\sqrt[3]{x+h})^2 + \sqrt[3]{x+h}\,\sqrt[3]{x} + (\sqrt[3]{x})^2}{(\sqrt[3]{x+h})^2 + \sqrt[3]{x+h}\,\sqrt[3]{x} + (\sqrt[3]{x})^2}$$

$$= \frac{(x+h) - x}{h\{(\sqrt[3]{x+h})^2 + \sqrt[3]{x+h}\,\sqrt[3]{x} + (\sqrt[3]{x})^2\}}$$

Since $x - x = 0$ and then $h/h = 1$ for $h \neq 0$ we have

(15) $$Qf(x)(h) = \frac{1}{(\sqrt[3]{x+h})^2 + \sqrt[3]{x+h}\,\sqrt[3]{x} + (\sqrt[3]{x})^2}.$$

Now in particular, upon setting $x = 0$ we have

$$Qf(0)(h) = \frac{1}{(\sqrt[3]{h})^2 + \sqrt[3]{h}\,\sqrt[3]{0} + (\sqrt[3]{0})^2} = \frac{1}{(\sqrt[3]{h})^2}.$$

Since this fraction can be made as large as we please merely by choosing h small enough, we see that $x = 0$ is not (for this particular function f) in the domain of f'. Now for $x \neq 0$, we shall say (although we have not actually proved it) that the function defined by the denominator of (15) has limit at 0 equal to

$$(\sqrt[3]{x})^2 + \sqrt[3]{x}\,\sqrt[3]{x} + (\sqrt[3]{x})^2 = 3(\sqrt[3]{x})^2.$$

Thus:

For the function f defined by $f(x) = \sqrt[3]{x}$, the derived function f' is defined by

$$f'(x) = \frac{1}{3(\sqrt[3]{x})^2}, \qquad x \neq 0.$$

This fact will be seen to fit a general formula discussed later if we write $f(x)$ and $f'(x)$ in the forms

$$f(x) = x^{1/3} \quad \text{and} \quad f'(x) = \tfrac{1}{3}x^{1/3-1}.$$

As still another example, for the function I defined by $I(x) = x$, it follows that I' is defined by $I'(x) = 1$. For $I(x + h) = x + h$ and

$$QI(x)(h) = \frac{(x + h) - x}{h} = \frac{h}{h} = 1 \text{ since } h \neq 0.$$

Hence also $\lim_0 QI(x) = 1$ so that $I'(x) = 1$.

The reader should prove for himself that if c is any number and C is the function defined by $C(x) = c$, then C' is the function defined by $C'(x) = 0$. Geometrically interpreted, this means that any line parallel to the x-axis has slope zero.

THEOREM 15. *Let f be a function and x a number such that $f'(x)$ exists. Then $\lim_0 \Delta f(x) = 0$.*

Proof. Let I be the function such that $I(h) = h$ for each number h and note that $\lim_0 I = 0$. Now

$$Qf(x)(h) = \frac{f(x + h) - f(x)}{h} = \frac{f(x + h) - f(x)}{I(h)}$$

so that $I(h) \cdot Qf(x)(h) = f(x + h) - f(x) = \Delta f(x)(h)$. Now

$$0 = 0 f'(x) = \lim_0 I \cdot \lim_0 Qf(x) = \lim_0 \{IQf(x)\} = \lim_0 \Delta f(x)$$

as we wished to prove. (This really shows that f is continuous at x).

We are now ready to derive some general formulas.

THEOREM 16. *Let f and g be functions with a common domain and let x be a number such that $f'(x)$ and $g'(x)$ both exist. Also let k be a number. Then the functions $kf, f + g,$ and $f \cdot g$ also have derivatives at x and*

(i) $$(kf)'(x) = kf'(x)$$

(ii) $$(f + g)'(x) = f'(x) + g'(x)$$

(iii) $$(f \cdot g)'(x) = f(x) \cdot g'(x) + g(x) \cdot f'(x).$$

Under the additional restriction that $g(x) \neq 0$ the derivative of the function $f \div g$ also exists at x and

(iv) $$(f \div g)'(x) = \frac{g(x)f'(x) - f(x)g'(x)}{g^2(x)}$$

where $g^2(x) = g(x) \cdot g(x)$.

Proof of (ii). Since $(f + g)(x + h) = f(x + h) + g(x + h)$ we have

$$Q(f + g)(x)(h) = \frac{\{f(x + h) + g(x + h)\} - \{f(x) + g(x)\}}{h}$$

$$= \frac{f(x + h) - f(x)}{h} + \frac{g(x + h) - g(x)}{h}$$

$$= Qf(x)(h) + Qg(x)(h).$$

We may thus write, knowing the existence of each limit written,

$$f'(x) + g'(x) = \lim_0 Qf(x) + \lim_0 Qg(x)$$

$$= \lim_0 \{Qf(x) + Qg(x)\} \qquad \text{by Theorem 11 (ii)}$$

$$= \lim_0 Q(f + g)(x) = (f + g)'(x).$$

Proof of (iii). Since $\Delta f(x)(h) = f(x + h) - f(x)$ we have

$$f(x + h) = \Delta f(x)(h) + f(x)$$

with the same relation for g. Thus

$$\Delta(f \cdot g)(x)(h) = (f \cdot g)(x + h) - (f \cdot g)(x) = f(x + h)g(x + h) - f(x)g(x)$$

$$= \{\Delta f(x)(h) + f(x)\}\{\Delta g(x)(h) + g(x)\} - f(x)g(x)$$

$$= \Delta f(x)(h) \cdot \Delta g(x)(h) + f(x) \cdot \Delta g(x)(h) + g(x) \cdot \Delta f(x)(h).$$

Hence upon dividing by h and then using the definition of Q

$$\frac{\Delta(f \cdot g)(x)(h)}{h} = \Delta f(x)(h) \cdot \frac{\Delta g(x)(h)}{h} + f(x) \cdot \frac{\Delta g(x)(h)}{h} + g(x) \cdot \frac{\Delta f(x)(h)}{h}$$

$$Q(f \cdot g)(x)(h) = \Delta f(x)(h) \cdot Qg(x)(h) + f(x) \cdot Qg(x)(h) + g(x) \cdot Qf(x)(h).$$

Now $\lim_0 \Delta f(x) = 0$ by Theorem 15, so by using several parts of Theorem 11, we obtain that $\lim_0 Q(f \cdot g)(x)$ exists and is equal to

$$0 \cdot g'(x) + f(x)g'(x) + g(x)f'(x).$$

By omitting the term equal to zero, we thus see that (iii) holds.

The proofs of (i) and (iv) are left for the reader to check his understanding.

We shall now show how these formulas may be used in bypassing the definition of a derivative to find the derived function for each function in a large class.

For each positive integer n, let f_n be the function defined for each number x by

$$f_n(x) = x^n.$$

In this notation a result proved earlier is:

$$f_1'(x) = 1.$$

We may now write $f_2 = f_1 \cdot f_1$, apply (ii) with f and g both replaced by f_1 to obtain

$$f_2'(x) = f_1(x)f_1'(x) + f_1'(x)f_1(x) = x \cdot 1 + x \cdot 1 = 2x.$$

Now with $f_3 = f_1 f_2$ we apply (ii) with f replaced by f_1 and g replaced by f_2 to obtain

$$f_3'(x) = f_1(x)f_2'(x) + f_2(x)f_1'(x)$$
$$= x(2x) + x^2(1) = 3x^2.$$

For the pair of equations and (the way they were obtained)

$$f_2(x) = x^2, \qquad f_2'(x) = 2x^{2-1} \qquad\qquad \text{and}$$
$$f_3(x) = x^3, \qquad f_3'(x) = 3x^{3-1}$$

one might suspect for any integer n whatever that $f_n'(x) = nx^{n-1}$. Assume that n is an integer for which this relation holds. Then write

$$f_{n+1}(x) = x^{n+1} = x \cdot x^n = f_1(x)f_n(x)$$

and from (ii) obtain

$$f_{n+1}'(x) = f_1(x) \cdot f_n'(x) + f_n(x) \cdot f_1'(x)$$
$$= x(nx^{n-1}) + x^n \cdot 1 = nx^n + x^n$$
$$= (n+1)x^n = (n+1)x^{(n+1)-1}$$

which has the same form as $f_n'(x) = nx^{n-1}$ with n replaced by $n+1$. Thus knowing that $f_3'(x) = 3x^{3-1}$ we know immediately that $f_4'(x) = 4x^{4-1}$, and then immediately that $f_5'(x) = 5x^{5-1}$ and so on to as many integers as we have the patience to go. By the principle of Mathematical Induction the formula $f_n'(x) = nx^{n-1}$ holds for any positive integer n.

We have already shown that for f defined by $f(x) = x^{1/2}$, then $f'(x) = \frac{1}{2}x^{1/2-1}$, for $x \neq 0$. Now let g be defined by $g(x) = x^{3/2}$. Then $g(x) = f_1(x)f(x)$ when $f_1(x) = x$ and

$$g'(x) = f_1(x)f'(x) + f(x)f_1'(x)$$
$$= x(\tfrac{1}{2}x^{1/2-1}) + x^{1/2} \cdot 1 = \tfrac{1}{2}x^{1/2} + x^{1/2} = \tfrac{3}{2}x^{1/2} = \tfrac{3}{2}x^{3/2-1}.$$

We shall pursue this line of reasoning no further, but shall state without proof that:

If p is any number and for each number x for which x^p is defined, we set $f(x) = x^p$, then the function f so defined has its derived function f' given by $f'(x) = px^{p-1}$ for each number x such that x^{p-1} is defined.

For c a number, two functions f and g related by $g(x) = f(x) + c$ have the same derived function; *i.e.* $g' = f'$. For upon letting C be the function defined by $C(x) = c$ we have

$$g'(x) = (f + C)'(x) = f'(x) + C'(x) \qquad \text{by (ii)}$$

and since (as mentioned earlier) $C'(x) = 0$ we have $g'(x) = f'(x)$.

In particular for A a number such that $A > 0$ the function defined by $f(x) = x^2 - A$ is such that $f'(x) = 2x$. We now use this fact to give an application of derivatives. Draw the graph of the function f and note that this graph is a parabola crossing the x-axis at the points $(-\sqrt{A}, 0)$ and $(\sqrt{A}, 0)$. Let a_1 be any number such that $a_1 > \sqrt{A}$. The point $(a_1, a_1^2 - A)$ is on the graph and the tangent to the graph at this point has slope

$$f'(a_1) = 2a_1 .$$

The reader should draw the figure described. The point where this tangent cuts the x-axis we label $(a_2, 0)$. The points $(a_2, 0)$ and $(a_1, a_1^2 - A)$ being on the tangent, the tangent has slope

$$\frac{(a_1^2 - A) - 0}{a_1 - a_2} = \frac{a_1^2 - A}{a_1 - a_2}$$

We now have two expressions for the slope of this tangent and hence set

$$2a_1 = \frac{a_1^2 - A}{a_1 - a_2} .$$

Consequently $2a_1^2 - 2a_1a_2 = a_1^2 - A$, $a_1^2 - 2a_1a_2 = -A$, $2a_1a_2 = a_1^2 + A$ so that

$$a_2 = \tfrac{1}{2}(a_1 + A/a_1).$$

Now by proceeding with a_2 exactly as we did with a_1 we obtain

$$a_3 = \tfrac{1}{2}(a_2 + A/a_2).$$

By continuing this process we obtain the sequence first mentioned on page 203 and later on page 215 proved to converge to $\sqrt{A}$. If the figure were drawn (as suggested above), the geometry should suggest why the sequence converges to $\sqrt{A}$.

The reader may now obtain a sequence which converges to $\sqrt[3]{A}$ by starting with the function defined by $f(x) = x^3 - A$ and proceeding as above.

The above arguments were made for the function f defined by $f(x) = x^2 - A$ to approximate a solution of $f(x) = 0$.

Now let f be any function whose derived function f' exists and also such that $f(x) = 0$ has a solution. Let a_1 be an approximation of a solution of $f(x) = 0$. In any specific case a_1 is obtained by guessing, probably with the aid of a graph. By following the above arguments, numbers a_2, a_3, $\cdots$, where

$$a_2 = a_1 - \frac{f(a_1)}{f'(a_1)}, \qquad a_3 = a_2 - \frac{f(a_2)}{f'(a_2)}, \cdots,$$

will be obtained. If a sufficiently good choice of a_1 is made the sequence a_1, a_2, a_3, $\cdots$ will converge to a solution of the equation $f(x) = 0$. This method of approximating roots of an equation is known as *Newton's Method*.

Integration. We shall pre-illustrate the general theory with the following example.

Example. Let f be the function defined by $f(x) = 2\sqrt{x}$, $x \geq 0$. Let S be the point set bounded by the graph of f, the x-axis and the vertical lines through the points $(1, 0)$ and $(9, 0)$. We shall show that the area of S exists and shall find its value.

For x a number such that $1 \leq x \leq 9$, let $\bar{F}(x)$ be the outer area of the portion of S between the vertical lines through the points $(1, 0)$ and $(x, 0)$. Hence $\bar{F}$ is a function whose domain consists of the numbers from 1 to 9 inclusive. Note that

(16) $\bar{F}(1) = 0$

and that $\bar{F}(9)$ is the (as yet unknown) outer area of S. For $1 < x \leq 9$ we do not know the value of $\bar{F}(x)$ but, paradoxical as it may seem, we can find the derivative $\bar{F}'(x)$. We shall show how to find $\bar{F}'(x)$ and then show how to use this knowledge to find $\bar{F}(x)$ itself.

Let x be a number such that $1 \leq x < 9$ and let h be a number such that $h > 0$ and $x + h \leq 9$. Then $\bar{F}(x + h)$ is the outer area of the portion of S out to the vertical line through the point $(x + h, 0)$. $\bar{F}(x + h) - \bar{F}(x)$ is thus (by Theorem 9) the outer area of the vertical strip with base of length h, lengths of sides $2\sqrt{x}$ and $2\sqrt{x + h}$, but with curved top which is a portion of the graph of f. This strip contains a rectangle with base h and altitude $2\sqrt{x}$, but is contained in a rectangle of base h and altitude $2\sqrt{x + h}$ so that (by Theorem 10)

$$h2\sqrt{x} \leq \bar{F}(x + h) - \bar{F}(x) \leq h2\sqrt{x + h}.$$

Thus upon dividing by the positive number h,

$$2\sqrt{x} \leqq \frac{\bar{F}(x + h) - \bar{F}(x)}{h} \leqq 2\sqrt{x + h}$$

or in different notation $2\sqrt{x} \leqq Q\bar{F}(x)(h) \leqq 2\sqrt{x + h}$. The number x having been chosen, the function defined by $2\sqrt{x + h}$ has $2\sqrt{x}$ as its limit at 0. Thus $\lim\limits_{0+} Q\bar{F}(x)$ exists and

$$2\sqrt{x} \leqq \lim\limits_{0+} Q\bar{F}(x) \leqq 2\sqrt{x}, \qquad\qquad 1 \leqq x < 9.$$

If now we take $1 < x \leqq 9$ and $h < 0$, and make appropriate adjustments in the argument, we obtain $2\sqrt{x} \leqq \lim\limits_{0-} Q\bar{F}(x) \leqq 2\sqrt{x}$. Thus $\lim\limits_{0} Q\bar{F}(x)$ exists and is equal to $2\sqrt{x}$; (see Theorem 13); that is,

$$\bar{F}'(x) = 2\sqrt{x}, \qquad\qquad 1 \leqq x \leqq 9.$$

Now for c a number the function g defined by $g(x) = \frac{4}{3}x^{3/2} + c$ is such that

$$g'(x) = \frac{4}{3}(\frac{3}{2}x^{3/2-1}) = 2x^{1/2}.$$

Since[3] also $\bar{F}'(x) = 2x^{1/2}$, is it possible to determine a number c such that $\bar{F}(x) = (\frac{4}{3})x^{3/2} + c$? If this relation holds then, in particular,

$$\bar{F}(1) = \frac{4}{3} + c.$$

Since we know (see (16)) that $\bar{F}(1) = 0$ we set $\frac{4}{3} + c = 0$ and obtain $c = -\frac{4}{3}$. Hence we set

$$\bar{F}(x) = \frac{4}{3}x^{3/2} - \frac{4}{3}. \qquad\qquad \text{for } 1 \leqq x \leqq 9.$$

Consequently $\bar{F}(9) = (\frac{4}{3}) 9^{3/2} - \frac{4}{3} = (\frac{4}{3}) 3^3 - \frac{4}{3} = (104)/3$ is the number of square units for the outer area of S.

Now by going over the whole argument again with the word "outer" replaced by "inner" (and of course $\bar{F}$ by $\underline{F}$) we obtain that also

$$\underline{F}(x) = \frac{4}{3}x^{3/2} - \frac{4}{3}, \qquad\qquad \text{for } 1 \leqq x \leqq 9.$$

Consequently $\underline{F}(x) = \bar{F}(x)$ for each number x such that $1 \leqq x \leqq 9$ and the area of the portion of S above the interval from $(1, 0)$ to $(x, 0)$ exists. In particular the area of S exists and is equal to $(104)/3$ sq. units.

---

[3] We have shown that if c is a number and f and g are functions such that $f = g + c$, then $f' = g'$. The converse is also true (although we do not prove it); the converse being: If f and g are functions such that $f' = g'$, then there is a number c such that $f = g + c$.

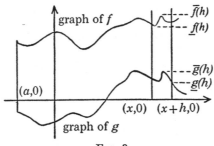

graph of f

$\bar{f}(h)$

$\underline{f}(h)$

$\bar{g}(h)$

$\underline{g}(h)$

(a,0)

(x,0)   (x+h,0)

graph of g

FIG. 3

THEOREM 17. *Let $a < b$ be numbers and f and g continuous functions such that $g(x) \leq f(x)$ for $a \leq x \leq b$. Let S be the point set between the graphs of f and g and the vertical lines through the points $(a, 0)$ and $(b, 0)$. For $a \leq x \leq b$ let $\bar{F}(x)$ and $\underline{F}(x)$ be the outer and inner areas of the portion of S between the vertical lines through the points $(a, 0)$ and $(x, 0)$. Then the functions so defined have derived functions and*

$$\bar{F}'(x) = \underline{F}'(x) = f(x) - g(x) \qquad \text{for } a \leq x \leq b.$$

Proof. For x such that $a \leq x < b$ let $h > 0$ be such that $x + h \leq b$. Then $\bar{F}(x + h) - \bar{F}(x)$ and $\underline{F}(x + h) - \underline{F}(x)$ are the outer and inner areas of the portion of S between the vertical lines through the points $(x, 0)$ and $(x + h, 0)$. On the interval between these points the graphs of f and g have maximum and minimum points (since f and g are continuous) and we call the ordinates of these points $\bar{f}(h)$, $\underline{f}(h)$, $\bar{g}(h)$, $\underline{g}(h)$, respectively. (See Fig. 3.) The portion of S being considered contains a rectangle of base h and altitude $\underline{f}(h) - \bar{g}(h)$ but is contained in a rectangle of base h and altitude $\bar{f}(h) - \underline{g}(h)$ so that both

$$\{\underline{f}(h) - \bar{g}(h)\}h \leq \bar{F}(x + h) - \bar{F}(x) \leq \{\bar{f}(h) - \underline{g}(h)\}h \qquad \text{and}$$

$$\{\underline{f}(h) - \bar{g}(h)\}h \leq \underline{F}(x + h) - \underline{F}(x) \leq \{\bar{f}(h) - \underline{g}(h)\}h$$

Upon dividing by the positive number h we thus obtain

$$\underline{f}(h) - \bar{g}(h) \leq Q\bar{F}(x)(h) \leq \bar{f}(h) - \underline{g}(h)$$

and the same relation with $\underline{F}$ in place of $\bar{F}$. By making the proper adjustments, these same relations hold for $a < x \leq b$ and $h < 0$ such that $a \leq x + h$. Since f and g are both continuous it follows that

$$\lim_{0} \underline{f} = f(x), \qquad \lim_{0} \bar{f} = f(x), \qquad \lim_{0} \underline{g} = g(x), \qquad \text{and} \quad \lim_{0} \bar{g} = g(x).$$

Thus we have, from the existence of these limits, the equalities

$$\lim_{0} Q\bar{F}(x) = f(x) - g(x) \quad \text{and} \quad \lim_{0} Q\underline{F}(x) = f(x) - g(x);$$

that is, as we wished to prove $\bar{F}'(x) = f(x) - g(x)$ and $\underline{F}'(x) = f(x) - g(x)$.

In the following corollary we use the same notation and assumptions as in the theorem.

COROLLARY. $\bar{F}(x) = \underline{F}(x)$ for $a \leq x \leq b$.

Proof. Since $\bar{F}'(x) = \underline{F}'(x)$ for $a \leq x \leq b$, there is a number c such that $\bar{F}(x) = \underline{F}(x) + c$ for $a \leq x \leq b$. (See footnote p. 234.) But $\bar{F}(a) = 0$ and $\underline{F}(a) = 0$ and consequently $c = 0$. Therefore $\bar{F}(x) = \underline{F}(x)$ for $a \leq x \leq b$.

We now give a theorem which is pertinent to the next definition.

THEOREM 18. *Let* $a < b$ *be numbers and* F *and* G *functions such that* $F'(x) = G'(x)$ *for* $a \leq x \leq b$. *Then* $F(x) - F(a) = G(x) - G(a)$ *for* $a \leq x \leq b$ *and in particular*

$$F(b) - F(a) = G(b) - G(a).$$

Proof. Since $F'(x) = G'(x)$ we let c be the number (asserted in footnote page 234 to exist) such that $F(x) = G(x) + c$ for $a \leq x \leq b$. Consequently $F(a) = G(a) + c$ so that $c = F(a) - G(a)$. Hence for $a \leq x \leq b$,

$$F(x) = G(x) + \{F(a) - G(a)\}$$

so that $F(x) - F(a) = G(x) - G(a)$.

We now give the definition of the definite integral.

DEFINITION 9. For $a < b$ numbers and f a function, if there is a function F such that

$$F'(x) = f(x) \quad \text{for} \quad a \leq x \leq b,$$

then the difference $F(b) - F(a)$ is called the *definite integral* of f from a to b and is represented[4] by $\int_a^b f$ so that

$$\int_a^b f = F(b) - F(a).$$

Note from Theorem 18 that if G is any other function such that $G'(x) = f(x)$ for $a \leq x \leq b$, then $F(b) - F(a) = G(b) - G(a)$ so that also

$$\int_a^b f = G(b) - G(a).$$

Example. Given the function f defined by $f(x) = 5x^2$, find $\int_{-1}^2 f$.

---

[4] It will be noted that this is not the traditional symbolism. One should refer to page 220 for this author's definition of function.

Solution. We merely look for a function whose derived function is f. Since the function F defined by $F(x) = (\frac{5}{3})x^3$ is such that $F'(x) = (\frac{5}{3}) \cdot 3\, x^{3-1} = 5x^2 = f(x)$ we see that

$$\int_{-1}^{2} f = F(2) - F(-1) = \frac{5}{3}(2)^3 - \frac{5}{3}(-1)^3 = \frac{40 + 5}{3} = 15.$$

Note that the definition of an integral contains the phrase " if there is a function F such that $F'(x) = f(x)$ for $a \leq x \leq b$." This means that if a given function f happens to be such that there is no function whose derived function is f, then this function f has no definite integral. The following theorem shows, however, that at least each continuous function has a definite integral.

THEOREM 19. *Let $a < b$ be numbers and let f be a function which is continuous at each number x such that $a \leq x \leq b$. Then the definite integral of f from a to b exists.*

Proof. Since, where we are considering it, f is continuous it has a minimum value there and we let m be this minimum value, *i.e.* m is a number and

$$f(x) \geq m \qquad\qquad \text{for } a \leq x \leq b.$$

For convenience we let g be the function defined by

$$g(x) = m \qquad\qquad \text{for } a \leq x \leq b.$$

Hence g is also a continuous function and in addition

$$f(x) \geq g(x) \qquad\qquad \text{for } a \leq x \leq b.$$

Let S denote the point set lying between the graphs of f and g and the vertical lines at the points $(a, 0)$ and $(b, 0)$. Now for each number x such that $a \leq x \leq b$ the portion of S lying between the verticals at the points $(a, 0)$ and $(x, 0)$ is known (since f and g are both continuous) to have an area (*i.e.* its outer and inner areas are equal) and we let $\mathcal{C}(x)$ be this area. We also know (by Theorem 17 and its corollary) that the function $\mathcal{C}$ so defined has a derived function and that $\mathcal{C}'(x) = f(x) - g(x)$.

Now, merely because it will do what we want, we let F be the function defined by

$$F(x) = \mathcal{C}(x) + m \cdot x \qquad\qquad \text{for } a \leq x \leq b.$$

Note for this function F that for $a \leq x \leq b$

$$F'(x) = \mathcal{C}'(x) + m$$
$$= f(x) - g(x) + m$$
$$= f(x) \qquad\qquad (\text{since } g(x) = m).$$

Thus, no matter how we got it, we have a function F such that $F'(x) = f(x)$ for $a \leqq x \leqq b$. Hence, by definition, the definite integral of f exists (and equals $F(b) - F(a)$) so the theorem is proved.

There are, of course, many more theorems about derivatives and integrals, but we must bring this chapter to a close. Also we regret that we did not have space to discuss some of the many physical applications of these concepts. Our concluding remarks depend upon two facts (which we do not prove):

1. If f and g have definite integrals from a to b, then the definite integral from a to b of $f - g$ also exists and

$$\int_a^b (f - g) = \int_a^b f - \int_a^b g.$$

2. The function θ defined by $\theta(x) =$ Arctan x (this is the principal value mentioned in most trigonometry books) is such that $\theta'(x)$ exists and

$$\theta'(x) = \frac{1}{1 + x^2}.$$

If the functions f and g in 1 are continuous and such that $f(x) \geqq g(x)$ for $a \leqq x \leqq b$, then

(17) $$\int_a^b f \geqq \int_a^b g.$$

This inequality follows from 1 since $\int_a^b (f - g)$ is the area of the point set lying between the two graphs and two uprights and this area is certainly $\geqq 0$.

From 2 we see that

(18) $$\int_0^1 \theta' = \text{Arctan } 1 - \text{Arctan } 0 = \frac{\pi}{4} - 0 = \frac{\pi}{4}.$$

Now by dividing 1 by $1 + x^2$ to obtain a quotient and a remainder we have, for example,

$$\frac{1}{1 + x^2} = 1 - x^2 + x^4 - x^6 + \frac{x^8}{1 + x^2}$$

or by carrying the division process one step further

$$\frac{1}{1 + x^2} = 1 - x^2 + x^4 - x^6 + x^8 - \frac{x^{10}}{1 + x^2}.$$

Since $0 \leq x^8/(1 + x^2)$ and $0 \geq -x^{10}/(1 + x^2)$ we thus have

(19) $1 - x^2 + x^4 - x^6 \leq \dfrac{1}{1 + x^2} \leq 1 - x^2 + x^4 - x^6 + x^8.$

Upon letting f and g be the functions defined by $f(x) = 1 - x^2 + x^4 - x^6$ and $g(x) = 1 - x^2 + x^4 - x^6 + x^8$ we thus have

$$f(x) \leq \theta'(x) \leq g(x)$$

so that by using the general principle expressed by (17)

$$\int_0^1 f \leq \int_0^1 \theta' \leq \int_0^1 g.$$

Note that the functions F and G defined by $F(x) = x - x^3/3 + x^5/5 - x^7/7$ and $G(x) = x - x^3/3 + x^5/5 - x^7/7 + x^9/9$ are such that $F'(x) = f(x)$ and $G'(x) = g(x)$. Thus

$$\int_0^1 f = F(1) - F(0) = 1 - \frac{1}{3} + \frac{1}{5} - \frac{1}{7},$$

$$\int_0^1 g = G(1) - G(0) = 1 - \frac{1}{3} + \frac{1}{5} - \frac{1}{7} + \frac{1}{9}$$

so that, remembering (18)

$$1 - \frac{1}{3} + \frac{1}{5} - \frac{1}{7} \leq \frac{\pi}{4} \leq 1 - \frac{1}{3} + \frac{1}{5} - \frac{1}{7} + \frac{1}{9}$$

Thus $1 - \frac{1}{3} + \frac{1}{5} - \frac{1}{7}$ approximates $\pi/4$ to within $\frac{1}{9}$. Actually this is not a very good approximation of $\pi/4$, but note that by expressing $1/(1 + x^2)$ with more and more terms in the quotient, approximations of $\pi/4$ (and thus of π) may be obtained to any desired degree of accuracy.

As established by the Greeks, for any unit of length and any circle the circumference divided by the diameter is the same number as for any other unit and any other circle. By actual measurements the approximations 3.1 and $2\frac{2}{7}$ of this number called π were obtained quite early, but it is known today that π is an irrational number.

The procedure described above for obtaining approximations of π to any desired degree of accuracy was one of the earliest and most striking triumphs of calculus. Since π was defined in terms of lengths it may seem incredible that π could be approximated by any means other than by actual measurements, but we now see that no instruments whatever were necessary to obtain any desired degree of approximation.

At present there is no known reason for wanting 150 decimal places of

π (although even more have been computed). There is, however, good reason for being sure of the approximation of π to one more place than any instrument can measure, for then this value can be used to check the accuracy of the instrument.

The method of approximating π is an illustration of how mathematics is sometimes used in practical applications. For if a physical constant is sufficiently well defined, then by using no measurements whatever, but only mathematical methods, the constant can be approximated and the result used to check the accuracy of subsequent observations.

As an exercise, check the trigonometric relation

$$\frac{\pi}{4} = 4 \operatorname{Arctan} \frac{1}{5} - \operatorname{Arctan} \frac{1}{239}.$$

Then note that this equality may be written as

$$\frac{\pi}{4} = 4 \int_0^{1/5} \theta' + \int_0^{1/239} \theta'.$$

Now by using (19) to approximate the first integral and

$$1 - x^2 \leqq \frac{1}{1 + x^2} \leqq 1 - x^2 + x^3$$

to approximate the second integral, obtain the approximation of π to five decimal places.

BIBLIOGRAPHY

1. BOYER, CARL B. *The Concepts of the Calculus.* New York: Columbia University Press, 1939.
2. COURANT, RICHARD, and ROBBINS, HERBERT. *What is Mathematics?* New York: Oxford University Press, 1941. P. 272–452.
3. KASNER, EDWARD, and NEWMAN, JAMES. *Mathematics and the Imagination.* New York: Simon & Schuster, 1940. P. 299–356.
4. RANDOLPH, JOHN F. *Calculus.* New York: Macmillan Co., 1952.
5. WADE, THOMAS L. *Calculus.* Boston: Ginn and Co., 1953.

Functions

RUDOLPH E. LANGER

THE notion of the function is central to mathematics. The relationship of one variable to another, which is the essence of the definition of the function, is discernible in all the manifestations of nature, and no less in the workings of such man-made disciplines as economics, education, and politics. When studies in any field are pressed beyond the rudimentary stage, the attention has to be focused upon quantitative relationships. There mathematics as a tool enters the field, and from there on thinking frames itself in terms of functions. Precise and sustained analysis is made possible by means of mathematics, and the studies are thereby lifted from the empirical level where the preoccupation is with observations, to the intellectual level where the search for underlying patterns of order is the primary concern. By the discovery of the governing relationships of order, understanding is won and the power to predict results from causes known or supposed follows.

It is because of the ubiquitous prevalence of functional relationships that mathematics insinuates itself into all fields of human thought. It is because of this that the "Queen of the Sciences" continually finds herself also, and at the same time, to be the "Handmaiden."

The variety of functional relationships—which is to say, the variety of functions—is, of course, endless. This fact could easily have proved to be a discouraging one, and would have been so, if every such relationship were utterly dissimilar from every other one, and therefore required its own analysis in every detail. But fortunately orderliness presents itself here too, and experience has shown that certain special functions, relatively few in number, are come upon in a great variety of connections, and govern phenomena which at first glance often seem wholly disparate. The study of these functions has therefore paid off again and again. And their properties, which in the course of time have become well known, are often appealed to for an understanding of other functions which, though different, approximate them in some sense.

241

It is our intention to analyse in this chapter a few of these special functions. Although these are certainly among the simpler ones, and can be discussed quite fully in elementary terms, they have secure holds upon the palm of importance. In making these analyses our thought is not upon the presentation of new or unfamiliar results, but rather to show with the use of familiar material something in the way of method by which, in modern mathematics, facts about functions can be obtained.

MODES OF REPRESENTING FUNCTIONS

A function, namely, a relation between variables, may be amenable to representation through several different modes of expression. In the elementary calculus consideration is given almost exclusively to functions that are representable by mathematical formulas. Examples of such are

$$y = e^x + 1, \qquad v = \frac{\log z}{1 + z^2}, \qquad w = t \sin 3t.$$

It is, however, nowhere implied in the definition of a function that it need be expressible in that way, and a number of other modes of representation are, in fact, familiar and in extensive use. As instances of a few of these we mention the following ones:

(i) *The graph.* This is familiar. At any weather station the temperature as a function of the time is recorded as a graph by an automatism which governs a moving pen in contact with paper on a rotating cylindrical drum. Whenever data of any kind are plotted as points upon coordinate paper, it is with the purpose of envisaging the graph through them, or nearly through them, as representing a function to which they actually or approximately conform.

(ii) *The table.* This also is familiar. In effect it is a convenient scheme for displaying the values of a function at selected values of the variable, and therefore differs no more than in form from a plot of points on coordinate paper. Our dependence upon this mode of representation is in many instances much more extensive than appears upon the surface. For while we use such symbols as $\log x$, e^{-t^2}, $\sin z$, etc. in mathematical formulas, we must ordinarily have recourse to tables for the values to which these symbols refer whenever such values are called for.

(iii) *The infinite series.* This is, in the end, a type of formula, though a special one. One recognizes that, for instance, when the right-hand member of a formula

$$y = (1 + x)^a,$$

is replaced by its binomial expansion to make the relation appear as

$$y = 1 + ax + \frac{a(a - 1)}{2!}\, x^2 + \frac{a(a - 1)(a - 2)}{3!}\, x^3 + \cdots .$$

For many purposes the representation by infinite series has peculiar advantages. It is especially well adapted when the numerical calculation of the functional values is an end in view.

(iv) *The differential equation.* This mode expresses the function in terms of the inter-relationship between it and its derivatives. Many functions that express laws of nature are discovered through their representations in this form, and for many functions this representation remains the most revealing or the only practicable mode of expression.

The formula, the graph, and the table are the most commonly used modes of functional representation at the elementary level. This is not to say that drawbacks in these modes are not recognized even there. The graph and the table are conspicuously subject to the faults of incompleteness or inaccuracy, whereas the formula lacks the flexibility that a universally applicable mode of representation would have to have, since it depends upon standard combinations of the rather limited number of functional elements to which the familiar symbols are assigned. Other modes of representation are therefore far from extraneous. In the desire to give some orientation as to the use and usefulness of these less familiar media, we propose, in the following discussion, to lean heavily upon the differential equation and the infinite series. It will appear that these seemingly more sophisticated modes can often be called upon extensively to yield an abundance of information, and to do so readily and in quite uncomplicated terms.

THE EXPONENTIAL FUNCTION

If the multiplication rate (excess of birth rate over death rate) of a unit group of individuals, say of bacteria or insects, or of farm animals or human beings, is fixed, the rate at which a whole population of them increases is proportional to the size of that population. The rate of increase is small for a population that is small, and is proportionally larger for a larger population. This is a law of growth. There is a corresponding law of decay. The rate at which a unit amount of any radioactive substance dissipates itself by radiation is fixed. Therefore a certain lump of such a substance diminishes, or decays, at a rate which is continually proportional to its surviving amount.

Whether of growth or decay, the instance here is one in which an amount of something is a function whose derivative (its rate of change) maintains a fixed ratio to the function itself. With k as the constant of proportionality the law is thus

(1) $y' = ky,$

with y' denoting the derivative dy/dx. With $k = 0$ the law would signify a vanishing rate of change, namely a function that is constant. With such we are not concerned. We shall suppose, therefore, that $k \neq 0$. A positive y is then an increasing function (a growth function) under this law if $k > 0$, and a decreasing function (a decay function) if $k < 0$.

The differential equation (1) is met with in every course in the elementary calculus. The procedure to solve it formally is to divide it by y. With the recognition that y'/y is the derivative of $\log y$, one is then led to integrate it into the form

$$\log y = kx + \log c,$$

with $\log c$ as the arbitrary constant of integration. Finally, using the fact that $\log y$ is the inverse of the function e^y, one puts the result into the form $y = ce^{kx}$.

Aside from the questionable step of dividing by y, before ever having disposed of the possibility that the value of this might be zero, the indicated "integration" is no more than a pure formalism which yields no information whatever, if the functions to which the symbols $\log y$ and e^{kx} refer have not previously and otherwise been identified and characterized as to their properties—the properties of logarithms and exponents. In the absence of that, the designation e^{kx} can only be looked upon as a convenient symbol to be assigned to a certain solution of the differential equation (1). How this solution is identified as a function of x, and how it behaves, would remain as obscure as at the start.

Now the function e^{kx} is among the best known functions. Characterizations of it are possible by various methods. In this instance, therefore, the omission can be filled, the procedure can be meaningful, the calculus is not to be impugned. However, that is not generally the case with functions that are defined by differential equations, and many functions that are of prime importance in modern mathematics are defined in precisely that way. It is therefore fortunate that a differential equation can be made to yield a fill of information about the functions that solve it, their characteristics and their laws of behavior. We shall show how that can be done, by carrying out a program of investigation upon the differential equation (1). Without ever drawing upon any other source, we shall deduce from this differential equation the nature of the exponential function and all the familiar laws of exponents. The reasoning we shall employ is typical of much that is constantly being applied to the analysis of functions that are as yet less well understood. By means of it many functions have been investigated, as we shall have some occasion to show in what follows.

Let us, to begin with, consider the differential equation in which $k = 1$, namely

$$(2) \qquad\qquad y' = y.$$

Any function y which fulfills this has a derivative. For y' occurs in the equation, and if y had no derivative the equation could not apply to it. We shall prove that y has derivatives of all orders. For this the method of mathematical induction is convenient. (See pages 9–11.) As the first step, we assume that y has an nth derivative for some integer n. Then, by virtue of the differential equation, y' has an nth derivative, which is to say that y has an $(n + 1)$th derivative. As the second step, we observe that the initial assumption is fulfilled for $n = 1$. Hence it is fulfilled for all n.

The fact that y is indefinitely differentiable insures that its graph is a smooth continuous curve, namely one that has neither breaks (discontinuities) nor corners. It does not of itself guarantee that y is representable by a power series, although it does suggest that that may be so. Following the suggestion, therefore, we consider the possibility of a relation

$$(3) \qquad y = 1 + c_1 x + c_2 x^2 + \cdots + c_n x^n + \cdots .$$

From the calculus we know that within its interval of convergence, if any, a power series can be differentiated by differentiating each term. A convergent relation (3) would thus imply that

$$y' = c_1 + 2c_2 x + 3c_3 x^2 + \cdots + (n + 1)c_{n+1} x^n + \cdots .$$

and this would formally fulfill the differential equation if

$$c_1 = 1,$$

$$2c_2 = c_1 ,$$

$$3c_3 = c_2 ,$$

$$\cdots\cdots$$

$$(n + 1)c_{n+1} = c_n ,$$

$$\cdots\cdots .$$

These equations give the values

$$c_1 = 1,$$

$$c_2 = \frac{c_1}{2} = \frac{1}{2},$$

$$c_3 = \frac{c_2}{3} = \frac{1}{3!},$$

$$\cdots$$

$$c_n = \frac{c_{n-1}}{n} = \frac{1}{n!},$$

$$\cdots.$$

Our attention is thus especially drawn to the explicit relation

(4) $$y(x) = 1 + x + \frac{x^2}{2!} + \frac{x^3}{3!} + \cdots .$$

We have thus far proved nothing concerning this relation, since our formal derivation of it was based upon a number of assumptions the justifications of which we did not stop to establish. Proofs are now, however, possible. We consider the relation (4) afresh, without any regard to the suggestion and chain of manipulations by means of which it was reached. The infinite series which it contains has a test-ratio (ratio of the nth term to the one preceding it) whose absolute value is $|x|/n$. This approaches zero as a limit, irrespective of what the (fixed) value of x may be, when n is taken larger and larger without limit. Therefore the series converges for all x, and (4) is accordingly a formula for a function. The derivative of this function, as it is obtainable from the formula, is the same as the function itself. Therefore (4) is a formula for a solution of the differential equation (2). At $x = 0$ this formula gives y the value of 1, namely

(5) $$y(0) = 1.$$

With k as any constant (positive or negative) let $y_k(x)$ designate the function which at x has the value $y(kx)$, namely

(6) $$y_k(x) \equiv y(kx).$$

This relation is in fact of the form $y_k(x) = y(s)$, with $s = kx$. By the rules of the calculus, therefore

$$y_k'(x) = \frac{dy(s)}{dx} = \frac{dy(s)}{ds} \cdot \frac{ds}{dx} = \frac{dy(s)}{ds} \cdot k$$

and since the differential equation asserts that $dy(s)/ds = y(s)$, we have further

$$y_k'(x) = y(s)k = ky_k(x).$$

We see thus that the function $y_k(x)$ fulfills the relation

(7) $$y_k' = ky_k,$$

namely that it is a solution of the differential equation (1). Because at $x = 0$ the values of y_k and of y are the same, by (6), it follows from (5) that

(8) $$y_k(0) = 1.$$

Since the value of k was unrestricted, we may assign to it successively any prescribed values a and b. The respective functions $y_a(x)$ and $y_b(x)$ fulfill the relations

$$y_a' = ay_a,$$

$$y_b' = by_b.$$

On multiplying these equations respectively by y_b and y_a and then adding them, we obtain the result that

$$y_b y_a' + y_a y_b' = (a + b)y_a y_b.$$

In this equation the left-hand member is the derivative of the product $y_a y_b$. This product is therefore a solution of the differential equation

(9) $$y' = (a + b)y.$$

We consider this result first when $a = 1$ and $b = -1$, observing that $y_1 \equiv y$. The equation (9) is in this instance simply $y' = 0$. Since any product yy_{-1} is a solution of this, and so has a vanishing derivative, it is constant. In symbols $yy_{-1} \equiv c$. This yields us an important fact, namely, that every solution y of the differential equation (2) that is different from zero at any single x is different from zero at every x. Inasmuch as every solution is continuous, it accordingly maintains its sign; is either positive for all x or negative for all x. In the instance of the particular solution (4) we have $y_{-1} = y(-x)$, and, since the relation (5) applies, the constant value of the product $y(x)y(-x)$ is 1. This we may write as

(10) $$y(-x) = \frac{1}{y(x)}.$$

The function $y(x)$ is positive for every x. Therefore, by the differential equation, its derivative is positive—the function is increasing—at every

x. Because of this, $y(x)$ has different values at every two different values of x.

We shall show now that every *positive* value A is the value of $y(x)$ at some x. Whatever A may be, there is a positive integer n such that A has a value between those of $1/(n + 1)$ and $(n + 1)$. Now at $x = n$ the terms of the infinite series in (4) are all positive, and, after the first term, each of the first $(n + 1)$ terms is greater than 1 since it has a numerator that exceeds its denominator. The value of $y(n)$, as given by the series, is thus greater than $(n + 1)$, which is greater than A. By (10), therefore, the value of $y(-n)$ is less than $1/(n + 1)$ which is less than A. The graph of $y(x)$ is accordingly below the line $y = A$ at $x = -n$ and above this line at $x = n$. Since it is continuous it intersects the line $y = A$. The abscissa of the point of intersection is the x for which $y(x)$ has the value A.

Suppose now that y_k is any solution of the differential equation (1) that is not 0, and that Y_k is any solution whatever of the same equation. Then $Y_k' = kY_k$, and after this and the equation (7) have been multiplied respectively by y_k and Y_k their difference is

$$y_k Y_k' - Y_k y_k' = 0.$$

After division by y_k^2, which is legitimate since y_k is not zero for any x, the left-hand member of this relation appears as the derivative of the ratio Y_k/y_k. This ratio is accordingly a constant, say c, namely

(11) $$Y_k(x) = cy_k(x).$$

Every solution of the differential equation (1) is thus some constant multiple of $y_k(x)$.

We can apply this result directly. The product $y_a(x)y_b(x)$ was observed above to be a solution of the differential equation (9). It is therefore a constant multiple of $y_{a+b}(x)$, and since these solutions both have the value 1 at $x = 0$, it follows that $y_a(x)y_b(x) = y_{a+b}(x)$. We may write this

(12) $$y(ax)y(bx) = y(ax + bx).$$

Let us multiply the equation (2) by ky^{k-1}, where k is any constant. The equation then appears as

$$ky^{k-1}y' = ky^k.$$

Since the left-hand member of this is the derivative of y^k, we see that $y^k(x)$ is a solution of the equation (1). Its value at $x = 0$ is 1. Therefore we may conclude that $y^k(x) = y_k(x)$, namely that

(13) $$y(kx) = y^k(x).$$

This effectively completes our characterization of the solution (4) of the differential equation (2). The relations (12) and (13) are the "functional equations" which the solution $y(x)$ fulfills. We may now draw consequences from these. The value $y(k)$ for any k is shown by the relation (13), with $x = 1$, to be the kth power of the constant $y(1)$. We shall denote this constant by e. Its value is given by the formula (4) to be

$$(14) \qquad e = 1 + 1 + \frac{1}{2!} + \frac{1}{3!} + \frac{1}{4!} + \cdots .$$

By (10) the reciprocal of e is obtainable from the formula (4) by giving x the value -1. Thus

$$(15) \qquad \frac{1}{e} = \frac{1}{2!} - \frac{1}{3!} + \frac{1}{4!} - \frac{1}{5!} + \cdots .$$

For the calculation of e, to any prescribed number of significant figures, the formula (15) is handier than (14). The infinite series in (15) is one whose terms have alternate signs and have numerical values that decrease steadily to zero. It is shown in the calculus that the value of such a series differs from the sum of its first n terms by a fraction of the $(n + 1)$th term. Thus e^{-1} differs from the value

$$\frac{1}{2!} - \frac{1}{3!} + \frac{1}{4!} - \frac{1}{5!} + \frac{1}{6!} - \frac{1}{7!},$$

which, to four figures, is found to be 0.3679, by a fraction of $\frac{1}{8}!$, namely by less than 0.00003. To four figures, therefore, $e^{-1} = 0.3679$, whence

$$(16) \qquad e = 2.718 \cdots .$$

From the equation (13) now, with $x = 1$, we have $y(k) = e^k$. Since k is arbitrary, this may be written as

$$y(x) = e^x.$$

The function e^x is the exponential function. Its laws, by (12) and (13) are

(17) (a)
$$e^{ax} \cdot e^{bx} = e^{ax+bx},$$
   (b)
$$e^{kx} = (e^x)^k.$$

We recall now that every positive constant is the value of $y(x)$ at some x. Hence if any such constants A, B, and C are given there are corresponding constants a, b, and c such that

$$A = e^a, \qquad B = e^b, \qquad C = e^c.$$

The evaluations

$$AB = e^{a+b}, \qquad e^{-c} = \frac{1}{C}, \qquad e^{ax} = A^x,$$

follow from the relations (17a) with $x = 1$, and from (17b) with $x = c$ and $k = -1$, and with $k = a$. The relation (17a) with $x = c$ now appears alternatively as the first two laws of the set

$$A^c B^c = (AB)^c,$$

(18) $$\qquad C^a C^b = C^{a+b},$$

$$C^o = 1,$$

$$A^{bc} = (A^b)^c.$$

The third one of these laws follows from the second one for $b = -a$, and the fourth one is obtained from (17b) by setting $x = ab$ and $k = c$. In (18) we have the laws of exponents for entirely unrestricted exponents a, b, and c. We have extracted these as well as the laws (17) of the exponential function, wholly from the differential equation (2).

For use in plotting the graph of the exponential relation $y = e^{kx}$, we may draw from the differential equation (1) the fact that for a positive y the derivative has the sign of k at all x. The graph is thus a rising one if $k > 0$, and a falling one if $k < 0$. Also from the differential equation $y'' = ky'$, whence $y'' = k^2 y$. For a positive y the second derivative is thus everywhere positive, the graph is concave upward at all its points. The graphs for $k = 1$ and $k = -1$ are as shown in Figure 1.

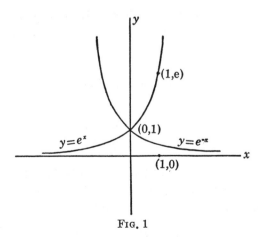

FIG. 1

Since $e^{kx} = y_k(x)$, whereas $y_k(x) = y_{\pm 1}(\pm kx)$, it follows that the graph of $y = e^{kx}$ is obtainable from that of $y = e^x$ or that of $y = e^{-x}$, according as $k > 0$ or $k < 0$, by replacing x by $|k| x$. In geometrical terms this means by suitably contracting or stretching the (x, y) plane in the direction of the horizontal.

THE LOGARITHMIC FUNCTION

It was observed on page 249 that every positive constant A has corresponding to it a constant a such that $A = e^a$. The formula $y = A^x$ can therefore be expressed as $y = e^{ax}$. It thus gives the solution of the differential equation $dy/dx = ay$, whose value is 1 at $x = 0$. This solution was found to be positive for all x, and, if $a \neq 0$, namely if $A \neq 1$, to have any given positive value y for some associated (single) x. That is to say, however, that every positive y determines an x, namely that on the domain of positive y values the relation $y = A^x$ defines x as a function of y. This function $x(y)$ is a solution of the differential equation $dx/dy = 1/ay$, for by the calculus the derivative dx/dy is the reciprocal of dy/dx. The fact that $y = 1$ is associated with $x = 0$, is re-expressed in the relation $x(1) = 0$.

Let us re-state this summary of facts with the letters x and y interchanged. It appears then as follows: The relation

$$(19) \qquad\qquad x = A^y,$$

in which A is any positive number other than 1, determines y as a function of x over the domain of positive x values. This function solves the differential equation

$$(20) \qquad\qquad y' = \frac{1}{ax},$$

in which a is the constant for which

$$(21) \qquad\qquad A = e^a.$$

It is the solution for which

$$(22) \qquad\qquad y(1) = 0.$$

We shall denote this function by the symbol $\log_A x$, and shall call it "the logarithm of x to the base A." To give it a name and symbol, as we have now done, does not, of course, tell us anything about it. We are therefore confronted with the matter of determining the properties of this "logarithmic function"—the laws to which it conforms. To accom-

plish that we shall again refer to the differential equation, and to no
other source.

The differential equation (20) defines y as a function whose derivative
is inversely proportional to the variable. Inasmuch as the right-hand
member of this equation is an explicit function of x that is indefinitely
differentiable, the function y has derivatives of all orders for $x > 0$.
This suggests the possibility of a power series representation. Such a
series, however, could not proceed by powers of x. For a series in powers
of x, if it represents a function at all, does so at $x = 0$. This point is not
in our specified domain, and could not even be added to the domain
because the differential equation would be meaningless there. Because
the function we are concerned with has an assigned value at $x = 1$, the
suggestion next in order is that the power series proceed by powers of
$(x - 1)$. On account of (22) the relation proposed would therefore be

$$y = c_1(x - 1) + c_2(x - 1)^2 + c_3(x - 1)^3 + \cdots + c_n(x - 1)^n + \cdots.$$

This would yield

$$y' = c_1 + 2c_2(x - 1) + 3c_3(x - 1)^2 + \cdots + nc_n(x - 1)^{n-1} + \cdots.$$

To equate this to $1/ax$, as the differential equation requires, we must
first express $1/ax$ in comparable terms, namely also in powers of $(x - 1)$.
This can be done readily. After writing $1/ax$ in the form

$$\frac{1}{a}\left[\frac{1}{1 + (x - 1)}\right],$$

we recall that $1/(1 + s)$ is the sum of the infinite geometric series $1 -
s + s^2 - s^3 + \cdots$. Therefore

$$\frac{1}{ax} = \frac{1}{a}[1 - (x - 1) + (x - 1)^2 - (x - 1)^3 + \cdots].$$

This formally equals y' if

$$c_1 = \frac{1}{a},$$

$$2c_2 = -\frac{1}{a},$$

$$3c_3 = \frac{1}{a},$$

$$\cdots\cdots.$$

The relation to which our attention is thus especially drawn is

$$y(x) = \frac{1}{a}\left[(x-1) - \frac{1}{2}(x-1)^2 + \frac{1}{3}(x-1)^3\right.$$

(23)

$$\left. - \frac{1}{4}(x-1)^4 + \cdots\right].$$

As yet we have no assurance that this is a formula. For it was derived upon the assumption that a power series might represent $y(x)$, whereas conceivably no such representation may be possible.

Let us therefore consider the relation (23) afresh, that is, without any regard to the source from which it was obtained. The infinite series in it has a test-ratio whose absolute value is $(1 - 1/n)\,|\,x - 1\,|$. As n is taken larger and larger without limit this value approaches the limit $|\,x - 1\,|$. Inasmuch as a series converges when its test-ratio limit is less than 1, the series here in question does so when $|\,x - 1\,| < 1$, that is to say, for

(24) $0 < x < 2.$

For these values of x the relation (23) is therefore a formula, and represents a function. The derivative of this function, as it is given by the series (23) is identical with the one that was obtained above for $1/ax$. Hence the function (23) is a solution of the differential equation (20). At $x = 1$ each of its terms vanishes. The solution represented therefore conforms to (22). It should, perhaps, be observed that the formula (23) does not represent $y(x)$ when $x \geq 2$. That, however, is the fault of the formula, which then diverges, and does not reflect upon $y(x)$ itself. The differential equation, which remains our source of reference, applies to $y(x)$ for all positive x values.

Suppose that, along with $y(x)$, the function $Y(x)$ is any solution of the differential equation (20). Then $Y' = 1/ax$, and thus $Y'(x)$ and $y'(x)$ are equal. Therefore the difference $(Y(x) - y(x))$ has a vanishing derivative and so is constant. At $x = 1$ the difference has the value $Y(1)$, by (22). Thus

(25) $Y(x) = y(x) + Y(1),$

namely, every solution of the differential equation differs from the one given by the formula (23) only by a constant.

Let $y_k(x)$ be the function whose value at x is $y(kx)$, where k is any positive constant. This is to say that $y_k(x) = y(s)$, with $s = kx$, and by the rules of the calculus, therefore

$$\frac{dy_k(x)}{dx} = \frac{dy(s)}{ds} \cdot \frac{ds}{dx}.$$

But $ds/dx = k$, whereas, by (20), $dy(s)/ds = 1/as$. Thus $y_k' = k/as$ and since the right-hand member of this equals $1/ax$, it follows that $y_k(x)$ is a solution of the differential equation (20). Therefore, by (25), we have $y_k(x) = y(x) + y_k(1)$, namely

(26) $y(kx) = y(x) + y(k)$.

With any constant c (positive or negative) let $Y(x) = y(x^c)$. The relationship indicated is $Y(x) = y(s)$, with $s = x^c$, and from it we draw

$$Y'(x) = \frac{dy(s)}{ds}\frac{ds}{dx} = \frac{1}{as} \cdot cx^{c-1} = \frac{c}{ax}.$$

So the function $Y(x)$ is a solution of the differential equation $Y' = c/ax$. The function $cy(x)$ is also a solution of this same differential equation. Therefore these two functions differ only by a constant. Inasmuch as both functions have the value 0 at $x = 1$, it follows that $Y(x) = cy(x)$. Thus

(27) $y(x^c) = cy(x)$.

Consider now the function $y(e^x)$. On denoting it by $u(x)$, we have $u(x) = y(s)$, with $s = e^x$. Therefore

$$u'(x) = \frac{dy(s)}{ds} \cdot \frac{ds}{dx} = \frac{1}{as}\frac{ds}{dx}.$$

But by the previous section the function s, as it is here in question, is a solution of the equation $ds/dx = s$. We see thus that $u'(x) = 1/a$, in accordance with which $u(x)$ equals x/a plus a constant. The constant is zero, for $x = 0$ corresponds to $s = 1$, wherefore $u(0) = y(1) = 0$. Thus $u(x) = x/a$, namely $y(e^x) = x/a$. At $x = a$ this gives $y(e^a) = 1$, and recalling the relation (21) we may thus conclude that

(28) $y(A) = 1$.

The formula

(29) $z(x) = \frac{a}{b} y(x),$

in which b is any constant other than zero, defines a function which fulfills the differential equation $z' = 1/bx$, and the relation $z(1) = 0$. This is the function to which we have assigned the designation $\log_B x$, where $B = e^b$. At $x = A$ the formula (29) yields, by virtue of (28), the evaluation $z(A) = a/b$. With the use of this we may replace the formula (29) by the equivalent one

$$(30) \qquad\qquad z(x) = z(A)y(x).$$

Herewith we have effectively accomplished the complete characterization of the logarithmic function. On giving to $y(x)$ and $z(x)$ their proper designations $\log_A x$ and $\log_B x$, the relations (22), (26), (27), (28), and (30) appear in the forms

$$\log_A 1 = 0,$$

$$\log_A (kx) = \log_A x + \log_A k,$$

$$(31) \qquad\qquad \log_A (x^c) = c \log_A x,$$

$$\log_A A = 1,$$

$$\log_B x = \log_B A \, \log_A x.$$

These are the laws of logarithms.

The last one of the laws (31) permits two observations. At $x = B$ it reduces to $1 = \log_B A \, \log_A B$, and thus shows that

$$(32) \qquad\qquad \log_B A = \frac{1}{\log_A B}.$$

It also shows that logarithms with any one base, B, are proportional to those with any other base, A, the factor of proportionality being $\log_B A$. To accomplish the calculation of logarithms to a base B when those to a base A are known, it is therefore only necessary to determine this factor of proportionality, which, by (32), is also $1/\log_A B$. Thus if $A = e$, and $B = 10$, the relationships are

$$(33) \qquad \begin{aligned} \log_{10} x &= \frac{\log_e x}{\log_e 10}, \\[2mm] \log_e x &= \frac{\log_{10} x}{\log_{10} e}. \end{aligned}$$

These are the conversion formulas that relate the logarithms to the two widely used bases 10 and e. Every positive number other than 1 is eligible for use as a base.

To determine the character of the graph of the logarithmic relation

$$y = \log_A x$$

we may again resort to the differential equation. Since $\log_A x$ has been defined only for $x > 0$, the graph lies entirely to the right of the y axis in the (x, y) plane. The first and fourth relations (31) show that it crosses the x-axis at $x = 1$, and has at $x = A$ the height 1. Beyond this, we shall distinguish between the case in which $A > 1$ and that in which $A < 1$.

If $A > 1$ the constant a that corresponds to it by (21) is positive. In accordance with the differential equation (20), the slope of the graph is thus positive at every positive x. The graph is therefore a rising one. Since it crosses the axis at $x = 1$, it is below the axis for $0 < x < 1$, and above the axis for $x > 1$. The logarithms of numbers less than 1 are therefore negative, and those of numbers greater than 1 are positive. By the differential equation also $y'' = -1/ax^2$. The second derivative of y is therefore negative. By the calculus, this signifies that the graph is concave down. The graph is shown in Figure 2.

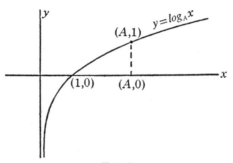

FIG. 2

If, in the third law (31), we take $c = -1$, it follows that the negative of the logarithm of a number is the logarithm of the reciprocal number. Thus, for instance, $-4\frac{7}{8}$ (namely -4.8750) is the logarithm of the number whose reciprocal has the logarithm 4.8750, and similarly the number $8.1432 - 10$, (by which we mean $-2 + 0.1432$) is the logarithm of the number whose reciprocal has the logarithm $2 - 0.1432$ (namely 1.8568). Numbers which are reciprocals of each other thus have logarithms that are numerically equal but have opposite signs.

We can use this principle in an interesting way in plotting the graph. Suppose x is any abscissa greater than 1. From the point $(x, 0)$ draw a tangent line to the circle with center at the origin and the radius 1. The abscissa of the point of contact is $1/x$. The proof of this is a simple exercise in plane geometry. At x and $1/x$ the graph is equidistant from the x-axis and on opposite sides of it.

If $A < 1$ the constant a that is associated with it by the relation (21) is negative. Therefore, by the differential equation, y' is negative and y'' is positive; the graph is a descending one that is concave up. It is as shown in Figure 3. This graph may seem unusual. If so, that is because the logarithms that are most familiar are those to the bases 10 and e, and these bases are greater than 1. Curiously enough, this was not originally so. Napier, who brought out the first table of logarithms, in 1614, used as his base the number 0.9999999, which is less than 1. He had good reason for doing so. The purpose he had mainly in mind was the solution of triangles by trigonometry. The numbers that are most prominent in that are sines and cosines of angles, and these are numerically never greater than 1. Now a glance at the graph of Figure 3

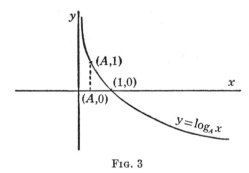

Fig. 3

shows that, when this graph applies, the numbers x that are less than 1 have positive logarithms. Thus Napier had chosen his base with an eye to avoiding negative logarithms in the work for which his table was primarily designed.

THE TRIGONOMETRIC FUNCTIONS

Trigonometric analysis, in which the sine and cosine functions play the basic role, is an extensive branch of modern mathematics. It is of profound mathematical interest, has been immensely influential upon the course of mathematical growth and development, and is a mainstay among the mathematical methods that have been devised for the investigation of problems of science and engineering. It is therefore responsible, often quite directly so, for much of our present-day understanding of physical phenomena and their control, and has made possible much of contemporary technology.

The functions that are here in question are, of course, those that enter into the solution of triangles in elementary trigonometry. However, our interest here is to be focused upon the functions themselves, not upon triangles. We shall seek out the laws to which the functions conform, namely the laws that are expressed in their addition and differentiation formulas, in the characters of their graphs etc. It should be especially noted that we shall not be concerned in any way with such relations as the so-called "law of sines" or "law of cosines," etc. These are prominent, to be sure, in trigonometry. They are, however, in every sense laws of the triangle, and in no sense laws through which the properties of the functions themselves come to expression. Nor, when we write sin x or cos x shall we think of x as an angle as is customary in elementary trigonometry. It will be a numerical variable, precisely as we think of it in the case of functions like e^x and log x. The angle is the natural variable when the triangle is concerned. However, in the whole compass of trigonometric analysis the solution of triangles is only a very minor part.

The medium through which we shall investigate the functions before us will again be the differential equation. This will, in this instance, be the differential equation of the second order

$$(34) \qquad\qquad y'' + y = 0.$$

The domain of x over which we shall consider the functions will be the whole range of real numbers, positive and negative.

Any solution y of the differential equation (34) is indefinitely differentiable. For, when the equation is written as $y'' = -y$, its right-hand member has a derivative. Otherwise y'' would not exist, and the equation could not apply. Now if y has an nth derivative the equation shows that it has an $(n + 2)$th derivative. These facts can be arranged into an orderly proof by mathematical induction that y has derivatives of all orders.

By this fact the suggestion that a solution may permit representation by means of a power series is again brought forth. The relation

$$y = c_0 + c_1 x + c_2 x^2 + \cdots + c_n x^n + \cdots,$$

yields formally

$$y'' = 2c_2 + 2 \cdot 3 c_3 x + 3 \cdot 4 c_4 x^2 + \cdots + (n + 1)(n + 2)c_{n+2} x^n + \cdots,$$

and this, again, formally, is the negative of y itself if

$$2c_2 = -c_0,$$

$$2 \cdot 3 c_3 = -c_1,$$

$$3 \cdot 4 c_4 = -c_2,$$

$$\cdots$$

$$(n + 1)(n + 2)c_{n+2} = -c_n ,$$

$$\cdots$$

$$\cdots .$$

These equations yield the evaluations

$$c_2 = -\frac{c_0}{2} ,$$

$$c_3 = -\frac{c_1}{3!} ,$$

$$c_4 = -\frac{c_2}{3 \cdot 4} = \frac{c_0}{4!} ,$$

$$c_5 = -\frac{c_3}{4 \cdot 5} = \frac{c_1}{5!} ,$$

$$c_6 = -\frac{c_4}{5 \cdot 6} = -\frac{c_0}{6!} ,$$

$$\cdots$$

$$\cdots ,$$

wherewith the relationship assumes the form

$$y = c_0 \left[1 - \frac{x^2}{2!} + \frac{x^4}{4!} - \frac{x^6}{6!} + \cdots \right] + c_1 \left[x - \frac{x^3}{3!} + \frac{x^5}{5!} - \frac{x^7}{7!} + \cdots \right].$$

In this deduction, no specification as to the value of either one of the coefficients c_0 and c_1 appears. A free choice of these values therefore seems permissible. The choices $c_0 = 0$, $c_1 = 1$, and $c_0 = 1$, $c_1 = 0$ are particular ones. As a consequence of them we have our attention drawn especially to the respective relations

(35)

$$y_1(x) = x - \frac{x^3}{3!} + \frac{x^5}{5!} - \frac{x^7}{7!} + \cdots ,$$

$$y_2(x) = 1 - \frac{x^2}{2!} + \frac{x^4}{4!} - \frac{x^6}{6!} + \cdots .$$

We may start with these relations afresh, as we did in the analogous circumstances in the two previous sections. The test-ratios show that the infinite series both converge for all x. The relations (35) are therefore formulas. Their second derivatives differ from them only in sign. Therefore the functions $y_1(x)$ and $y_2(x)$ given by these formulas are solutions of the differential equation (34). They are seen to conform to the relations

(36)
$$y_1(0) = 0, \qquad y_2(0) = 1.$$
$$y_1'(0) = 1, \qquad y_2'(0) = 0.$$

Inasmuch as each term of the formula for $y_1(x)$ is replaced by its negative, and each term of $y_2(x)$ is replaced by its equal when x is replaced by $-x$, it is seen that

(37)
$$y_1(-x) = -y_1(x),$$
$$y_2(-x) = y_2(x).$$

Because of these relations $y_1(x)$ is said to be an *odd* function of x, and $y_2(x)$ an *even* function. Finally, a comparison of the formulas (35) with their derivatives shows that

(38)
$$y_1'(x) = y_2(x),$$
$$y_2'(x) = -y_1(x).$$

On being multiplied by $2y'$, the differential equation (34) takes on the form $2y'y'' + 2yy' = 0$. Its left-hand member is now the derivative of the quantity $(y'^2 + y^2)$. This quantity is therefore a constant, and if y refers to $y_2(x)$, as it may, the constant is shown by (36) to be 1. Thus $y_2'^2(x) + y_2^2(x) \equiv 1$, and by the use of the relations (38) this is seen to imply both that

(39)
$$y_1^2(x) + y_2^2(x) \equiv 1,$$

and that

(40)
$$y_1(x)y_2'(x) - y_1'(x)y_2(x) \equiv -1.$$

Suppose now that y and y_3 are any two solutions of the differential equation, so that the equations $y_3'' + y_3 = 0$ and (34) both maintain. If these equations are multiplied respectively by y and y_3, their difference appears as $yy_3'' - y''y_3 = 0$. The left-hand member of this is the derivative of the quantity $(yy_3' - y'y_3)$. For, when this quantity is differentiated, two of the resulting four terms cancel. The quantity is therefore constant, namely

(41)
$$yy_3' - y'y_3 = \text{a constant}.$$

What the constant is, depends upon the solutions y and y_3 that are involved. When y is $y_1(x)$ and y_3 is $y_2(x)$ the constant is -1, as is shown by (40).

Consider the equation

$$y_1 y_2' y_3 - y_1' y_2 y_3 + y_2 y_3' y_1 - y_2' y_3 y_1 + y_3 y_1' y_2 - y_3' y_1 y_2 = 0.$$

It is fulfilled; vacuously so, because its terms all cancel. However by the insertion of parentheses we may write it

$$(y_1 y_2' - y_1' y_2) y_3 + (y_2 y_3' - y_2' y_3) y_1 + (y_3 y_1' - y_3' y_1) y_2 = 0.$$

Now by (41) each parenthesis encloses a constant, and the constant that multiplies y_3 is -1. Thus with appropriate constants k_1 and k_2,

$$(42) \qquad\qquad y_3 = k_1 y_1(x) + k_2 y_2(x).$$

This is important. It asserts that every solution of the differential equation (34) is a sum of constant multiples of the particular solutions given by the formulas (35).

We can apply this result immediately. Suppose $Y_1(x) = y_1(x + a)$, and $Y_2(x) = y_2(x + a)$ where a is any constant. Then with either subscript attached we have

$$Y''(x) = y''(x + a) = -y(x + a) = -Y(x),$$

from which it is seen that $Y_1(x)$ and $Y_2(x)$ are both solutions of the differential equation (34). Therefore, with appropriate constant coefficients,

$$(43) \qquad \begin{aligned} Y_1(x) &= k_{1,1} y_1(x) + k_{1,2} y_2(x), \\ Y_2(x) &= k_{2,1} y_1(x) + k_{2,2} y_2(x), \end{aligned}$$

whereas the derivatives of these relations are

$$\begin{aligned} Y_1'(x) &= k_{1,1} y_1'(x) + k_{1,2} y_2'(x), \\ Y_2'(x) &= k_{2,1} y_1'(x) + k_{2,2} y_2'(x). \end{aligned}$$

These equations are, in particular, all valid at $x = 0$, where the evaluations (36) apply. Therefore they yield

$$Y_1(0) = k_{1,2}, \qquad Y_2(0) = k_{2,2},$$
$$Y_1'(0) = k_{1,1}, \qquad Y_2'(0) = k_{2,1}.$$

We thus have $k_{1,2} = y_1(a)$, $k_{2,2} = y_2(a)$, and $k_{1,1} = y_1'(a) = y_2(a)$, $k_{2,1} = y_2'(a) = -y_1(a)$. The relations (43) are thus explicitly

$$(44) \qquad \begin{aligned} y_1(x + a) &= y_2(a) y_1(x) + y_1(a) y_2(x), \\ y_2(x + a) &= -y_1(a) y_1(x) + y_2(a) y_2(x). \end{aligned}$$

We turn now to a study of the graphs. By the differential equation, y'' has at each point the sign opposite to that of y. The graph of any solution of this equation is, accordingly, concave downward wherever it is above the x-axis, and concave upward wherever it is below the axis. Consider first the graph of $y_2(x)$. Because $y_2(x)$ has at $x = 0$ the values (36), its graph issues from the y-axis toward the right from a point at the height 1 and in the horizontal direction. Since it is concave downward it falls, and since it remains concave downward it comes to an intersection with the x-axis. Let us assign to the abscissa of this intersection point the symbol $\pi/2$, leaving entirely aside, for the moment, what the value of this may be.

The graph of $y_1(x)$, by (36), issues from the origin toward the right as a rising curve also concave downward. Since its ordinate y_1 is bound to y_2, the ordinate of the graph of $y_2(x)$, by the relation (39), it continues to rise while the graph of $y_2(x)$ falls, until at the value $x = \pi/2$ it attains the height 1. We have therefore

(45) $$y_1\left(\frac{\pi}{2}\right) = 1, \qquad y_2\left(\frac{\pi}{2}\right) = 0.$$

By virtue of the relations (38) we infer from these evaluations the further ones $y_1'(\pi/2) = 0$, $y_2'(\pi/2) = -1$. The graph of $y_1(x)$ thus has at $x = \pi/2$ a horizontal tangent line, whereas the graph of $y_2(x)$ has there the slope -1. On the interval from $x = 0$ to $x = \pi/2$ the graphs therefore appear as in Figure 4.

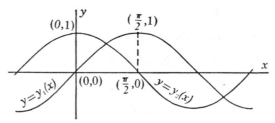

FIG. 4

To obtain the continuations of the graphs we draw from the relations (44) and (45) the facts that

$$y_1\left(x + \frac{\pi}{2}\right) = y_2(x),$$

(46)

$$y_2\left(x + \frac{\pi}{2}\right) = -y_1(x).$$

These show that the graph of $y_1(x)$ is obtainable from that of $y_2(x)$ by merely displacing the latter through the distance $\pi/2$ to the right, and that the graph of $y_2(x)$ is similarly obtainable from that of $-y_1(x)$. The relations (37) serve to extend the graphs to the left.

Herewith we have effectively completed the analysis of the solutions $y_1(x)$ and $y_2(x)$. If we now assign these solutions the respective proper names "sine of x" and "cosine of x" and symbolize them accordingly by $\sin x$ and $\cos x$, the results (36), (37), (39), (44), (45) and (38) appear in the forms

$$\sin 0 = 0,$$

$$\cos 0 = 1,$$

$$\sin (-x) = -\sin x,$$

$$\cos (-x) = \cos x,$$

$$\sin^2 x + \cos^2 x = 1,$$

$$\sin (x + a) = \sin x \cos a + \cos x \sin a,$$

(47) $$\cos (x + a) = \cos x \cos a - \sin x \sin a,$$

$$\sin \frac{\pi}{2} = 1,$$

$$\cos \frac{\pi}{2} = 0,$$

$$\frac{d}{dx} \sin x = \cos x,$$

$$\frac{d}{dx} \cos x = -\sin x.$$

From these many other familiar relations can be drawn. Thus with $x = a$ the addition formulas give the results

$$\sin 2a = 2 \sin a \cos a,$$

$$\cos 2a = \cos^2 a - \sin^2 a.$$

In the second one of these we may replace $\cos^2 a$ by $1 - \sin^2 a$ and solve for $\sin a$, or alternatively replace $\sin^2 a$ by $1 - \cos^2 a$ and solve for $\cos^2 a$. In that way one obtains the results

$$\sin^2 a = \tfrac{1}{2}(1 - \cos 2a),$$

$$\cos^2 a = \tfrac{1}{2}(1 + \cos 2a).$$

With $a = \pi/2$ the addition formulas yield

$$\sin\left(x + \frac{\pi}{2}\right) = \cos x,$$

$$\cos\left(x + \frac{\pi}{2}\right) = -\sin x,$$

and the first one of these becomes, when x is replaced by $-x$,

$$\sin\left(\frac{\pi}{2} - x\right) = \cos x,$$

etc.

The relations (47) are the laws to which the trigonometric functions conform. We emphasize again that they have been deduced here without any reference to either an angle or a triangle.

Before concluding this section let us return to the constant π. This was introduced as a symbol, namely for twice the least positive abscissa for which the solution $y_2(x)$ vanishes. From that characterization, however, we may calculate its value, at least approximately to any desired degree of accuracy. The second formula (35) expresses $y_2(x)$ as an infinite series whose terms (from a certain one on) decrease steadily to the limiting value 0 and have alternate signs. The sum of the first n terms of the series therefore gives the value of y_2 to within an amount which is some fraction of the $(n + 1)$th term (provided n is suitably large). Applying this at $x = 1.5$ and $x = 1.6$ we have

$$y_2(1.5) = 1 - \frac{(1.5)^2}{2!} + \frac{(1.5)^4}{4!} - \left(\text{some fraction of } \frac{(1.5)^6}{6!}\right),$$

$$y_2(1.6) = 1 - \frac{(1.6)^2}{2!} + \left(\text{some fraction of } \frac{(1.6)^4}{4!}\right).$$

These values are respectively positive and negative, irrespective of the fractions that are involved. Hence $y_2(x)$ has opposite signs at $x = 1.5$ and at $x = 1.6$. The abscissa $\pi/2$ at which it crosses the axis is between these, namely π is between 3.0 and 3.2. The method can be applied with increasing refinement. Thus y_2 (1.56) is found to be positive. Hence π is between 3.12 and 3.2, etc.

ANALYSIS OF VIBRATIONS

A phenomenon that is almost ever-present in mechanical systems, and in electrical ones no less, is that of vibration. The mathematical analysis of vibrations, upon which designs to obviate or reduce or modify them must be based, is therefore a matter of central concern in technology. The method by which such an analysis can be made often depends upon the trigonometric functions. Let us, for the sake of such orientation as it may give, consider the formal outline of such an analysis. The rounding out of this into a rigorous mathematical theory would require the supply of a considerable amount of technical proof, and also of much careful probing into the conditions under which the various steps may be taken. That precise analysis, all important as it is, we shall leave aside, since our purpose is only one of attaining to a perspective of the methodical program.

A steel wire drawn taut and then fastened at two of its points is in equilibrium as a straight stretched wire. If such a wire is drawn aside from its equilibrium position into the shape of some curve and is thereupon released, it vibrates. A plucked harp-string is an instance in point; one in which the vibration produces a musical sound that is known to be composed of a fundamental tone and a collection of overtones. This stretched wire is, to be sure, an ideally simple mechanical system. That, however, is as it should be, inasmuch as we seek only to pursue the central ideas of a method, not its adaptation to encumbering complications.

By the choice of an appropriate system of plane rectangular coördinates and unit of length, we may take the wire to be fastened on the x-axis at the points $x = 0$ and $x = \pi$, and to vibrate in the (x, u) plane. We shall suppose the vibration to be transverse, namely, such that any point of the wire maintains a constant abscissa. The ordinate u varies, and so depends upon the time. The displacement of the wire from its equilibrium position is thus given by the function $u(x, t)$. The time derivatives $\partial u/\partial t$ and $\partial^2 u/\partial t^2$ represent respectively the wire's velocity and acceleration. Now Newton's law of motion says, that the product of the acceleration by the mass of any small piece of the wire is equal to the force acting upon this piece. Thus if the piece, when in equilibrium, is of density w and length Δx, so that its mass is $w\Delta x$, the formula for the force F acting upon it is

$$(48) \qquad\qquad F = w\Delta x \, \frac{\partial^2 u}{\partial t^2} .$$

Consider the wire now in any displaced position, such as is shown in Figure 5. The force of tension T at any point is directed along the tangent line, and has the vertical and horizontal components $T \sin z$ and $T \cos z$. The total transverse (vertical) force acting upon the piece of wire is the difference of the value of $T \sin z$ at its two ends, namely the "increment"

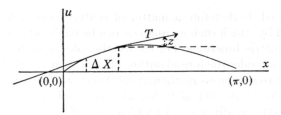

$\Delta(T \sin z)$. Thus

(49) $\Delta(T \sin z) = F.$

The horizontal force $T \cos z$ has the same intensity at both ends, for otherwise the piece would move longitudinally. Thus $T \cos z$ is a constant, say T_0. Since

$$\frac{1}{T_0} \Delta(T \sin z) = \Delta\left(\frac{T \sin z}{T_0}\right),$$

whereas $T \sin z / T_0 = T \sin z / T \cos z = \tan z$, and whereas further $\tan z$ is the slope of the wire, $\partial u/\partial x$, the effect of dividing the equation (49) by $T_0 \Delta x$ is to give it the form

$$\frac{\Delta\left(\dfrac{\partial u}{\partial x}\right)}{\Delta x} = \frac{F}{T_0 \Delta x} .$$

By (48) this is

(50) $$\frac{\Delta\left(\dfrac{\partial u}{\partial x}\right)}{\Delta x} = \frac{1}{a^2} \frac{\partial^2 u}{\partial t^2} .$$

with $a^2 = T_0/w$.

In the manner familiar from the calculus we now think of the piece of wire as taken smaller and smaller, namely of Δx as approaching 0. Since

$$\lim_{\Delta x \to 0} \frac{\Delta\left(\dfrac{\partial u}{\partial x}\right)}{\Delta x} = \frac{\partial}{\partial x}\left(\frac{\partial u}{\partial x}\right) = \frac{\partial^2 u}{\partial x^2},$$

this leads us to

(51) $$\frac{\partial^2 u}{\partial x^2} = \frac{1}{a^2} \frac{\partial^2 u}{\partial t^2},$$

as the limiting form of the relation (50). This partial differential equation (51) is now adopted as the one to which the motion of the wire conforms. The motion is also taken to conform to the "boundary conditions"

(52)
$$u(0, t) = 0,$$
$$u(\pi, t) = 0,$$

and the "initial conditions"

(53)
$$\frac{\partial u}{\partial t}\bigg]_{t=0} = 0,$$
$$u(x, 0) = f(x).$$

The conditions (52) merely assert that the displacements at the points where the wire is fastened are zero at all times. The conditions (53) assert that the wire is without motion at the time $t = 0$ when it is released, and that its shape then is along the curve $u = f(x)$. The constant a^2 in the equation (51) is positive and depends upon the wire and its adjustment. It is inversely proportional to the wire's density (weight) and directly proportional to the tension under which it is stretched.

A method by which solutions of the differential equation (51) may be sought is by a "separation of variables." This is centered upon possible solutions which factor into functions of x and t separately, thus

(54) $$u(x, t) = y(x)v(t).$$

By substituting this form into it, the equation (51) is made to appear as

$$y''(x)v(t) = \frac{1}{a^2} y(x) \frac{d^2v}{dt^2}.$$

Because this must hold when either t or x is fixed, it requires that y'' be proportional to $-y(x)$, and $(1/a^2)(d^2v/dt^2)$ be accordingly proportional to $-v(t)$ with the same factor of proportionality. The factor itself must be a constant. Thus we are led to consider the ordinary differential equations

(55)
$$y'' = -ky,$$
$$\frac{d^2v}{dt^2} = -ka^2v.$$

We may now consider these differential equations afresh. If $y(x)$ fulfills the first one and also the relations

(56) $$y(0) = 0, \qquad y(\pi) = 0,$$

and $v(t)$ fulfills the second one as well as the relation

(57) $$\left.\frac{dv}{dt}\right]_{t=0} = 0,$$

it can be verified by direct substitution that the product (54) is a solution of the differential equation (51) and that it conforms with the conditions (52) and the first one of the conditions (53).

The first differential equation (55) is closely akin to the equation (34). If $y(x)$ is a solution of it, the function $y_3(x) = y(x/\sqrt{k})$ is a solution of the equation (48). Therefore, by (42) it has the form $k_1 \sin x + k_2 \cos x$. This is to say that

$$y(x) = k_1 \sin (\sqrt{k}x) + k_2 \cos (\sqrt{k}x).$$

Upon this relation the conditions (56) impose the requirements $k_2 = 0$, $k_1 \sin (\sqrt{k}\pi) = 0$. These are fulfilled by $k_1 = 1$, $k_2 = 0$, $\sqrt{k} = n$, where n is any positive integer. The respective functions $y(x)$ are

$$y_n(x) = \sin nx, \qquad \text{with } n = 1, 2, 3, \cdots.$$

The second differential equation (55) can be similarly dealt with (with t in the place of x). Its solutions are of the form

$$v(t) = k_1 \sin (\sqrt{k}at) + k_2 \cos (\sqrt{k}at),$$

and upon this the condition (57) imposes the requirement $k_1\sqrt{k}a = 0$. This is fulfilled if $k_1 = 0$, $k_2 = 1$. The respective functions $v(t)$ are

$$v_n(t) = \cos (nat).$$

The corresponding products (54) are accordingly

(58) $$\sin nx \cos nat, \qquad n = 1, 2, 3, \cdots.$$

Each of these, it was observed, fulfills the differential equation (51), the conditions (52) and the first one of the conditions (53).

We can go further than this. By direct substitution it can be verified that any sum of constant multiples of products (58), say the sum

$$c_1 \sin x \cos (at) + c_2 \sin 2x \cos (2at) + \cdots + c_n \sin nx \cos (nat),$$

fulfills the differential equation (51) and the first three of the conditions (52) and (53). Such a sum has flexibility, to the extent that its coefficients are arbitrary and can therefore be adjusted to some remaining purpose. We have a remaining purpose, namely to conform with the second condition (53), in which $f(x)$ is a specifically given function. Inasmuch as this calls for conformity of a sort of infinite kind, namely at all points x of the interval from $x = 0$ to $x = \pi$, the suggestion presents itself that a solution may, perhaps, be obtainable through an infinite series representation

(59) $u(x, t) = c_1 \sin x \cos (at)$

$$+ c_2 \sin 2x \cos (2at) + \cdots + c_n \sin nx \cos (nat) + \cdots .$$

By (53) this could be so only if the formula yields $f(x)$ when $t = 0$, namely if

(60) $f(x) = c_1 \sin x + c_2 \sin 2x + \cdots + c_n \sin nx + \cdots .$

Historically it seemed, even to some great mathematicians, that a representation (60) would be impossible unless the function $f(x)$ were of some quite special type. That that conviction was erroneous, was shown by Fourier in about 1810. The trigonometric functions are in fact capable of representing, in the manner (60), the functions of a very wide category. Granting this fact, the question remains, how are the coefficients c_n to be determined when the function $f(x)$ is at hand? To this we can give a definite answer.

The functions $y_n(x)$ and $y_m(x)$, namely $\sin nx$ and $\sin mx$, are such that

$$y_n'' = -n^2 y_n ,$$
$$y_m'' = -m^2 y_m .$$

After these equations have been multiplied by y_m and y_n respectively, their difference is

$$y_m y_n'' - y_m'' y_n = (m^2 - n^2) y_m y_n .$$

The left-hand member of this relation is the derivative of the quantity $(y_m y_n' - y_m' y_n)$. An integration from $x = 0$ to $x = \pi$ therefore gives

$$\left[y_m y_n' - y_m' y_n \right]_{x=0}^{x=\pi} = (m^2 - n^2) \int_0^\pi y_m(x) y_n(x) \, dx.$$

In this the left-hand member is zero, since y_m and y_n both vanish at $x = 0$ and at $x = \pi$. Thus

$$0 = (m^2 - n^2) \int_0^\pi \sin mx \sin nx \, dx,$$

namely

(61) $\int_0^\pi \sin mx \sin nx \, dx = 0,$ if $m \neq n.$

Let the relation (60) therefore be multiplied by sin mx with any integer m. If it is then integrated term by term, each resulting term on the right, except the mth one, is zero, and the relation obtained is

$$\int_0^\pi f(x) \sin mx \, dx = c_m \int_0^\pi \sin mx \sin mx \, dx.$$

Because

$$\int_0^\pi \sin^2 mx \, dx = \frac{1}{2} \int_0^\pi (1 - \cos 2mx) \, dx = \frac{1}{2} \left(x - \frac{1}{2m} \sin 2mx \right)_{x=0}^{x=\pi} = \frac{\pi}{2},$$

the result obtained is

$$(62) \qquad\qquad c_m = \frac{2}{\pi} \int_0^\pi f(x) \sin mx \, dx.$$

These values c_m are called the *Fourier coefficients* of the function $f(x)$.

If the infinite series in (59), with the coefficients (62), converges suitably, the relation (59) is a formula, and the function $u(x, t)$ it represents fulfills the differential equation (51) and the conditions (52) and (53). Thus $u(x, t)$ gives the position of the vibrating wire at any abscissa x and at any time t. Through it the analysis of the vibration has been accomplished. From the standpoint of acoustics, the first term of the formula (59) represents the harp string's fundamental tone, the intensity of which is indicated by the coefficient c_1 . The nth term of the formula and its coefficient c_n , correspondingly represent the $(n - 1)$th overtone and its intensity. From (62) it is clear that these various intensities depend upon the shape of the curve $u = f(x)$ from which the string is released.

CONCLUSION

The method of trigonometric analysis which we have outlined in the previous section is a powerful mathematical tool. However it is only one among many mathematical methods whose applicabilities extend beyond the field of vibration phenomena over the field of flow manifestations, as they take place with fluids, electricity or heat, over chemical diffusion phenomena, and over those of wave motions, such as are displayed by sound, light, radio, etc. In many of these methods the trigonometric functions are important. There are, however, also other functions that have proved themselves so widely applicable that much study has been given to them. In some cases proper names have been assigned these, and their values have been computed and tabulated. As examples of such the following few will bear mention:

(i) *The hypergeometric functions*, which appear as solutions of differential equations

$$x(1 - x)y'' + \{a - (b + c + 1)x\}y' - bcy = 0,$$

in which a, b, and c are constants. For different sets of these constants different sets of hypergeometric functions are obtained, and these sets are inter-related in many interesting ways.

(ii) *The Bessel functions*, which appear as solutions of a differential equation

$$x^2y'' + xy' + (x^2 - a^2)y = 0.$$

Different constants a again yield different sets of functions.

(iii) *The Legendre functions*, whose differential equation is

$$(1 - x^2)y'' - 2xy' + a(a + 1)y = 0.$$

(iv) *The Mathieu functions*, that are obtained as solutions of the differential equation

$$y'' + (a + b \cos 2x)y = 0.$$

Diverse as the instances of applicability of these functions, and of others of similar category, are, it remains true that most mathematical problems depend upon functions that are peculiar to themselves. To solve such problems mathematical researchers must study these functions to determine their properties. To do that they must often depend almost solely upon the differential equation. It is hoped that the deductions of this chapter may have illustrated the types of reasoning and procedure by which much of this is accomplished.

BIBLIOGRAPHY

1. Bôcher, M. "Introduction to the Theory of Fourier's Series." *Annals of Mathematics*, (2) Vol. 7, p. 81–152.
2. Churchill, R. V. *Fourier Series and Boundary Value Problems*. New York: McGraw-Hill Book Co., 1941.
3. Jackson, D. *Fourier Series and Orthogonal Polynomials*. The Mathematical Association of America: Carus Monograph No. 6, 1941.
4. Klein, F. *Elementary Mathematics from an Advanced Standpoint*. Vol. I. New York: Dover Publications.
5. Langer, R. E. "Fourier's Series, The Genesis and Evolution of a Theory." *The American Mathematical Monthly*, Herbert Ellsworth Slaught Memorial Paper, No. 1, 1947.

6. LANGER, R. E. *A First Course in Differential Equations.* New York: John Wiley and Sons, 1954.

7. OSGOOD, W. F. "Lehrbuch der Funktionentheorie," Chap. 12. *Die elementaren Funktionen.* Leipzig: B. G. Teubner, 1928.

8. VAN VLECK, E. B. "The Influence of Fourier's Series upon the Development of Mathematics." *Science,* N. S. Vol. 39, p. 113–124, 1914.

IX

Origins and Development of Concepts of Geometry

S. H. GOULD

THE CONCEPT OF MATHEMATICAL PROOF

SINCE the time of Euclid (300 B.C.), a "mathematical" proof, based on precise axioms stated in advance, has been regarded as the model of perfection for all other reasoning. For example, the philosopher Spinoza (1632–1677) felt that he could argue most persuasively for his system of ethics by presenting it *more geometrico*, in the geometric manner with axioms and theorems. On the other hand, a carpenter makes use of geometric results without desiring any "proof" other than that they "work." Who first felt the need of such a proof? And why was the axiomatic method, which today pervades all mathematics, associated for so many centuries particularly with geometry? To answer these two questions let us look at the history of Greek geometry.

THE PYTHAGOREANS AND EUCLID

Many of the details in Greek writings on geometry come from Babylonian sources. For instance, the little book *Metrics*, written by Heron of Alexandria about 60 A.D., deals with the same numerical lengths, areas and volumes as the cuneiform texts of two thousand years earlier, and these examples are handed down in an unbroken line through Roman times into the Middle Ages. Such a collection of practical rules for surveyors and architects shows little trace of the axioms and proofs of Euclid, so that we find two streams of tradition in Greek mathematics, the one, empirical, concrete, typically Oriental; the other, scientific, abstract, typically Greek. This second, or Greek, approach to geometry, of immense importance for all subsequent mathematics, must have been developed between 500 B.C., when the Pythagoreans first became acquainted with Babylonian mathematics, and 300 B.C., when Euclid wrote the *Elements*. We can trace its growth by examining the theorem of Pythagoras, namely,

> EUCLID I, 47: *the square on the hypotenuse of a right triangle is equal to the sum of the squares on the other two sides.*

Although many specific examples of this theorem (e.g., triangles with sides 3, 4, 5; 5, 12, 13; etc.) are mentioned by the Babylonians and it is known that they were able to find all such triplets of integers, still the Pythagorean school seems to have been the first to give what would now be regarded as anything like a mathematical proof. The neo-Platonic philosopher Proclus (410–485 A.D.) tells us: "The ancients refer this theorem to Pythagoras himself and say that he sacrificed an ox in honor of its discovery." But then he goes on: "For my part, while I admire those who first observed the truth of the theorem, I marvel more at the writer of the *Elements* . . . because he made it certain by a most lucid demonstration,"—a statement that naturally leads us to ask: How was the theorem first proved by the Pythagoreans and what was the change introduced by Euclid?

To judge from later discussions, the original proof of the Pythagoreans must have been as follows (see Fig. 1):

Since $\triangle ABC \sim \triangle DBA$, $\dfrac{CB}{BA} = \dfrac{BA}{BD}$, or $CB \cdot BD = BA^2$;

also $\triangle ABC \sim \triangle DAC$, so that $CB \cdot CD = CA^2$,

and by addition $CB^2 = BA^2 + CA^2$, as desired.

But the proof given in Euclid I, 47 — the one praised so highly by Proclus — is based on actually constructing the squares (see Fig. 1):

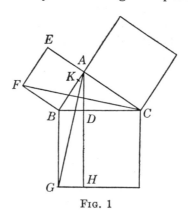

FIG. 1

$$BA^2 = 2 \triangle FBC; \; BG \cdot BD = 2 \triangle ABG;$$

while $\triangle FBC = \triangle ABG,$

so that $BA^2 = $ rect. BH. Similarly, $CA^2 = $ rect. CH,

so that $BA^2 + CA^2 = BC^2$.

The first proof is seen to depend on the property of similar triangles stated in

EUCLID VI, 4: *in equiangular triangles, the sides containing the equal angles are proportional,*

which is proved thus (see Fig. 2): if $\triangle ADE \sim \triangle ABC$, then $DE \parallel BC$ by

EUCLID I, 27: *if a transversal cutting two lines makes alternate angles equal, the lines are parallel* (for Euclid's proof, see below),

so that the desired result is given by

EUCLID VI, 2: *a line parallel to one side of a triangle cuts the other sides proportionally,*

for which the proof is (see Fig. 2):

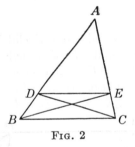

FIG. 2

$\triangle DEB = \triangle DEC$, by Eucl. I, 38 (see below),

so that, by adding $\triangle ADE$, we get $\triangle ABE = \triangle ADC$.

Also $\dfrac{AD}{AB} = \dfrac{\triangle ADE}{\triangle ABE}$, by Eucl. VI. 1 (see below).

Thus $\dfrac{AD}{AB} = \dfrac{\triangle ADE}{\triangle ADC} = \dfrac{AE}{AC}$, as desired.

Here we have used the two theorems:

EUCLID VI, 1: *triangles with the same height are to each other as their bases,*

EUCLID I, 38: *triangles on equal bases and between the same parallels are equal,*

whereas Euclid's proof uses only the second of these two theorems.

So Euclid's change consists of making the proof of the Pythagorean theorem depend directly on Eucl. I, 38, without the intervention of Eucl. VI, 1. Why did Proclus give so much praise to such a slight modification?

For answer, let us look at the Pythagorean definition of proportionality given in

EUCLID VII, DEF. 20: *four magnitudes are proportional when the first is the same multiple, submultiple, or multiple of submultiple, of the second that the third is of the fourth:*

or, as we would express it,

$$a \text{ is to } b \text{ as } c \text{ is to } d \quad \text{if} \quad a = \frac{m}{n} b \quad \text{and} \quad c = \frac{m}{n} d$$

for some integers m and n.

Thus the Pythagoreans would prove Euclid VI, 1 as follows (see Fig. 3): Measure the bases BC and EF of triangles ABC and DEF with a unit of measurement $BB_2 = EE_2$ which is contained an integral number of times, say m, in BC, and also, say n, in EF. Mark off $BB_2 = B_2B_3 = \cdots = B_mC$ and $EE_2 = E_2E_3 = \cdots = E_nF$, as indicated in Figure 3.

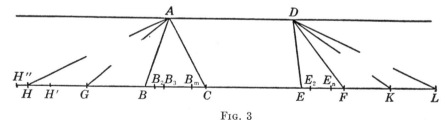

FIG. 3

Then

$$BC = mBB_2 = mEE_2 = \frac{m}{n} EF,$$

and, from Eucl. I, 38:

$$\triangle ABC = m\triangle ABB_2 = m\triangle DEF_2 = \frac{m}{n}\triangle DEF.$$

so that

$$\frac{\triangle ABC}{\triangle DEF} = \frac{BC}{EF}$$

by the Pythagorean definition of proportionality. In their words, $\triangle ABC$ is the same multiple, namely, the mth, of the same submultiple, namely, the nth, of $\triangle DEF$, as BC is of DE.

Here we must note that for the Pythagoreans it was evident, up to the time of the "scandal" mentioned below, that such a common measure $BB_2 = EE_2$ for the lines BC and EF must exist; that is, BC and EF must each be a whole number of times as long as some segment BB_2, provided BB_2 is sufficiently short. For this belief they had many reasons. For example, being the first to discover the role played by small integers in the theory of music, they made a religion of such numbers, asserting that "God works in whole numbers" or even that "God is number." So it did not at first occur to them that "whole numbers" might not also measure any two line segments, and for a time they were happy with their proof of the Pythagorean theorem.

THE SCANDAL IN GEOMETRY

But what about the simplest case of all, the isosceles right triangle with $AB = AC$ (see Fig. 1) so that AB is the side and BC the diagonal of a square? Do the lines AB and BC have a common measure? If so, let it be AK, so that $AB = nAK$ and $BC = mAK$. Then, since $BC^2 = AB^2 + AC^2 = 2AB^2$, we have $m^2 AK^2 = 2n^2 AK^2$, or $m^2 = 2n^2$, with integers m and n from which we may suppose that common factors have been cancelled. But this is impossible, since m^2, being equal to $2n^2$, must be even, so that m itself is even and therefore n is odd. Setting $m = 2p$, we get $2n^2 = 4p^2$ or $n^2 = 2p^2$, so that n is even, and as Aristotle (384–322 B.C.) expressed it, "the same number is both even and odd." This contradiction showed that the hypotenuse and side of an isosceles right triangle are incommensurable, so that nobody knew whether for such a triangle the Pythagorean theorem is true or not.

Let us try to realize the magnitude of the resulting scandal, which lasted perhaps a hundred years and formed one of the great turning points in the history of science. The feelings of the Pythagoreans themselves are enshrined in the following legend.

At first the disastrous discovery, demolishing their favorite dogma that "God is number," was known only to a few members of the cult in southern Italy, and Pythagoras himself warned them not to divulge to the vulgar world the terrible blow that God had received. But one of their number, whose name was Hippasus, was so weighed down by the dread secret that he let it slip. Shortly after, on a voyage to Greece, he was drowned in a shipwreck in which everyone else was saved. The details of such legends are wholly untrustworthy and, in fact, we know neither the date of the discovery of incommensurables nor whether Pythagoras himself, an almost mythical character, took any share in it. But the story of Hippasus well illustrates the bewilderment of geometers who thought they had built up a body of theorems about similar triangles and now realized that even their *definition* of proportionality applied only to commensurable segments. (The nth submultiple of b must also measure a, namely, m times.)

The fundamental difficulty was removed about 360 B.C. by one of the great mathematicians of all time, Eudoxus, the pupil of Plato. Following a suggestion of De Morgan's, we may imagine him thinking along these lines:

Imagine a picket fence, with distance P between pickets, in front of a colonnade with distance C between columns, the first picket being directly in front of the first column. Now make a drawing (the reader is advised to do this) with respective distances p and c. What do we mean by saying that the drawing is *similar* to the original, or, in other words, that P is to C as p is to c? If P and C are *commensurable*, a picket and a column will coincide at regular intervals and the original definition of proportionality will apply. But if P and C are *incommensurable*, then no picket after the first will coincide with any column. In other words, if we consider a particular picket, say the mth, where m is any integer, and a particular column, say the nth, then the mth picket will come either before ($mP < nC$) or after ($mP > nC$) the nth column, and to say that the drawing is similar to the original means that, whichever is true in the original, the same will be true in the drawing, for every choice of m and n. Thus, including the commensurable case, we may say:

P is to C as p is to c if all integers m and n which make $mP \gtreqless nC$ also make $mp \gtreqless nc$, or, in the original words of Eudoxus:

EUCLID V, DEF. 5: *four magnitudes are in the same ratio, the first to the second as the third to the fourth, when, if any multiple whatever is taken of the first and the third (the same multiple of each) and also any multiple*

*of the second and the fourth (the same multiple of
each) then the multiple of the third exceeds, is equal
to, or falls short of the multiple of the fourth, according as the multiple of the first exceeds, is equal
to, or falls short of the multiple of the second.*

From our familiar notions of similarity in the world around us, we see
at once that this definition of Eudoxus, which has sometimes been criticized for its complexity, is in fact the natural definition of proportionality
in view of the existence of incommensurables. Moreover, it is extremely
useful and easy to apply. Thus Euclid VI, 1 now follows immediately
from Eucl. I, 38. For (see Fig. 3) marking off $BC = GB = HG = \cdots$
any number of times, say m, and $EF = FK = KL = \cdots$, say n times,
we have:

$$\text{if } HC = EL, \qquad \text{then} \qquad \triangle AHC = \triangle DEL$$

by Eucl. I, 38; so that

$$\text{if } mBC = nEF, \qquad \text{then} \qquad m\triangle ABC = n\triangle DEL.$$

Also, if $HC > EL$, take $H'C = EL$, so that $\triangle AHC = \triangle AHH' + \triangle AH'C > \triangle AH'C = \triangle DEL$. Then $mBC > nEF$ and $m\triangle ABC > n\triangle DEF$. Again, if $HC < EL$, take $H''C = EL$, etc., whereupon $mBC < nEF$ and $m\triangle ABC < n\triangle DEF$. Thus

$$\frac{\triangle ABC}{\triangle DEF} = \frac{BC}{EF},$$

by the definition of Eudoxus.

GEOMETRIC ALGEBRA

In this way Eudoxus restored the prestige of geometry and enabled
it to go forward again. Then, in the hands of Archimedes (287–212 B.C.)
and Apollonius (born about 260 B.C.) mathematics enjoyed a brief period of brilliance that can be compared only with the explosive growth
of modern science. Yet it had a certain bias, toward geometry and away
from algebra, that was not removed until the time of Descartes and Fermat (about 1650). For example, the Babylonians had devoted much attention to the solution, by algebraic means, of quadratic equations
in two unknowns, which they would first reduce to the two standard
forms

$$\text{(i) } xy = c^2, \qquad x + y = 2p,$$

$$\text{(ii) } xy = c^2, \qquad x - y = 2p,$$

and then, by eliminating y, to the familiar quadratic equation, say $x^2 - 2px - c^2 = 0$. For the Babylonians, the c^2 and p were known positive numbers, and there was no necessary connection with geometry. If the answers happened to be irrational, the Babylonians remained unaware of that fact and were content with an approximating fraction.

But in Euclid the same two problems are stated in a purely geometric way:

EUCLID II, 5, 6:   *to apply to a given line-segment a rectangle with*
VI, 28, 29: *given area and falling short (or exceeding) by a*
square (or more generally, by a rectangle similar to a
given rectangle).

Thus, in Figure 4, the problem is to apply rectangle AO with area c^2 to the given line AB in such a way as to exceed (*i.e.*, exceed rectangle AE) by a square BO, which is called the *excess* of AO. In Figure 4, let

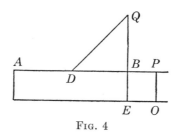

FIG. 4

AP be taken as x, AB as $2p$, and AB be bisected at D. Then

$$AD = DB = p \quad \text{and} \quad DP = x - p.$$

Since $PO = BP = x - 2p$, it follows that

$$c^2 = AP \cdot PO = x(x - 2p).$$

Furthermore

$$(x - p)^2 = x^2 - 2px + p^2 = x(x - 2p) + p^2 = c^2 + p^2.$$

Thus we see that the desired line length DP can be obtained (Euclid's construction was slightly different) as the hypotenuse of a right triangle with the given sides $DB = p$ and $BQ = c$. (See Fig. 4.)

This is the same as the Babylonian problem with $AB = 2p$ and with x and y for the sides of the required rectangle AO. Euclid gave a ruler-and-compass construction that began with the bisecting of AB at D and must have been at least as familiar to Greek schoolboys as the solution of the quadratic to modern ones. In fact, they both depend on completing the square.

For dealing with any quadratic problem, the *geometric algebra* based on this type of application of areas was far superior in Greek eyes to the numerical work of the Babylonians, since the latter gave no meaning to an irrational answer; and it became standard Greek practice after the discovery by Menaechmus (about 360 B.C.) of conic sections.

There are many legends about the early history of conics, subsequently so important in art and science. For example, when Glaucus, the five-year-old son of Minos, king of Crete, met his death by getting his head stuck, like Pooh, in a pot of honey, the king was dissatisfied with the cubical tomb planned by his architects and ordered it doubled in size. This demand raised the problem (proved impossible only in the nineteenth century) of solving with ruler and compass the equation $x^3 = 2p^3$. Menaechmus pointed out that it could be solved "graphically" by finding the intersection of the two curves $x^2 = py$ and $y^2 = 2px$. But, although Menaechmus had very clear notions about ordinates and abscissas, he felt that until these curves could be given some geometric interpretation, their very existence was doubtful. (The first geometer to realize that an equation *defines* a curve was Fermat (1601–1665), who thereby became one of the inventors of analytic geometry.) Neugebauer has pointed out, however, that Menaechmus, by watching the shadow on a sundial, was able to "legitimize" these curves as *sections of a cone.* As a result, his successors were able to build up their wonderful system of theorems about parabolas, ellipses and hyperbolas, from which followed, for example, the astronomical developments of Kepler and Newton.

But these theorems are *planimetric*, that is, they deal only with one plane. So, except to provide a definition, the parent cone is superfluous. To study these conic sections conveniently, we may bypass the cone by finding some planar property of each conic from which we can deduce our other theorems. This property will most naturally take the form of some relation between the ordinate QV (see Fig. 5) and the abscissa PV of any point Q on the conic. (In Fig. 5, PP' is an ellipse cut from the cone ABC.) Thus, Apollonius asserts that if we draw $AF \parallel PP'$ and take DE such that

$$\frac{DE}{PP'} = \frac{BF \cdot FC}{AF},$$

and then apply to DE a rectangle (Fig. 6) with area equal to QV^2 and defect similar to DG (see Fig. 5, where $EG = PP'$), then the width of the applied rectangle is equal to the abscissa PV. The length of DE,

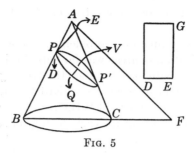

FIG. 5

call it p, is the *parameter* of the ellipse and PP', of length say d, is its *diameter*. The position of DE is of no importance but, in proving the above statement, it was convenient for Apollonius to erect DE at P perpendicular to PP', for which reason he calls DE the *erect segment* (in Latin, *latus erectum*, or later *latus rectum*).

Since the curve in Figure 5 is defined by application with *defect* it is called an *ellipse* (cf. *ellipsis*); similarly the *hyperbola* will show *excess* (cf. *hyperbole*), while the *parabola* will be defined by exact application of one area to another (cf. *parable*).

Now all this is straight algebra in geometric dress. For example, the equation of the above ellipse may be shown, in the following manner to be $y^2 = px - (p/d)x^2$. In Figure 6, the rectangle DN is the rectangle

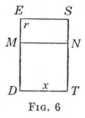

FIG. 6

applied to DE with the defect MS which is similar to rectangle DG of Figure 5. Whence,

$$QV^2 = DE \cdot DT - EM \cdot MN,$$

$$y^2 = px - rx.$$

Since $r{:}x = p{:}d$, it follows that

$$y^2 = px - (p/d)x^2.$$

But the discovery of the irrational made the Greeks distrustful of algebra, while the definition of proportionality by Eudoxus enabled them to put geometry on that firm axiomatic basis to which we shall now turn. Not until Dedekind (1872) applied the Eudoxian idea to the domain of numbers, and thereby gave the first definition of $\sqrt{2}$, can this domination by geometry be said to have ended.

Modern science is based on manipulations with numbers that are a natural development of the old Babylonian mathematics. The Greek "detour through geometry," though rich in results of the very greatest importance, nevertheless hindered this development by failing, for example, to consider powers of numbers higher than the third, and thereby, for almost two thousand years, retarded the whole progress of science.

THE BIRTH OF AXIOMATICS

The Pythagorean discovery of incommensurables showed that the old method of piling up examples, the argument of the "practical" man, for whom Heron's rules were intended, was not sufficient. The disturbing feature was not so much that practical experience had never suggested incommensurables, which therefore came as a shock, but rather, that experience could not possibly have suggested them. No practical carpenter could ever have observed them. It was the first example of what has often happened since; theoretical reasoning had led to results that experience could not possibly have foreseen. The value of logical proof is not so much that it corroborates experience as that it raises new questions.

As a result, Greek thinkers of the fourth century B.C. took the view that, in geometry, theoretical reasoning must play as large a role as possible. In other words, the list of "self-evident" facts taken from experience must be reduced to a minimum and written down in advance, where it can be scrutinized.

So Euclid and his predecessors set up the famous system of ten axioms found at the beginning of the *Elements*. They consist of five *postulates* and five *common notions*. For example,

POSTULATE I: *any two points can be joined by a straight line.*

COMMON NOTION I: *things which are equal to the same thing are equal to each other.*

Euclid himself appears to have believed that with these ten axioms he had provided a complete basis for the familiar geometry of the *Elements*, but in modern times it has been seen that this is not so. For instance, some of Euclid's proofs depend on the fact that a straight line containing an interior point of a triangle must also contain an exterior point (*axiom of Pasch*) or again that, given two line-segments, some multiple of the shorter exceeds the greater (*axiom of Eudoxus*, often called *axiom of Archimedes*). Again, the "proof by superposition" of

> EUCLID I, 4: *if two triangles have two sides and the contained angle respectively equal, the triangles are congruent,*

and

> EUCLID I, 8: *if three sides are equal to three sides, the triangles are congruent,*

really amounts to a postulate based on experience and is nowadays so regarded. There is plenty of evidence that Euclid himself disliked "superposition"; for example, he does not use it in

> EUCLID I, 25: *if two triangles have two sides respectively equal, then the greater base is opposite the greater contained angle.*

As for these missing axioms, some of which we shall need below, we content ourselves here with the remark that they have been supplied by modern writers.

THE PYTHAGOREAN THEOREM ON THE BASIS OF EUCLID'S AXIOMS

We saw above that the proof of the Pythagorean theorem depends on Eucl. I, 38, or, since a triangle is half of a parallelogram, on

> EUCLID I, 35: *parallelograms on the same base and between the same parallels are equal in area.*

But here certain questions arise. Parallel lines are defined as lines in the plane which do not meet, no matter how far extended, and we begin to feel that we have introduced a concept which is somehow not on quite so firm a basis as, *e.g.*, the two axioms quoted above. Certainly it does not make quite as strong an appeal to our experience. For we can actually observe or measure only a finite part of the universe, so that we may very well wonder what will happen to straight lines if they are produced into the reaches of space beyond our experience. Do any two parallel lines exist? Or is a parallelogram a contradiction in terms?

Fortunately, the existence of parallel lines can readily be proved. For we have:

> EUCLID I, 27: *if a transversal on two straight lines makes the alternate angles equal, then the straight lines are parallel.*

For if, say, $\angle DEB = \angle EBC$ in Figure 7, and if DF and BC were to meet at K, we would have a contradiction to

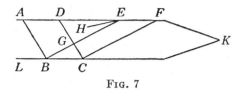

FIG. 7

EUCLID I, 16: *an exterior angle of a triangle is greater than either of the interior, non-adjacent angles,*

which is proved as follows (see Fig. 8): Let E be the midpoint of AC and

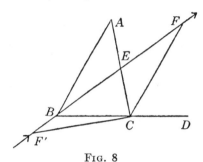

FIG. 8

take $BE = EF$. Then $\angle ACD > \angle FCE$ and $\angle FCE = \angle BAC$ (Euclid I, 4), so that $\angle ACD > \angle BAC$, as desired. (We shall return below, in connection with elliptic geometry and the theory of relativity, to the crucial statement here, namely, that $\angle ACD > \angle FCE$.)

Coming back to the proof of Eucl. I, 35 (see Fig. 7) we need only prove $\triangle ABE = \triangle DCF$, since then the result will follow by subtracting $\triangle DGE$ and adding $\triangle BGC$. But for this purpose we must prove $\angle AEB = \angle DFC$. In other words, we need the converse of Euclid I, 27, namely:

EUCLID I, 29: *if parallel lines are cut by a transversal, the alternate angles are equal.*

Now we proved Eucl. I, 27 without difficulty, but how are we going to prove Eucl. I, 29? It is natural to try another indirect proof (see Fig. 7): suppose $\angle AEB > \angle DFC$ and construct $\angle HEB = \angle DFC$. Then HE and BC would be parallel by Eucl. I, 27, so that we would have two distinct lines through E, namely AE and HE, both parallel to BC. This looks hopeful. We had no trouble proving (Eucl. I, 27) that there must be at least one such parallel. Surely we can prove equally well that there cannot be more than one?

But, before we begin, let us note that Euclid himself gave up the attempt, which, as we shall prove below, was in fact a hopeless one. He observed that the lines HE and LB cut BE in such a way that the angles HEB and LBE add up to less than two right angles and then took it as an axiom that such lines cannot be parallel. This is the famous Fifth Postulate of Euclid, the one sentence in the history of science that has given rise to more publication than any other, namely:

POSTULATE 5: *if a straight line falling on two straight lines make the interior angles on the same side less than two right angles, the two straight lines, if produced indefinitely, meet on that side on which are the angles less than the two right angles.*

Accepting this statement as a "self-evident" truth, that cannot be proved on the basis of other simpler or more acceptable axioms, we see that our proof of the theorem of Pythagoras is at last complete.

ATTEMPTS TO PROVE THE FIFTH POSTULATE

It was, indeed, inevitable that this postulate would be attacked as soon as it was set up. In comparison with the others it is very lengthy and has all the appearance of being a theorem. In fact, as we have just seen it is the converse of an easily proved theorem, namely Euclid I, 27. Taking it for an axiom was regarded as a blemish in Euclid's work, and attempts, lasting over two thousand years, were made to prove it on the basis alone of Euclid's other axioms. (Note that in these attempts we may use any theorem up through Euclid I, 28.)

Consider, for example, the supposed proof of Aganis, quoted in the Aristotelian commentator Simplicius (6th century, A.D.).

Let AB and DZ cut BZ as in Figure 9, with $\angle B + \angle Z < 2$ rt. $\angle$'s.

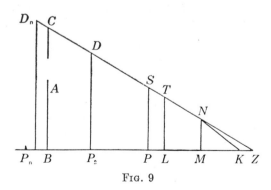

FIG. 9

Then, to find the supposed point C where ZD cuts BA, let T be any point on ZD and draw $TL \perp BZ$. (For convenience, we take $\angle B$ to be a right angle; otherwise, draw $TL \parallel AB$.) Bisect BZ at P, PZ at M, etc., often enough so that by the axiom of Eudoxus M falls between L and Z. Draw $MN \perp BZ$. (Note that MN will meet SZ by the axiom of Pasch.) Let ZC be the same multiple, say the nth, of ZN that ZB is of ZM. Then $ZC = nZN$ and ZM is the projection of ZN on BZ, and assuming that equal segments have equal projections, we get (proj. of ZC) $= nZM = ZB$, so that C is on BA.

But the objection to this work of Aganis is obvious. How can we prove that equal segments have equal projections? So we turn next to the Persian Nasiraddin (1201–1274), who introduced a figure destined to become famous, namely, the *birectangular isosceles quadrilateral* obtained by making $CA \perp AB$, $DB \perp AB$ and $CA = DB$ and then joining CD (see Fig. 10). It is usually referred to as the *Saccheri quadrilateral*, AB

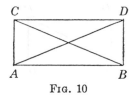

FIG. 10

being the *base*, CD the *summit* and the angles at C and D the *summit angles*.

Then, with or without the Fifth Postulate, we can prove that the summit angles are equal. For join AD and BC. Then $\triangle ABC \equiv \triangle ABD$, (Eucl. I, 4), so that $AD = BC$ and therefore $\triangle ACD \equiv \triangle BDC$ (Eucl. I, 8), so that $\angle ACD = \angle BDC$, as desired.

We shall refer to this theorem as

SACCH. I, 1: *the summit angles of a Saccheri quadrilateral are equal.*

Nasiraddin then said that $\angle ACD$ must be a right angle; for if it were acute the lines AB and CD would be approaching each other, so that DB would be shorter than CA, contrary to construction; and if $\angle ACD = \angle BDC = 90°$, he proved (see below) that the Fifth Postulate must be true. But he failed to give a satisfactory argument that the lines AB and CD will approach each other if $\angle ACD$ is acute.

The study of the quadrilateral in Figure 10 was carried much deeper by the Jesuit priest Saccheri, the first person of really high mathematical gifts to give his attention to the problem of proving Euclid's Fifth Postulate. In Book One of his *Euclid Freed from Every Blemish* (1733), he rightly rejects Nasiraddin's argument that the summit angles at C

and D must be right angles. Instead, he considers the three possible hypotheses, abbreviated as *Hyp. Rect.*, *Hyp. Obt.* and *Hyp. Acut.*

Hyp. Rect.: the summit angles are right-angles;
Hyp. Obt.:   they are obtuse-angles;
Hyp. Acut.: they are acute-angles,

and finds the striking result:

SACCH. I, 9: *if* $\triangle ABC$ (see Fig. 11) *is right-angled (any triangle can be divided into right triangles), then*

$$\angle A + \angle B + \angle C \gtreqless 2 \text{ rt. } \angle\text{'s} \quad \begin{matrix} Hyp. \ Obt. \\ Hyp. \ Rect. \\ Hyp. \ Acut. \end{matrix}$$

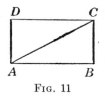

FIG. 11

(The amount by which the angle-sum of any triangle differs from two right angles is called the *defect*, or the *excess*, of the triangle.)

For the proof of this fundamental theorem, we need two lemmas:

SACCH. I, 2: *the line joining the midpoints of base and summit of a Saccheri quadrilateral is perpendicular to both of them,*

the proof of which (see Fig. 12) runs: $\triangle ACM \equiv \triangle BDM$, by Euclid I,

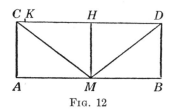

FIG. 12

4. Thus $CM = MD$ and $\triangle MCH \equiv \triangle MHD$, by Eucl. I, 8, so that $\angle CHM = \angle DHM = $ rt. $\angle$, and similarly for the angles at M. The other necessary lemma is:

SACCH. I, 3: *summit $CD \lesseqgtr$ base AB* $\begin{matrix} Hyp. \ Obt. \\ Hyp. \ Rect. \\ Hyp. \ Acut., \end{matrix}$

the proof of which for *Hyp. Obt.* (the other two cases proceed in the same way) is this: bisect AB at M and CD at H and join MH. Then, supposing $CD \geq AB$, we take $HK = MA$, so that $MHKA$ is a Saccheri quadrilateral (Sacch. I, 2) and $\angle HKA = \angle MAK = $ rt. $\angle$ (Sacch. I, 1); but $\angle HKA > \angle HCA = $ obt. $\angle$. (*Hyp. Obt.* and Eucl. I, 16), a contradiction from which we get $CD < AB$, as desired.

The proof of Sacch. I, 9 is now immediate. For, if we construct the Saccheri quadrilateral on base AB (see Fig. 11), we have

$$CD \lesseqgtr AB \quad \begin{matrix} Hyp.\ Obt. \\ Hyp.\ Rect. \\ Hyp.\ Acut. \end{matrix}$$

so that $\angle DAC \lesseqgtr \angle ACB$, by Euclid I, 25. Thus in $\triangle ABC$ it follows that $\angle A + \angle B + \angle C \gtreqless \angle A + \angle DAC + \angle B = 2$ rt. $\angle$'s, as desired. Also, for any quadrilateral $ABCD$, by dividing it into two triangles, we get a theorem which we will call:

$$\text{Sacch. I, 9b:}\ \angle A + \angle B + \angle C + \angle D \gtreqless 4\ \text{rt.}\ \angle\text{'s.}\quad \begin{matrix} Hyp.\ Obt. \\ Hyp.\ Rect. \\ Hyp.\ Acut. \end{matrix}$$

The earlier work of Aganis and Nasiraddin can now be corrected and extended.

Concerning the projections PM and MZ (which Aganis had assumed to be equal) of the equal segments SN and NZ in Figure 9, we have

$$\text{Sacch. I, 11a, 12a:}\ PM \gtreqless MZ\quad \begin{matrix} Hyp.\ Obt. \\ Hyp.\ Rect. \\ Hyp.\ Acut. \end{matrix}$$

For join NP. Then $NP > NZ$ (again we consider only *Hyp. Obt.*). For if $NP \leq NZ = SN$, then $\angle NPS \geq \angle NSP$ (cf. Eucl. I, 18, quoted below). Thus $\angle NPS + \angle SZP > \angle NSP + \angle SZP >$ rt. $\angle$, by Sacch. I, 9. But $\angle NPS + \angle NPZ = $ rt. $\angle$, so that $\angle NZP = \angle SZP > \angle NPZ$, from which $NP < NZ$, by Eucl. I, 18. But if $NP > NZ$, then $PM > MZ$. For if $PM \leq MZ$, take $MK = PM$ (see Fig. 9) and join NK. Then $\angle NPK = \angle NKP$ (Eucl. I, 4), and $\angle NKP > \angle NZP$ (Eucl. I, 16), so that $\angle NPK > \angle NZP$, leading to the contradition $NZ > NP$, by reason of

Euclid I, 18: *in any triangle a greater side subtends a greater angle.*

Passing on to Nasiraddin, we noted above that he showed how the Fifth Postulate follows from the *Hyp. Rect.* We now prove and extend this result as follows (main theorem of Saccheri):

SACCH. I, 13: *the Fifth Postulate follows from Hyp. Rect. and also from Hyp. Obt.*

To prove this, we need

SACCH. I, 11, 12 (*Hyp. Rect. or Hyp. Obt.*): *if the line BA* (see Fig. 9) *cuts BZ at right angles and ZD cuts BZ at an acute angle, then BA and ZD must intersect.*

For proof, we cut off on ZD an arbitrary number n of equal segments $ZN = NS = SD = \cdots$ and draw NM, SP, $DP_2 \cdots \perp BZ$. Then by Sacch. I, 11a, 12a, we have, for *Hyp. Obt.* and *Hyp. Rect.* $PM \geqq MZ$; $PZ \geqq 2MZ$; $P_2Z \geqq 4MZ$, $\cdots \geqq 2^nMZ$. Suppose n taken great enough, by the axiom of Eudoxus, so that P_n falls outside B. Then in $\triangle ZD_nP_n$, the line BC cannot cut P_nD_n (Eucl. I, 27) and so, by the axiom of Pasch, must cut ZD_n, as desired.

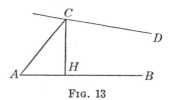

FIG. 13

The proof of the main theorem Sacch. I, 13, now runs (see Fig. 13): Suppose $\angle A + \angle ACD < 2$ rt. $\angle$'s, with $\angle BAC$ acute. Draw $CH \perp AB$. Then in $\triangle ACH$, $\angle A + \angle ACH + \angle H \geqq 2$ rt. $\angle$'s (Sacch. I, 9). But $\angle A + \angle ACH + \angle HCD < 2$ rt. $\angle$'s, by hypothesis. Thus $\angle H > \angle HCD$ = acute, so that, by Sacch. I, 11, 12, the lines AB and CD intersect as demanded by the Fifth Postulate.

From this main theorem of Saccheri we can at once dismiss *Hyp. Obt.* as self-contradictory. For it implies the Fifth Postulate, which in turn implies the whole of Euclid's geometry, including *Hyp. Rect.*, which is inconsistent with *Hyp. Obt.*

In order to prove the Fifth Postulate true in every case, Saccheri now turns in vigorous fashion to the (hopeless) task of demolishing *Hyp. Acut.* Though he finds many interesting theorems like the one proved above about the defect of a triangle, still none of them produce the desired self-contradiction. They are, in fact, theorems of non-Euclidean (plane) geometry, to which we now proceed.

NON-EUCLIDEAN (HYPERBOLIC) GEOMETRY

The search for a proof of the Fifth Postulate, with publications on the subject by several hundred mathematicians, had now gone on for more than two thousand years. The notion of its necessarily being hopeless occurred, at about the same time, to three mathematicians, Gauss (1777–1855), Bolyai (1802–1860) and Lobachevski (1793–1856). They systematically developed the consequences of introducing a postulate equivalent to Saccheri's *Hyp. Acut.*, namely,

NON-EUCLIDEAN POSTULATE: *through a given point, not on a given line, more than one line can be drawn not intersecting the given line,*

a postulate which they were the first to regard, not as leading to a contradiction, but as enabling them to build up a consistent geometry. Thus in Figure 14, let *PD* and *PB* be two lines through *P* not intersecting the

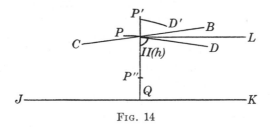

FIG. 14

line *JK*. Draw $PQ \perp JK$ and suppose *PD* rotated, if necessary, about *P* until the $\angle QPD$ is as small as possible (that is, so that if rotated further, *PD* will intersect *JK*). Similarly let $\angle QPB$ be as large as possible, and produce *BP* to *C*. Then no line, *e.g.* $PL \perp PQ$, in the angle *DPB* can intersect *JK*, so that there are infinitely many lines through *P* not intersecting *JK*. For convenience we restrict the name "parallel" to the two lines *PD* and *PC* only.

Now we can see at once that our present non-Euclidean geometry (also called *hyperbolic* because of the similarity of the lines *PC* and *PD* to the asymptotes of a hyperbola) is much richer than Euclidean geometry, in the following respect, at least.

In both geometries there exists what may be called an "intrinsic" unit-angle, say a right angle, in terms of which every other angle can be measured. By this is meant that it is defined (as half a straight angle) by the very axioms. Nobody needs to preserve a right angle (or any other angle) at the Bureau of Standards in Washington, since a geometer or carpenter can easily construct one for himself. In fact, in both geometries, it is customary to take as unit-angle the $\pi/2$th part of a right angle, where $\pi = 3.14159 \cdots$ is the limit, as a circle shrinks to zero, of the ratio of circumference to diameter. (In hyperbolic geometry this ratio does not remain constant as the circle shrinks!)

But in Euclidean geometry there is no intrinsic *unit-length*. If the bronze yardstick at Westminster, England, is lost, the theorems of Euclidean geometry (though they tell us how to construct a right angle) cannot help us make a yardstick. The original of all yardsticks (was it the arm of an English king?) had nothing to do with the axioms of Euclid.

However, matters are far different in hyperbolic geometry. Here angles and line-segments are inseparably connected. In Figure 14, for example, the angle QPD, denoted by $II(h)$, is called the "angle of parallelism" for the segment $PQ = h$, and it is easy to prove that $II(h)$ decreases as h increases. In Figure 14, we need only show $\angle QPD > \angle QP'D'$, which is proved like Eucl. I, 16, since $\angle QPD$ is an exterior angle for the "improper" triangle $DPP'D'$. Thus to every angle $II(h)$ there is associated a line-segment h and conversely. Let us take as our "intrinsic" unit-length the segment associated, say, with half-a-right-angle. Then a unit-segment can be constructed at will. For, having bisected the right-angle LPP'' (see Fig. 14) with the line DP, we need only determine the point Q (a standard construction for ruler and compasses, given by both Bolyai and Lobachevski) such that, if KQ is drawn perpendicular to PQ, then KQ is parallel to PD. As a matter of fact, however, Bolyai and Lobachevski (Gauss did not publish on the subject) used for the angle $II(1)$, not $45°$, as we have done, but the slightly smaller angle defined by $\cot \frac{1}{2}II(1) = e$, where $e = \lim_{n\to\infty}\left(1 + \frac{1}{n}\right)^n = 2.71828$ $\cdots$ is the natural base for logarithms, so that $II(1) = 40°24'$ approximately. Then defining, *e.g.*, the *sine* of an angle as the limit of opposite-over-hypotenuse as the sides of the triangle shrink to zero (there are no similar triangles of unequal size in hyperbolic geometry), they showed that $\cot \frac{1}{2}II(h) = e^h$ for a segment of any length h.

As soon as we have defined an intrinsic unit-length in this way the interesting question arises: How many *yards* long is it? Since, to judge from daily experience, we would need to travel a very long way indeed before the angle of parallelism QPD in Figure 14 became perceptibly less than a right angle, to say nothing of its shrinking down to $40°24'$, we can expect the number of yards in the unit-segment to be very large. Let us examine the question more closely.

HYPERBOLIC GEOMETRY AND THE ACTUAL WORLD

Since it is possible in hyperbolic geometry to prove many striking theorems in which measurement is involved, *e.g.*, that the area of a triangle is proportional to its defect, we naturally wonder whether the physicists and astronomers can prove, or disprove, the new geometry by observation of the world around us. The two most famous attempts proceeded as follows.

Gauss measured the largest convenient triangle he could find, formed by the peaks of three German mountains, *Broken, Hohehagen* and *Inselberg*, whose distance from one another is of the order of 100 miles. But the defect of this triangle, if it existed at all, was small enough to fall within the limits of error of measurement, and Gauss was forced to conclude that the triangle was too small to give evidence.

The other attempt, made by Lobachevski, was on a larger scale.

In Figure 15 let S be a fixed star and let E_1 and E_2 be two positions

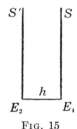

FIG. 15

of the earth six months apart. Then $h = E_1E_2 = 186,000,000$ miles is the diameter of the earth's orbit, and $\angle SE_2E_1 < \text{II}(h)$, with $\angle SE_2E_1$ approaching $\text{II}(h)$ for more distant stars. The apparent change of position of the star S, namely, $(\pi/2) - \angle SE_2E_1$, which we shall denote by $2p$, is called the (half-year) *parallax* of S. For many stars the angle $2p$ is directly measurable; *e.g.*, for Sirius, Lobachevski set $p = 1.24''$. So we have

$$\frac{\pi}{2} - 2p < \text{II}(h)$$

and therefore, by the preceding section,

$$\cot \frac{1}{2}\left(\frac{\pi}{2} - 2p\right) > \cot \frac{1}{2}\text{II}(h) = e^h.$$

But

$$\cot \frac{1}{2}\left(\frac{\pi}{2} - 2p\right) = \frac{1 + \tan p}{1 - \tan p}.$$

Thus

$$h < \log \frac{1 + \tan p}{1 - \tan p},$$

and, by Maclaurin's series,

$$\log \left(\frac{1 + \tan p}{1 - \tan p}\right) = 2(\tan p + \frac{1}{3} \tan^3 p + \frac{1}{5} \tan^5 p + \cdots).$$

Also

$$\tan 2p = \frac{2 \tan p}{1 - \tan^2 p} = 2(\tan p + \tan^3 p + \tan^5 p + \cdots)$$

so that $h < \tan 2p$. Thus, using the above information for Sirius, $h < \tan 2.48'' = 0.000006012 < \frac{1}{140,000}$, which means that our unit-length is at least 140,000 times as great as the diameter of the earth's orbit, a number which, with modern measurements of parallax, can be increased beyond 10,000,000. Or expressed in another way: To find a measurable defect, Gauss would have had to choose a triangle with sides many million times as great as the distance from the earth to the sun. So it is clear that for practical purposes the simpler geometry of Euclid will still serve on the surface of the earth.

Moreover, an apparently fundamental objection can be raised against applying hyperbolic geometry to astronomy or physics. It is most emphatically stated by Poincaré (Paris, 1903, Dover translation):

> If Lobachevski's geometry is true, the parallaxes of very distant stars will all be above a certain constant. If Riemann's is true [for Riemannian geometry, see below], they will be negative. These are results which seem within the reach of experiment, and it is hoped that astronomical observations may enable us to decide between the two geometries. But what we call a straight line in astronomy is simply the path of a ray of light. If, therefore, we were to discover negative parallaxes, or to prove that all parallaxes are higher than a certain limit, we should have a choice between two conclusions: we could give up Euclidean geometry, or modify the laws of optics, and suppose that light is not rigorously propagated in a straight line. It is needless to add that every one would look upon this solution as the more advantageous. Euclidean geometry, therefore, has nothing to fear from fresh experiments.

RIEMANN'S FINITE BUT UNBOUNDED SPACE.

So it would seem that non-Euclidean geometry deserved the neglect it received at the hands of physicists for a half-century or more after its discovery. Nevertheless, their attitude today is very different. It is now agreed, since the advent of Einstein's General Theory of Relativity (1915), that astronomical measurements support the non-Euclidean view of space; or, rather, that it is simpler to state the laws of physics in terms of non-Euclidean geometry than to make the changes in the laws of optics, gravitation, and so forth, necessary to rescue Euclidean geometry. Let us see how all this came about.

So far it would seem, from Saccheri's investigations, that there is only one type of non-Euclidean geometry, namely, the hyperbolic. This is true provided we accept all the other axioms of Euclid, denying only the Fifth Postulate. As we have seen above, these other axioms have a very cogent character; however, it was pointed out by Riemann, in his famous lecture *On the Hypotheses of Geometry* (1854) that one of them, namely, "a straight line is of infinite length," is in some ways of the same nature as the parallel postulate, there being again a reference to immense distances. It is true that we cannot imagine an "end" to a straight line, at a "boundary" with no space beyond it. As Riemann said: "The unboundedness of space possesses greater empirical certainty than any other external experience." But then he adds the significant remark: "However, its infinite extent by no means follows from this."

In other words, although we will not readily accept an axiom to the effect that a straight line cannot be produced beyond a certain point, still, may not the following situation arise? Let us proceed in any direction as far as we wish, by laying down yardsticks one after another. When we stop, how far will we be from home? In Euclidean or hyperbolic geometry, the answer is, as far as we have already proceeded. But in the course of our journey, may it not have happened that we passed very close to immense bodies of matter, and since after all our yardsticks are composed of matter, may it not be that these bodies have had such an effect on them that we may have arrived at a point from which we could get back by a shorter route? If we travel more than about twelve thousand miles along a great circle on the surface of the earth, the shortest way of returning home is to proceed straight ahead. In space also then, may there not be an upper bound to the distance we can get away from our starting point, no matter how often we lay down a yardstick? May it not be that a straight line, traversed in this way, can return upon itself? Or, putting it more vividly, may it not soon happen, with the improvements in modern telescopes, that an astronomer can see the back of his own head? If so, what will be the circumference, defined in this way, of the universe? It will depend, of course, on the actual distribution of matter in the universe, about which astronomers are able to supply some information. Provisional estimates indicate that the number of years taken for the light to get around from the back of the astronomer's head will be of the order of ten to fiftieth power.

Of course, in the older view, people would say that, in laying down our yardsticks, we were not "really" putting them in a straight line;

they were being warped by the neighboring matter, and as we saw above in the quotation from Poincaré, this is a possible interpretation. Certainly, however, the yardsticks would not appear warped to us, a phenomenon which the older people would interpret on the ground that we also were being warped, but which Riemann prefers to explain by saying that in the neighborhood of matter, the geometry of space is modified. In the absence of large bodies of matter (in this regard the earth is a microscopic speck), our measurements indicate that space is Euclidean, but as we approach a massive body, its presence is indicated by a growing non-Euclideanism of the geometry of the space.

By analogy with the geometry of the curved surface of the earth, Riemann spoke of this non-Euclideanism as "curvature," and considered the three-dimensional space of experience as having greater or smaller curvature at various points. The expression "curvature of space" is perhaps unfortunate, since it has given rise to a misunderstanding. The "curvature" of a line or surface is ordinarily thought of with reference to surrounding space, and some readers have therefore concluded that Riemann considered the space of experience to be immersed in a supposed surrounding space of four or more dimensions, their argument being that "if it is curved, it must have some place to curve in." But this notion is quite mistaken. Even if we were confined to the surface of the earth, we could still determine, by making measurements on it, that it is not a plane, and could then say, if we liked, that it is "curved." By saying that our three-dimensional space is "curved" Riemann means only that it is non-Euclidean.

CONSISTENCY OF THE NON-EUCLIDEAN GEOMETRIES

In the light of these remarks, let us return to Euclid's Fifth Postulate for the plane. Following Gauss, Lobachevski and Bolyai, we have seen that it can be replaced, with consistent results, by the postulate that through a point not on a given line there are at least two lines parallel to the given line. Let us now follow Riemann and replace Euclid's Fifth Postulate by the postulate that no two lines are ever parallel. What sort of geometry can we then build up? At first sight our new postulate is inconsistent with Eucl. I, 27, but, as we have seen above, the proof of Eucl. I, 27 depends on Eucl. I, 16, and reference to Figure 8 shows that if we admit the possibility of a straight line returning on itself, then when we mark off $EF = BE$, it is possible, if the triangle ABC is large enough, that F will come into a position (indicated by F' in the figure) which is close to B from the other side. Then $\angle ACD$ will no longer be greater than $\angle F'CE$ and the argument given by Euclid will no longer apply.

In some respects the new geometry is more complicated than hyperbolic geometry. For example, if two lines cannot be parallel, how many points of intersection will they have? Retaining Euclid's axiom (that there cannot be more than one) leads to consistent results, sometimes called *elliptic geometry proper*, while the assumption that there are two (natural in view of the behavior of great circles on the earth's surface, and in fact the only other possibility) leads to what is often called *spherical geometry* (of the plane). Again, in a space of "variable curvature" we would find that still more of the Euclidean axioms (*e.g.*, about congruent triangles) would become inapplicable, leading to a more general *Riemannian geometry*.

Instead of developing the details of these geometries, let us turn to a question of fundamental importance. The earlier investigators of an alternative for the Fifth Postulate were looking for a contradiction. But we have proceeded in quite a different spirit. We have produced a series of remarkable new theorems which, however startling they at first appear, have at least never contradicted one another. How can we prove to a person of the older persuasion that they never will?

One of the many possible answers runs as follows. If there is a contradiction in Riemann's spherical geometry on a plane, then there must be a contradiction in Euclidean solid geometry. For let us consider the points on the surface of a Euclidean sphere as the points on our Riemannian plane and the Euclidean great circles as Riemannian straight lines. This system then forms a model for Riemannian geometry in the sense that all theorems proved for the Riemannian plane will have their counterparts on the surface of the sphere, and any logical contradiction for the Riemannian lines will involve the corresponding contradiction for the Euclidean great circles.

In this way, and in many similar ways, we can answer the age-old question: Can Euclid's Fifth Postulate be proved on the basis of his other axioms? The answer is: No!

Nowadays, as a result of these proofs of its logical consistency and its usefulness in the description of the external world, non-Euclidean geometry has received a great deal of attention. The realization that we are not compelled to regard space as being Euclidean in order to discuss it in a rational and profitable way has had profound philosophical effects. For centuries, the axioms of Euclid were regarded as in some sense "absolutely true" and his theorems, established by mathematical reasoning, were beyond doubt. Whatever uncertainty men might feel about the nature of God and things, here at least they felt secure. Nowadays, however, Euclid's axioms are regarded as being based on a certain amount of experimental evidence, gathered by our forefathers from their observations in what we may call the "middle distances," *i.e.*, neither the great stretches of astronomy nor the minute dimensions of the atom. In these other reaches of experience, which we are only now beginning to examine, the relations among mutual distances of points appear to be different from what is possible in Euclid's geometry. There are no longer any "self-evident truths," the axioms of Euclid being now only convenient bases for a certain well-known mathematical system. In its effect on the minds of modern man, this realization that we can have no absolute knowledge of nature, no knowledge that may not be contradicted by more refined observation, is comparable to the theories of Copernicus and Darwin.

PROJECTIVE GEOMETRY

During the long dispute about what our measurements would disclose in space, and whether they would support one or another of the non-Euclidean geometries, another type of geometry was being built up, at first sight completely different from any "geometry of measurement," and yet one about which Cayley (1821–1895) could say, "projective geometry includes all geometry." Let us briefly examine its origin and modern form with a view to discovering the meaning, and limitations, of Cayley's remark.

Projective geometry is often called "pure geometry" for the following reason. In early life we acquire two basic mathematical notions, namely, of the number of objects in a set (the fundamental notion of arithmetic, analysis, etc.), and secondly, of certain spatial relations (the subject-matter of geometry) about points and lines, such as whether certain points are on a straight line, which points separate which other points, and so forth. Now measurement is closely associated with number, namely, the number of times the yardstick is contained in the measured distance, and the modern "pure" geometer will have none of that. He wants to build up a geometry without the idea of distance, on the basis of points and lines only, and not of line-segments. In Euclid, many of the axioms, *e.g.*, about congruent triangles, refer to the length of a segment. In projective geometry, on the other hand, all the axioms, and therefore the theorems, will be of the form: If certain points lie on certain lines, then certain other points will also lie on certain lines. Examples will be given below.

The fundamental idea for this pure geometry came from the desire of Renaissance painters to produce a "visual" geometry. How do things really look, and how can they be represented on the plane of the drawing? For example, there will be no parallel lines, since such lines appear to the eye to converge. In contrast with Euclidean geometry, this is the geometry of vision rather than measurement; it represents parallel rails as they appear from a fixed point of view in contrast to what we would find by measuring them at various points. Thus at first, in keeping with its avowed artistic purpose, it seemed much more subjective than the "scientific" geometry of Euclid, and in fact it remained on a scarcely mathematical basis until the time of Desargues (1593–1662) and particularly Poncelet (1788–1867).

The first painter to use and discuss the new method (see below) appears to have been Brunelleschi (1425), and the first treatise on it is by Alberti (1435). The most famous painter among its early proponents was Leonardo da Vinci, and the most influential in disseminating its doctrines was Albrecht Dürer (1528), whose famous work, *St. Jerome in His Study*, is a very perspicuous example.

To understand the method, let us consider the problem of representing on canvas a long hallway with its floor covered with square tiles. In Figure 16 the canvas is in the plane of the paper, imagined vertical. The

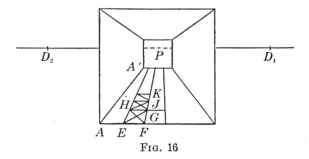

Fig. 16

eye of the artist is at a fixed point O, not shown in Figure 16 but such that PO extends at the right angles from the paper toward the eye of the reader. Then the drawing on the canvas is the *section of a projection* in the following sense. The set of lines drawn from the artist's eye to the various points of the object, which is in this case the hallway, constitute the *projection* of the object and are called the *Euclidean cone*. Then the *section* of this cone made by the canvas is the desired drawing. The horizontal plane through the artist's eye will intersect the canvas in the *vanishing-line* D_1PD_2 (see Fig. 16), so-called because the representation of points on the artist's horizon (imagining the hallway to extend infinitely far) will lie on this line. The point P, on the perpendicular from the eye to the canvas, is called the *principal vanishing point* and the points D_1 and D_2, at the same distance from P as the artist's eye, are the *diagonal vanishing points* (for their importance, see below). The point P, which may be chosen at will (the artist may stand where he likes), is usually the most important point in the picture. For example, in the *Last Supper* of da Vinci it is the head of Christ. We can then prove the

FUNDAMENTAL THEOREM: *the representation of any line perpendicular to the canvas must go through P.*

For consider any such line, *e.g.*, AA'. The planes OAA' and OAP must coincide, since both of them are perpendicular to the canvas and contain the two points O and A; and the representation of the line AA' is the line (passing through P) in which this plane cuts the canvas. Thus lines like AA', EE', DD', *etc.*, which in the hallway are parallel, are drawn to meet at P.

More generally: *parallel lines in the object converge in the picture to the point where the canvas is pierced by the line from the eye parallel to the given lines.*

How, then, are we to represent the square tiles of the floor? (See Fig. 16.) Once this question is settled, we can deal with any object, as Dürer explained, by covering it with a fine mesh of squares. The sides perpendicular to the canvas will give no trouble; they need only be connected to the principal vanishing point. But what about the spacing of the other two sides of the squares? In other words, how rapidly should lines like FG, GJ, JK in the object get shorter in the drawing? Alberti says that some people thought they ought to be in geometric progression, "which is a great mistake." He then gives the correct solution, which begins by joining ED_1. Then, by the above general theorem, with an angle of 45°, the point G is given by the intersection of ED_1 with FP.

THE GEOMETRY OF A GROUP

Now what is of interest here to the mathematician? Evidently the question: What geometric properties of the object are preserved in the drawing? Lengths and angles are changed, but something must be kept, since the drawing remains recognizable. Our new geometry will consist of the study of these invariant properties.

In order to have a plane geometry, let us confine our attention to two-dimensional objects, say the tiled floor of the hallway, and let us also project the drawing onto the floor. In other words, we make a sequence of drawings, each of them serving as the object for the next one, the position of the canvas and of the artist's eye being successively at our choice, except that in the last case the canvas must lie on the original floor. Imagining the floor to extend over the whole of its plane (extended by the addition of the so-called line-at-infinity, namely, the artist's horizon, which becomes the vanishing-line in the drawing), we may then interpret our final drawing as a point-by-point transformation of the (extended) plane into itself, each point being moved to a well-defined other point (or perhaps the same point). Such a transformation of its points is called a *projective transformation* (or *mapping*) of the plane; and the plane itself, consisting of a set of points and straight lines such that every two points determine exactly one line and every two lines intersect in exactly one point (by adding the line at infinity we have provided for the intersection of parallels) is called a *projective plane*.

Then the class of all projective transformations will have these three properties:

(i) *the product of two elements of the class (i.e., the transformation obtained by carrying out two projective transformations in succession) is itself an element of the class.*

(ii) *the inverse of every element of the class (i.e., the mapping of the drawing back into its original) is in the class.*

(iii) *the identity-mapping (in which every point is mapped into itself) is in the class.*

A class of transformations meeting these three conditions is called a *group* and any geometric properties of a figure that are preserved under all the transformations of a group are called *invariants* of the group or are said to *belong* to it.

Now consider any theorem *i.e.*, any statement of the form: If a given figure has property A, then it also has property B. If property A belongs (in the above sense) to a given group of transformations, then the theorem is also said to belong to the group, and the study of all such theorems is called the *geometry of the group*. Thus projective geometry is simply the geometry of the projective group, or the geometry of projective properties. Let us consider some examples.

Does the property of being a circle belong to projective geometry? No, because it is defined in terms of distance from a center, and distance is not a projective invariant. A circle will not in general remain a circle under projection. In fact, by considering it as the section of a cone, we see that it can be projected into any other conic section. What about an ellipse? No, because it can be projected into a parabola or an hyperbola. But a conic section? Yes, since any projection of it is also a conic section (compare the projection and section of the Euclidean cone mentioned above). What about, say, a hexagon? Yes. A regular hexagon? No.

A good example of a projective theorem about conics (proved by Pascal in 1640, at age 16) is

PASCAL'S THEOREM: *the necessary and sufficient condition that a conic can be drawn through the six vertices of a given hexagon is that the three points of intersection of pairs of opposite sides be collinear,*

and a good example of a theorem that is not a projective theorem is

EUCLIDEAN THEOREM: *the necessary and sufficient condition that a circle can be drawn through the four vertices of a given quadrilateral is that its opposite angles be supplementary.*

CLASSIFICATION OF GEOMETRIES

Let us examine again the projective group of transformations. It has *subgroups*, that is, subsets of transformations meeting all the above conditions for a group. As an example, consider the transformations (called *affine*) that leave one line invariant in the sense that the map of any point of this specialized line is also a point on the line. Visualizing the specialized line as the line-at-infinity on the floor, let us also say that two lines are *parallel* if their intersection lies on the specialized line. By thus restricting the projective group we produce a *subgeometry*, called *affine geometry*, which is richer in special concepts than its parent. For example, the three types of conic sections, ellipse, parabola, hyperbola, can now be distinguished as meeting the specialized line in no point, one point, or two points, respectively, and any theorem that deals with affinely invariant properties, *e.g.*, that all chords through the center of an ellipse are bisected there, is an *affine theorem*. Apollonius may be called the greatest affine geometer that ever lived.

Euclidean geometry is now obtained, as a subgeometry of affine geometry, by further restricting the group of projective transformations in such a way as to introduce *perpendicularity*. To do this, we define an *involution* as a non-identical transformation which, when performed twice in succession, sends every point back into itself. Consider the transformations induced on the specialized line by affine transformations of the plane. Among them will occur involutions without a fixed point, arranging the points on the specialized line in pairs such that each point of a pair is sent by the involution into the other point. We now select at will any such involution on the specialized line of an affine plane, and say that two lines in the Euclidean plane (*i.e.*, the affine plane with the specialized line omitted) are *perpendicular* if their points of intersection with the specialized line are a pair in the chosen involution.

Then a *Euclidean transformation* (also called a *rigid displacement*) is defined as an affine transformation preserving perpendicularity. If we say that a segment AB is *congruent* to a segment CD when there exists a Euclidean transformation sending AB into CD, the usual theorems about congruent triangles (*i.e.*, Eucl. I, 4, *etc.*) will be found to hold. Moreover, the general theory of projective geometry can now be used to prove Euclidean theorems. For example, the theorem that the altitudes of a triangle are concurrent becomes a theorem about involutions.

As a further illustration of subgeometries, let us choose, not a specialized straight line, but a specialized conic, say an ellipse, and consider only the subgroup of projective transformations that send points on the conic into points on the conic and send interior points into interior points. Let us call such transformations *hyperbolic displacements* (for the choice of name, see below) and call the interior of the conic (*i.e.*, part of the projective plane) a *hyperbolic plane*. If we then say that two segments AB and CD are congruent when there exists a hyperbolic displacement carrying AB into CD (compare the above superposition in Euclidean geometry), we thereby have a model of the hyperbolic geometry of Bolyai and Lobachevski (see Fig. 17). Note that its lines are of infinite length; *i.e.*, if PQ is a unit segment, we can take $PQ = QQ_1 = Q_1Q_2 = \ldots$ without ever reaching the edge of the conic; and note also that there are infinitely many lines through R not intersecting PQ.

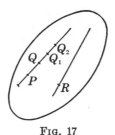

FIG. 17

These examples will illustrate how geometry as a whole has been unified by the notion, introduced by Klein in 1872, of the *geometry of a group*. Also we can see what was meant by Cayley's exclamation, quoted above, that "projective geometry is all geometry." But nowadays it is realized that Cayley was somewhat over-enthusiastic. It is true that the fundamental idea of modern geometry is that of a "mapping," of which projective transformations give an important illustration, but the concept of mapping is much more inclusive. For example, the maps of Mercator (1512–1594) and others are transformations of points from hemisphere to plane, and their properties, studied especially by Gauss in his important *General Investigations of Curved Surfaces* (1822), cannot profitably be fitted into the framework of projective geometry.

Again, the group of transformations of greatest importance in present-day mathematics, namely, the *group of topological transformations*, is far wider than the projective group. Here we are dealing with a *topological space;* that is, with a set of elements, call them *points*, for which the concept of a *neighborhood* is defined. In a Euclidean plane, for example, a neighborhood of a point may be taken to consist of any set of points containing a circle centered on the given point. Then any transformation preserving neighborhoods (*i.e.*, such that if A goes into A', then every neighborhood of A goes into a neighborhood of A') is called *topological*.

Topology may be visualized as rubber-sheet geometry, since a topological transformation permits any amount of stretching or compressing (without tearing) of the plane, and it might seem that under such freedom of transformation there could be no interesting theorems, *i.e.*, no interesting properties could be preserved. But consider, for example, the famous theorem, due to Euler (1707–1783) and belonging really to the pre-history of the subject, that it is impossible to *traverse* the seven Konigsberg bridges (*i.e.*, to cross each of them exactly once, see Fig. 18),

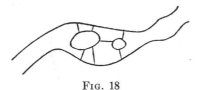

FIG. 18

which is certainly a topological theorem, since the figure will remain untraversable after any topological transformation. Modern topology, discussed in Chapter X, owes its great importance, especially in its applications to other branches of mathematics, to the fact that topological spaces can be of a very general nature. With this brief notice of the "most general of all geometries," if indeed topology may not be said to have outgrown even the name of geometry, we close the present review of some of the various possible types of geometry. Our purpose throughout has been to discuss those aspects of the subject that are of greatest interest for the study of Euclidean plane geometry in the high school.

BIBLIOGRAPHY

1. BONOLA, R. *Non-Euclidean Geometry*. Chicago: Open Court Publishing Co., 1912.
2. COOLIDGE, J. *A History of Geometrical Methods*. New York: Oxford University Press, 1940.
3. COURANT, RICHARD, and ROBBINS, HERBERT. *What Is Mathematics?* New York: Oxford University Press, 1941.
4. COXETER, H. S. M. *The Real Projective Plane*. New York: McGraw-Hill Book Co., 1949.
5. HALSTEAD, G. B. *Saccheri's Euclides Vindicatus*. Chicago: Open Court Publishing Co., 1920.
6. HEATH, T. L. *A History of Greek Mathematics*. Vols. I and II. New York: Oxford University Press, 1921.
7. HEATH, T. L. *Euclid's Elements*. Vols. I, II, III. Cambridge: Cambridge University Press, 1926.
8. KLINE, M. *Mathematics in Western Culture*. New York: Oxford University Press, 1953.
9. KLINE, M. "Projective Geometry." *Scientific American*, 192:80; January 1955.
10. LeCORBELLIER, P. "The Curvature of Space." *Scientific American*, 191:80; November 1954.
11. MESERVE, B. *Fundamental Concepts of Geometry*. Cambridge, Mass.: Addison-Wesley Publishing Company, 1955.
12. NEUGEBAUER, O. *The Exact Sciences in Antiquity*. Princeton, N. J.: Princeton University Press, 1952.
13. POINCARÉ, H. *Science and Hypothesis*. New York: Dover Publications, 1952.
14. ROBINSON, G. DEB. *The Foundations of Geometry*. Toronto: University of Toronto Press, 1940.
15. THOMAS, I. *Greek Mathematical Works*. Vols. I and II. Cambridge, Mass.: Loeb Classical Library, Harvard University, 1941.
16. VAN DER WAERDEN, B. L. *Science Awakening*. Groningen: P. Noordhoff, Ltd., 1954.
17. VEBLEN, O., AND YOUNG, J. W. *Projective Geometry*. Vols. I and II. Boston: Ginn and Company, 1910 and 1918.
18. WHITTAKER, E. T. *From Euclid to Eddington*. Cambridge: Cambridge University Press, 1949.
19. WOLFE, H. E. *Introduction to Non-Euclidean Geometry*. New York: The Dryden Press, 1948.
20. YOUNG, J. W. *Projective Geometry*. Carus Monograph No. 4, 1930. Chicago: Open Court Publishing Co.

X

Point Set Topology

R. H. BING

ONE who gains his knowledge of topology by listening to popular lectures and reading entertaining expository articles about it such as appear in some popular magazines and elementary texts may get the erroneous impression that topology is a brand of recreational mathematics. If such a person were to take a course in topology expecting it to consist of cutting out pretty figures and stretching rubber sheets, he would be in for a rude awakening. However, if he pursued the study further, he might be delighted to find that it is a subject of much substance and beauty. Of the references given at the end of this chapter, Newman's book, is the most pertinent to the material covered in this chapter. Some advanced texts on subjects other than topology have a chapter treating the point set topology used in their subject. Some of these treatments are very readable—for example, Chapter 1 of *Introduction to Measure and Integration* by M. E. Munroe.

Topology is a relatively new branch of mathematics. Those who took graduate training in mathematics thirty years ago did not have the opportunity to take a course in topology at many schools. Others had the opportunity but passed it by, thinking that topology was one of those "new-fangled" things that was not here to stay. In that respect it was like the automobile.

It is the purpose of this chapter to treat topology less rigorously than is done in most graduate texts on the subject. However, the treatment will not be entirely frivolous and the reader should on occasions get a glimpse of topology as it is frequently taught in many college courses.

Topology as an offshoot of geometry. In many respects topology may be considered as an offshoot of geometry.

From Euclid we have certain axioms for geometry. On the basis of some undefined terms, definitions, and axioms, certain theorems are proved. So it is with many treatments of topology. The undefined terms, definitions, and axioms that are used are different from those employed in geometry but there is an analogy in the synthetic methods used in the two courses. This schematic diagram shows the approach:

$$\left.\begin{array}{l} \textit{Undefined terms} \\ \quad \textit{(points, neighborhoods, ...)} \\ \textit{Definitions} \\ \quad \textit{(limit point, closed, ...)} \\ \textit{Axioms} \end{array}\right\} \Rightarrow \qquad \textit{Theorems.}$$

The prospective geometry teacher who wants to broaden his outlook on geometry by studying a related branch of mathematics might well choose an elementary course in topology. A teacher's knowledge of the subject should extend well beyond the material in the course. He does not want to be like the boy who fell out of bed—he went to sleep too near where he got in.

In studies of the foundations of geometry, one may consider modifying certain of the axioms and seeing how this affects the resulting theorems. Hilbert studied the effects of changing certain of Euclid's axioms. Veblen made further studies in this respect.

One fundamental idea from plane geometry is the notion of betweenness. We use it to develop the notion of straightness. If this notion were modified in a certain way, the two objects shown in Figure 1 would be alike. The first is a triangle and the second is a more general simple closed curve. In topology they are considered alike.

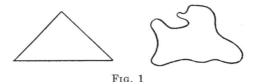

Fig. 1

In some respects there are as many, if not more, simple closed curves in the plane as there are triangles. In fact, if there were some way of selecting a simple closed curve at random from the plane, one might consider it odd indeed if the one chosen even contained a straight line segment. It does not take a great deal of imagination to see how a person that was seeking to learn about all kinds of figures in the plane would be led to a study of simple closed curves.

Studying this chapter. A person might be considered as having a superficial knowledge of geometry if he acquired it merely by reading about geometry but without actually proving theorems and working problems. Accordingly, only a shallow understanding of topology is gained by listening to discussions and reading alone, no matter how stimulating the discussions and reading might be.

Some readers may want merely to read the introduction and then skim through this chapter to get a scanty notion of some topological concepts. Others may be willing to read it all. It is hoped that there will be the occasional reader who will want to do some pondering and deep thinking such as is done by the students who get a joy out of solving hard problems and proving difficult theorems.

Some problems are included in this chapter to deepen the serious reader's understanding—it is not expected that these exercises will be used by most readers. However, one who sets for himself the task of mastering the chapter may find it advantageous to spend part of each period of study in working on the exercises. Answers in rather concise form are found at the end of the chapter.

TOPOLOGICAL EQUIVALENCE

We mentioned in the first section of this chapter that a triangle and a "wiggly" simple closed curve are alike in some respects. We say that they are topologically equivalent. A circle, a square, an ellipse, and a triangle are all called simple closed curves.

Topological equivalence. Two sets A, B are topologically equivalent if there is a 1–1 correspondence between them that is continuous both ways.

The curved arc A and the straight line segment B shown in Figure 2

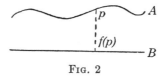

Fig. 2

are topologically equivalent. Although there are many 1–1 correspondences that could be chosen, the projection of A onto B provides a convenient one—that is, a point p of A corresponds to the point $f(p)$ of B directly beneath it.

The 1–1 correspondence from A to B above is continuous because points close together in A go into points that are close together in B. The 1–1 correspondence is continuous the other way because points close together in B correspond to points close together in A. The expression "close together" is not precise and we give a better explanation of continuity on pages 318–21 of this chapter.

A 1–1 correspondence of a set A onto a set B that is continuous both ways is called a *homeomorphism* of A onto B. The projection of A onto B is an example of a homeomorphism of A onto B. A homeomorphism of one set onto another set is a special type of transformation and will be discussed further in later sections.

Two sets are *topologically equivalent* if there is a homeomorphism of one onto the other. Accordingly, instead of saying that the two sets are topologically equivalent, we may say that they are *homeomorphic*. Each set is topologically equivalent or homeomorphic to itself because each point of the set may be made to correspond to itself.

It is possible to get a homeomorphism of a short segment I_1 onto a long segment I_2. (See Fig. 3.) The homeomorphism this time is given

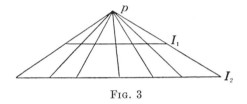

FIG. 3

by a projection from a point. This homeomorphism shows that in a certain sense, there are as many points on a short segment as on a long segment.

We mention some other objects that are topologically equivalent or homeomorphic—the surface of a tetrahedron and the surface of a sphere; a teacup and a doughnut; the x-axis and the graph of $y = x^2$; the half-open segment[1] $(0, 1]$ and the graph of $y = 1/x$ $(0 < x \leqq 1)$. (We suppose that a segment contains its endpoints unless we state otherwise.)

In Figure 4 we show some objects no two of which are homeomorphic. For example, the circle is not topologically equivalent to the segment because no point of it could be made to correspond (in the proper way) to the end of the segment.

A misconception about topological equivalence. One might suppose that if two sets are topologically equivalent, it is possible to deform one onto the other by "pulling and stretching but without breaking and tearing." We give some counterexamples to show that this is not always possible.

In E^3 (Euclidean 3-space—the familiar three-dimensional space of everyday experience) consider two sets each consisting of two tangent spheres; in the first set the spheres are tangent externally and in the

---

[1] By "half-open segment" is meant a segment which contains only one of its endpoints. For example, the half-open segment $(0, 1]$ contains the point 1 but not 0.

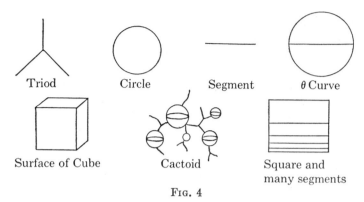

| Triod | Circle | Segment | θ Curve |

| Surface of Cube | Cactoid | Square and many segments |

Fig. 4

second set one sphere lies on the interior (except for the point of tangency) of the other. The two sets are topologically equivalent because there is a 1–1 correspondence of the proper sort between the two sets. However, it is not possible to deform one set onto the other in E^3 by "pulling and stretching but without breaking and tearing."

Let us consider another counterexample to the faulty "pulling and stretching" definition. One can form a cylinder or tube by sewing two opposite sides of a rectangular rubber sheet together. If the ends of the cylinder are sewn together, the resulting figure may resemble an inner tube. However, if a knot is tied in the cylinder before its ends are united, the exterior view of the inner tube will be changed but its interior structure will not be. If after the knot is tied and before the final sewing is done, one end of the cylinder is stretched and pulled over the knot before the ends are connected, another figure results. The three surfaces shown in Figure 5 are topologically equivalent but not one of them can be stretched to make it fit on any other one of them.

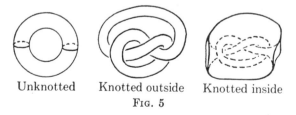

| Unknotted | Knotted outside | Knotted inside |

Fig. 5

If one figure can be deformed onto another by pulling and stretching but without breaking and tearing, the figures are topologically equivalent; the converse of this statement is not true. We repeat that two sets are topologically equivalent if there is a 1–1 correspondence between them that is continuous both ways. The definition makes no mention of pulling, stretching, and deforming.

The *complement* of a figure is the set of points not in the figure. (Complement of A in space $S = S - A$.) For example, on the surface of the earth, the surface of the oceans, seas, lakes, etc., is the complement of the surface of the land. In the plane, the complement of a circle is the sum of its interior and its exterior. The complement of a circle in E^3 would be all in one piece.

We now show that even though two figures in the same space are topologically equivalent, their complements may be topologically different.

In Figure 5, each of the surfaces has an interior and an exterior. It could be shown that the interior of the first is topologically equivalent to the interior of the second but not to the interior of the third. A small creature who knew about topological equivalence but not about straightness, length, etc., could not tell the difference between the inside of the first tube and the interior of the second. However, the exterior of the first tube in Figure 5 is like the exterior of the third but not like the exterior of the second. No two of the three figures have complements that are topologically equivalent.

J. W. Alexander described the *horned sphere* illustrated in Figure 6.

Fig. 6

It is topologically equivalent to the surface of a sphere but its exterior is not like the exterior of a sphere.

We might describe the horned sphere as follows. A long cylinder closed at both ends is folded until the ends are near each other and parallel. (In the figure we show only the ends of the closed cylinder.) Then tubes are pushed out each end until they hook as shown. The process is continued by pushing out additional tubes, etc. The resulting set has the property that although it is topologically equivalent to the surface of a sphere, there is a circle in the exterior of the horned sphere that cannot be shrunk in the exterior of the horned sphere to a point.

Although a horned sphere is topologically equivalent to a sphere, the complement of a horned sphere is not like the complement of a sphere because each simple closed curve in the complement of a sphere can be shrunk to a point without touching the sphere.

Exercise 1. Examine the keyboard of a typewriter and find which capital letters are topologically different from all others. On some keyboards the letters are formed without serifs as follows: A, B, C, D, E, F, G, H, I, J, K, L, M, N, O, P, Q, R, S, T, U, V, W, X, Y, Z.

THE JORDAN CURVE THEOREM

The interiors of circles and triangles appear to be somewhat alike. However, if one has a complicated simple closed curve such as shown in Figure 7, one might have difficulty in deciding if its interior was like

Fig. 7

the interior of a circle. If the curve had been even more complicated, it might have been impossible to tell by merely looking whether or not it even had an interior.

The Jordan curve theorem states that the complement of each simple closed curve J is the sum of two mutually separated connected pieces and is the boundary of each of them. (See pages 321–24 of this chapter for a discussion of the meaning of such terms as mutually separated, connected, boundary.) One extension of the Jordan curve theorem due to Schoenflies may be stated as follows:

An extension of the Jordan curve theorem. If J_1 and J_2 are simple closed curves in the plane, there is a homeomorphism of the plane onto itself that takes J_1 onto J_2.

In plane analytical geometry one studies translations and rotations of the plane onto itself—and perhaps even reflections through a line. These are examples of homeomorphisms of the plane onto itself. Under each of these particular homeomorphisms, a figure would go into a figure congruent to itself. However, the types of homeomorphisms used to show the truth of the above extension of the Jordan curve theorem would frequently change the shapes of objects.

To those who think the theorem needs no proof since it is intuitively obvious, it may come as a surprise that the analogous theorem is not true in 3-space. Although the horned sphere described in the last section is topologically like the surface of a sphere, there is no homeomorphism of 3-space onto itself that will take one onto the other.

The Jordan curve theorem is one of the more frequently used theorems of plane topology; it is used in such subjects as analysis or complex variables. Although several proofs of this theorem have been given—and some only recently—these proofs are still complicated.

In this section, we only treat the Jordan curve theorem where the simple closed curves involved are uncomplicated. However, if the reader works through even the first of the exercises, he will get some understanding of the meaning of the theorem.

Exercise 2. Describe a homeomorphism of the plane onto itself that takes a given triangle onto a given circle. See Figure 8 for a hint as how to proceed.

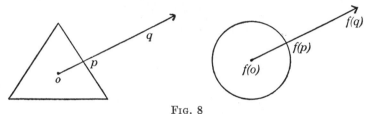

FIG. 8

Exercise 3. Describe a homeomorphism of the plane onto itself that takes a given triangle onto the graph of $(y - \sqrt{1 - x^2})(2y - \sqrt{1 - x^2}) = 0, (-1 \leq x \leq 1)$.

By using mathematical induction on the number of sides, it can be shown that for any polygon, there is a homeomorphism of the plane onto itself that takes the polygon onto a triangle. One without training in topology who obtained an airtight proof of such a theorem could be proud of his accomplishment. Indeed, it is no easy task for the beginner to even get Exercise 4. Some of the methods used in working the following exercise can be used to get the more general result.

Exercise 4. Describe a homeomorphism of the plane onto itself that takes polygon $ABCDEF$ of Figure 9 onto a triangle. See Figure 10 for a hint as how to proceed.

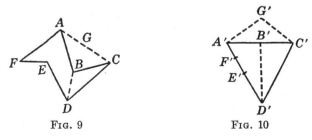

FIG. 9   FIG. 10

Some unsolved problems. There are many unsolved problems in mathematics. It was not known for many years whether or not the Jordan curve theorem was true—though during part of this time many people suspected it to be true. The four color problem is still unsolved. It asks if each map on the surface of a sphere can be colored with four colors if the boundaries of the states are simple closed curves. States whose boundaries meet at a point are allowed to have the same color, but states whose boundaries meet along an arc are not allowed to have the same color. If some of the states had boundaries that contained part of the graph of $y = \sin 1/x$, we might have a situation such as depicted in Figure 11 so it is essential to impose the condition that the boundaries of the states be simple closed curves.

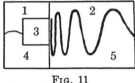

Fig. 11

Here is an unsolved problem of perhaps more importance and complexity. Research topologists have been learning many interesting things about 3-space in the last few years so the time may be ripe for its solution. The horned sphere provides us with a counterexample to the conjecture that if S is a set in 3-space topologically equivalent to the surface of a sphere, there is a homeomorphism of 3-space onto itself that takes S onto the surface of a sphere. What restrictions could be imposed on S to insure that there would be such a homeomorphism? Some topologists suspect that there is such a homeomorphism if the complement of S is simply connected and locally simply connected—but these are concepts that we shall not discuss here.

LIMIT POINTS

The notion of a limit is a very important one in mathematics. Chapter VII of this book discusses some aspects of limits. A student may do well in arithmetic, algebra, and even geometry (the notion of a limit was used more extensively here before the dilution of the curriculum) without understanding limits, but he must learn this concept in order to go far in mathematics. Although limits are used extensively in calculus, some students do the mechanical parts of this subject without gaining an insight into the concept. Limits are used so extensively in point set topology that it is inconceivable that a person could make much headway here without learning a considerable amount about limits. Learning about limit points is a good starting point.

We use the notion of a limit to tell us whether a point is "infinitely close" to a set or whether it "sticks onto" a set. In such a nice space as the Euclidean plane we can use the idea of distance to define limit points. In the more general spaces such as discussed on pages 327–31 of this chapter, we use neighborhoods to define limit points.

Neighborhood. A *neighborhood* in the Euclidean plane is the interior of a circle. If a point lies on the interior of a circle, this interior is called a neighborhood of the point. Each point has many neighborhoods—some small and some large. Each neighborhood is a neighborhood of many points—in fact, of each point in it.

A neighborhood in Euclidean 3-space is the interior of a sphere; in Euclidean 1-space (or a line) it is an open segment (interval without its end points)—in fact, in any space with a distance, it is the interior of a generalized sphere. In more general spaces the neighborhoods are certain designated point sets that satisfy certain conditions or axioms for these abstract spaces.

Neighborhood definition of limit point. The point p is a *limit point* of the set X if each neighborhood containing p contains a point other than p of X.

Distance definition of limit point. The point p is a *limit point* of the set X if for each positive number ϵ there is a point x of X such that $0 < \rho(p, x) < \epsilon$. (We use $\rho(p, x)$ to denote the distance from p to x.) This definition is not applicable in abstract spaces without a distance function.

Exercise 5. Suppose that, in the plane, X is the set of all points satisfying the conditions $0 < x^2 + y^2 < 1$ or $x = y = 1$.
Which of the following points are points or limit points of X? $p = (0, 0)$, $q = (0, 1), r = (1, 1), s = (2, 0), t = (0, \frac{1}{2})$.
Exercise 6. In an abstract space whose points are the points of the plane and whose neighborhoods are horizontal lines, which of the points p, q, r, s, t are limit points of the set X of Exercise 5?

Sets. Sets are discussed in Chapter III. A *point set* is a collection each of whose elements is a point. The collection of all points with a particular property P is designated by $\{p/p$ has property $P\}$. A point set with only one point p in it is indicated by $\{p\}$.

Subset. We say that A is a *subset* of B if each point of A is a point of B. We write $A \subset B$. If p is an element of A, we write $p \, \epsilon \, A$.

Sum or union. The *sum* of A and B $(A + B$ or $A \cup B)$ is the set of all points in either A or B—that is, $A + B = \{p/p \, \epsilon \, A$ or $p \, \epsilon \, B\}$.

Intersection, product or common part. The *intersection* of A and B

$(A \cdot B, A \cap B$, or AB) is the set of all points in both A and B—that is, $A \cdot B = \{p/p \; \epsilon \; A \text{ and } p \; \epsilon \; B\}$. If there is no point which is in both A and B we say that A does not intersect B or $A \cdot B = 0$.

Difference. We use $A - B$ to denote the collection of points of A that do not belong to B—that is, $A - B = \{p/p \; \epsilon \; A \text{ and } p \; \not\epsilon \; B\}$. If $A \subset B$, we write $A - B = 0$.

If a person were to squirt some black ink on the plane, the set of points A in the dark spot is an example of a point set. Suppose a set B is determined by squirting some red ink on the plane. Then $A + B$ designates the set of points covered by ink, $A \cdot B$ designates the set covered by both kinds of ink, and $A - B$ designates those covered by black but not red ink. (See Fig. 12.)

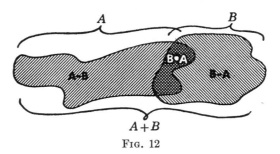

$$A + B$$

F_IG. 12

Some people use the artificial convention that a collection need have no elements. This frequently leads to complications in having to make exceptions for the null set, as this artificial thing is called. On the other hand, the convention may have some advantages in the algebra of sets (Boolean algebra) for then any two sets always have a product and difference. More topologists than algebraists criticize the convention.

Open set or domain. A point set D is an *open set* if for each point p of D there is a neighborhood of p that lies in D. An open set is also called a *domain.* The intersection of a set X and an open set is called an *open subset* of X.

An open set is the sum of neighborhoods, although they may be infinite in number. For the Euclidean plane, the interior of a square is a domain but a disk (circle plus its interior) is not.

Closed set. A *closed set* is one which contains all its limit points. Each of its limit points belongs to it. A line is closed in the plane.

The *closure* of a set X is the sum of X and all its limit points. This closure is designated by $\bar{X}$. The closure of the set X of Exercise 5 is the sum of a disk and $\{r\}$. The closure of the set X of Exercise 6 is the set of all the points above the line $y = -1$ and on or below the line $y = 1$.

Exercise 7. For the Euclidean plane give an example of (a) a closed set, (b) an open set, (c) a set neither open nor closed, and (d) a set that is both open and closed.

Exercise 8. What would need to be proved to show that the closure of each set is closed?

Exercise 9. If, in the plane, G is graph of $y = \sin 1/x, 0 < x < 1/\pi$, what is $\bar{G}$?

It is possible to prove each of the following theorems from the definitions alone without even knowing what a neighborhood looks like:

1. The sum of two open sets is open.
2. The product of two closed sets is closed.
3. The complement of a closed set is open.
4. The complement of an open set is closed.
5. The closure of each set is closed.

The reader may gain experience by proving these from the definitions. The first two results may be generalized to say that the sum of any collection (perhaps infinite) of open sets is open and the product of any collection of closed sets is closed.

TRANSFORMATIONS

Ordinarily, a person regards a transformation as a change. A seed becomes a tree, a baby becomes an adult, wood turns to smoke and ashes, a bomb changes to fragments, a wad of gum is squeezed, one object is transformed into another.

In mathematics a person may come to regard a transformation as a rule or law for associating points of one set X with points of another set Y. If f denotes the transformation and $x \in X$, we use $f(x)$ to denote the point of Y associated with x. (See Fig. 13.)

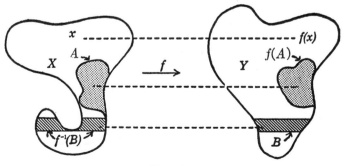

Fig. 13

We call $f(x)$ the image of x under f. If A is a subset of X, $f(A)$ denotes the collection of images of points of A. If B is a subset of Y, $f^{-1}(B)$ is the set of all points of X that go under f into a point of B; $f^{-1}(B)$ is called the *inverse* of B.

A useful and more sophisticated approach is to define a transformation of X into Y to be a collection of ordered pairs (x, y) such that $x \in X$, $y \in Y$, and each element of X is the first element of one and only one of the ordered pairs. For example, if one considers the equation $y = x^2$, the transformation might be regarded as the set of all ordered pairs of the form (a, a^2). The pairs would be the coordinates of points on a parabola.

We shall regard a transformation f of X into Y as a single valued function that assigns one and only one value of Y to each value of X. We leave it to those who define function to decide if it is a change, a rule, a collection of ordered pairs, or something else.

If each element of Y is the image of an element of X under f, we say that f takes X *onto* Y instead of merely into Y. Hence, of the two transformations of the reals into the reals given by $y = x^2$ and $y = x^3$, the second is an onto transformation but the other one is not.

If no element of Y is the image of two elements of X, we say that f is *one-to-one* or 1–1. Hence $y = \sin x$ is 1–1 for $0 < x < \pi/2$ but not for $0 < x < \pi$.

Intuitively, we say that a transformation f is continuous if points "close together" in X go into points "close together" in Y. The notion "close together" is not precise so we give the following definition:

Neighborhood definition of continuous. A transformation f of X into Y is *continuous at the point* x of X if for each neighborhood U of $f(x)$ there is a neighborhood V of x such that $f(V \cdot X) \subset U$. (See Fig. 14.)

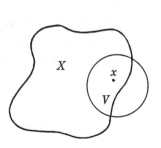
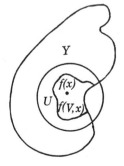

Fig. 14

The size of V in the above definition is a function of x and U—if U is taken to be small, V may need to be chosen as small. However, Example 8 below shows that V may not need to be small even if U is.

If the set X on which f is defined is all of space, the definition may be simplified by replacing "$f(V \cdot X) \subset U$" by "$f(V) \subset U$." We could make this change anyway if we called V an open subset of X containing x rather than a neighborhood of x.

A *function* is *continuous* if it is continuous at each point of its domain of definition X. (A domain of definition may not be a domain.)

For metric spaces (Euclidean spaces and other spaces with distance functions), we may use the following definition:

ϵ-δ *definition of continuity*. The transformation f of X into Y is *continuous at the point x of X* if for each positive number ϵ there is a positive number δ such that for each point x' of X within a distance δ of x, $f(x')$ is within ϵ of $f(x)$.

Let us consider some examples of transformations of the reals into the reals described by equations.

1. $f(x) = x^3 - 1$.
2. $f(x) = x^3 - x$.
3. $f(x) = x^2$.
4. $f(x) = \sin x$.
5. $f(x) = \text{Arc} \tan x$.
6. $f(x) = 0$ if x is irrational or 0, $f(p/q) = 1/q$ if $p/q \neq 0$ is rational and in lowest terms.
7. $f(x) = x$ or x^3 according as x is algebraic or nonalgebraic.
8. $f(x) = 1 - x^2$, 0, or $x^2 - 1$ according as $x < -1$, $-1 \leq x \leq 1$, $x > 1$.

Exercise 10. Which of the above examples is (a) onto, (b) 1–1, (c) continuous, (d) continuous at $x = 0$?

The homeomorphisms of the plane onto itself discussed on pages 312–17 were examples of transformations. No doubt the reader can think of many other transformations that can be described either geometrically or physically.

Here is an example of a homeomorphism f of the interior of a circle of radius 1 onto the plane. The center c of the circle goes into itself and if p is another point of the interior of the circle, $f(p)$ is on the ray from c through p and $\rho(c, f(p))$ is $\rho(c, p)/(1 - \rho(c, p))$.

Now let us consider a transformation of one segment AC onto a segment $C'B'$ as shown in Figure 15. The points between A and B go into

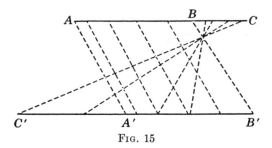

FIG. 15

points of $A'B'$ through parallel projections and the points of BC go into points of $B'C'$ by projections through a point.

This transformation may be represented by the graph shown in Figure 16. The interval AC is placed on the x-axis and the interval $C'B'$ on the

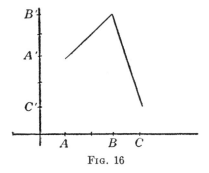

FIG. 16

y-axis. To find the image of a point of AC, project the point vertically onto the graph and then project this point horizontally onto the y-axis. The equation in this case is $f(x) = x + 2$ or $17 - 4x$ according as $0 \leq x \leq 3$ or $3 \leq x \leq 4$.

If X is a square plus its interior and Y is a segment, a transformation of X into Y may be represented by a graph in 3-space which may be a surface. However, if Y were also a square plus its interior, it would require a space of four dimensions to draw the graph—two dimensions for X and two others for Y.

The ϵ-δ definitions of continuity of functions given in analytics and calculus texts are puzzling to most students when they are encountered for the first time. In some cases, even an expert could not tell from the definitions given whether or not a function such as $f(x) = \sqrt{1 - x^2}$, $-1 \leq x \leq 1$, is continuous at $x = 1$ since mention of the domain of definition is sometimes omitted—in other cases the function would be called discontinuous because it is not defined for all values of x. From the definitions we have given, the corresponding transformation would be continuous.

Here is a geometric definition of the continuity of a graph which may be more meaningful to many students than the corresponding ϵ-δ definition. It is understood that the graph is in the plane.

Definition of continuity of a graph. The graph G is *continuous at the point p of G* if for each pair of horizontal lines, one above and one below p, there is a pair of vertical lines, one to the right and one to the left of p, such that each point of G between the vertical lines is between the horizontal lines. (See Fig. 17.)

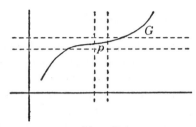

FIG. 17

Exercise 11. What in the above definition corresponds in the ϵ-δ definition to (a) x, (b) $f(x)$, (c) ϵ, (d) δ?

Exercise 12. Under what conditions would a graph in 3-space be continuous at one of its points p?

PROPERTIES PRESERVED BY MAPPINGS

A continuous transformation is called a *mapping*. In this section we point out that certain properties are preserved by mappings. The following exercise shows that in a certain sense, limit points are preserved.

Exercise 13. Show that if p is a limit point of X and if f is a map of $\bar{X}$ into Y, $f(p)$ is either a point or a limit point of $f(X)$.

Suppose x_1, x_2, $\cdots$ is an infinite sequence of points—we do not suppose that different elements of the sequence are different points. The sequence is said to *converge* to a point x_0 if for each neighborhood U of x_0 there is an integer n such that each point of x_n, x_{n+1}, $\cdots$ lies in U. A proof similar to that used to answer Exercise 13 shows that if f is a mapping of X, X contains x_0, x_1, x_2, $\cdots$, and x_1, x_2, $\cdots$ converges to x_0, then $f(x_1)$, $f(x_2)$, $\cdots$ converges to $f(x_0)$.

The preceding result may remind us of the definition of continuity of a function given in some calculus books. The function $f(x)$ is sometimes said to be continuous at $x = a$ provided the limit, as x approaches a, of $f(x)$ is $f(a)$.

Connectedness is another property that is preserved by mappings. Intuitively, we think that a thing is connected if it is "all in one piece." For example, the disk D shown in Figure 18 is connected but the Cantor

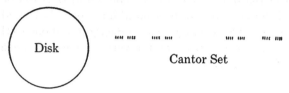

FIG. 18

set C is not. However, D is the sum of its interior and its boundary, so in a certain sense it is in two pieces. (The *boundary* of M is the intersection of $\bar{M}$ and the closure of the complement of M.) We shall give a more precise definition of connected. The Cantor set shown in Figure 18 is obtained by starting with a segment, removing the open middle third of it, removing each of the open middle thirds of each of the remaining pieces, removing the open middle thirds of the remaining pieces, etc.

Two sets are *mutually exclusive* if neither contains a point of the other. (A, B mutually exclusive means $A \cdot B = 0$.) The boundary of the disk D and the interior D are mutually exclusive.

Two sets are *mutually separated* if neither contains either a point or a limit point of the other. (A, B mutually separated means $\bar{A} \cdot B = A \cdot \bar{B} = 0$.) The interior of a disk and its boundary are not mutually separated since a point of the boundary is a limit point of the interior.

A set is *not connected* if it can be expressed as the sum of two mutually separated sets each of which contains a point. (This definition would be simpler if everyone adopted the convention that each point set contains a point.) The Cantor set is not connected because its right hand half and left hand half are mutually separated.

Definition of connected. A set is *connected* if it is not the sum of any two mutually separated (non-null) sets.

Exercise 14. Show that if f is a mapping of a connected set X, then $f(X)$ is connected also.

A closed and connected (non-null) point set is called a *continuum*. Being a continuum is not preserved under a mapping—we showed in the last section a homeomorphism of the interior of a circle (not a continuum) onto the plane (a continuum). The inverse of this homeomorphism is a mapping of a continuum onto something that is not closed.

Being open and closed is preserved in a certain sense by the inverses of mappings. In fact, rather than say that a transformation f of X into Y is continuous if it is continuous at each point of X, it is sometimes defined to be continuous if the inverse of each open subset in Y is an open subset of X.

Exercise 15. Show that if f is a continuous transformation of X into Y, the inverse of each open subset of Y is an open subset of X.

Exercise 16. Show that the transformation f of X into Y is continuous if the inverse of each open subset of Y is an open subset of X.

Exercises 15, 16 show that a transformation f of X into Y is continuous if and only if the inverse of each open subset of Y is an open subset of X. Similarly it might have been shown that f is continuous if and only if the inverse of each closed subset of Y is a closed subset of X. (A closed subset of X may not be closed but is merely the intersection of X with a closed set.)

As the population of the earth increases, it is inevitable that there be centers of dense population. In contrast, it is felt that if space travel becomes feasible and space is habitable, the population of the universe can increase without us having crowding. Even in the plane, it is possible to have infinitely many points such that no point is within 100 units of any two of them. (Put the points at $(0, 0)$, $(300, 0)$, $(600, 0)$, $\cdots$.) The limited property of the earth we have mentioned reminds us of a topological property called compactness. Compactness is preserved under a mapping if we use the following definition.

Definition of compactness. A set M is *compact* if each infinite subset of M has a limit point in M.

The plane is not compact but, as shown in the next section, a closed segment is. (Some definitions do not impose the condition that the limit point of M need belong to M. If this condition were omitted, the interior of a circle would be compact and Exercise 17 could not be done. Other definitions impose even more stringent conditions.)

Exercise 17. Show that if f is a mapping of X and X is compact, then $f(X)$ is compact also.

A set of tarpaulins is said to cover a football field if each spot of the field is under one of them. Similarly, a collection G of point sets is said to *cover* a point set X if each point of X belongs to an element of G. If each element of G is a domain, G is called an *open covering*.

The collection of all neighborhoods covers the plane but no finite sub-collection of them does. However, if a collection of neighborhoods covers the surface of a sphere or a segment, a finite number of them does also. This reminds us of the following topological property that is preserved under a mapping.

Definition of bicompactness. A set X is *bicompact* if each open covering of X contains a finite collection of elements which covers X.

Exercise 18. Show that bicompactness is preserved by a mapping.

The two properties, compactness and bicompactness, are not equivalent because there are topological spaces that are compact but not bicompact. However, any bicompact space is compact and any compact metric space is bicompact. Therefore, bicompactness is a stronger condition and a person who uses it in the hypothesis of a theorem is not proving as strong a result as one who uses mere compactness instead. Because of lack of agreement on terminology, some people use the term compact to mean bicompact as we have defined it here. Not everyone that mentions the word compact is using the weaker condition.

Exercise 19. Show that bicompactness implies compactness.

A set X is *homogeneous* if for each pair of points p, q of X there is a homeomorphism of X onto itself that takes p onto q. A circle is homogeneous because a rotation will take any point into any other point. A closed segment is not homogeneous because there is no homeomorphism of the segment onto itself that takes an end point onto a non-end point.

Unsolved problems. A difficulty in discussing unsolved problems in topology is that most of these problems involve complicated concepts meaningful only to the advanced worker. Most people would have neither the technical vocabulary nor the experience to appreciate the discussion. However, here is a problem which we have at least developed the vocabulary to present.

Is a simple closed curve the only homogeneous compact plane continuum other than a point? For many years it was conjectured that the answer was in the affirmative—in fact some papers appeared in scientific journals giving faulty proofs to this effect. However, in 1948, an example was given of a bounded plane continuum which was neither a simple closed curve nor a point but which was homogeneous nevertheless. In 1954 another example was announced. Are there other homogeneous bounded plane continua?

TOPOLOGY OF THE REAL LINE

Much of mathematics is devoted to a study of some aspects of the real number system. In analytics much use is made of the fact that there is a 1–1 correspondence between the real numbers and points on a line.

Certain of the topological properties of the real numbers can be developed from a study of the line. Consider the following axiom for a line.

AXIOM. *A line is connected.*

This axiom can be used to prove some very useful theorems used in calculus.

THEOREM. *A bounded (non-null) collection of numbers has a least upper bound.*

We shall give a proof of this theorem based on the above axiom. First we elaborate on its meaning.

A number x is an *upper bound* for a collection N if x is greater than or equal to each value of N. The *least upper bound* is the smallest of all the upper bounds. Similarly, a *lower bound* and a *greatest lower bound* are defined. A set of numbers with both an upper and a lower bound is called *bounded*.

Now for a proof of the theorem. Suppose N is a bounded collection of numbers. Let R be the set of all points p of the x-axis such that p is associated with an upper bound of N and L be the rest of the x-axis. (See Fig. 19.)

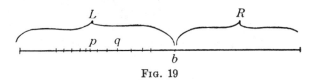

FIG. 19

There is a point in R because N was given to be bounded. There is a point in L because any number less than a number in N is not an upper bound of N.

No point p of L is a limit point of R because if p is in L, there is a point q associated with a number of N such that q is to the right of p—no point to the left of q is a point of R.

The axiom tells us that L, R cannot be mutually separated, so some point b of R is a limit point of L. This point will be associated with an upper bound x_b of N since b is in R. Since b is a limit point of L, for each positive number ϵ there is a point of N within ϵ of x_b. Then x_b is is the least of all the upper bounds of N.

We now mention some theorems of a related nature frequently used in calculus.

1. A collection (non-null) of numbers with a lower bound has a greatest lower bound.

2. A bounded monotone increasing sequence of numbers converges to the least upper bound of the sequence.

3. The Dedekind cut proposition holds. (The proposition, sometimes taken as an alternate axiom, states that if A, B are two (non-null) collections of numbers such that (a) $A \cdot B = 0$, (b) each number belongs to $A + B$ and (c) each number of B is less than every value of A, then either A has a minimum or B has a maximum.)

4. A bounded infinite collection of numbers has a greatest limit. (This limit is sometimes called the upper limit. If the numbers are regarded as points, it will be a limit point that is larger than any other limit point.)

Exercise 20. How would the x-axis be broken into two sets L, R to show that the above theorem (4) holds?

Exercise 21. Show from the axiom that a segment is connected.

A subset of the plane is said to be *bounded* if it lies on the interior of some circle. It can be shown that each closed and bounded subset of the plane is bicompact and hence compact. Also, each compact subset of the plane is closed and bounded.

Exercise 22. Show that each compact subset of a line is closed and bounded.

Exercise 23. In showing that each closed and bounded subset of the x-axis is bicompact, one must show that if G is an open covering of a closed and bounded subset S of the x-axis, then a finite number of elements of G cover S. How would the x-axis be broken into two sets L, R to show that the assumption that no finite subcollection of G covers S leads to the contradiction that the x-axis is not connected?

The following theorems used in calculus and concerning a continuous function $f(x)$ defined on a segment $(a \leq x \leq b)$ are special cases of more general theorems we have considered.

1. $f(x)$ is bounded.

2. $f(x)$ takes on its maximum and its minimum.

3. If $f(x)$ takes on two values, it takes on each value between them.

The first of these results follows from the facts that a segment is bicompact (Exercise 23), its image under a continuous transformation is bicompact (Exercise 18) and hence compact (Exercise 19), and a compact set of numbers is closed and bounded (Exercise 22).

The second of these results follows from the further facts that a closed and bounded set of numbers has a least upper bound and a greatest lower bound since it is bounded. It takes on these values since it is closed.

The third result follows from the facts that a segment is connected (Exercise 21) and therefore its image is connected (Exercise 14).

A set X is said to have the *fixed point property* if for each mapping f of X into itself, for some point p of X, $f(p) = p$. A circle does not have the fixed point property because a rotation can move each of its points. It can be shown that a segment, a disk, and a solid cube do have the fixed point property.

Exercise 24. Show that a segment has the fixed point property.

Some unsolved problems. It is conjectured that if C is a bounded plane continuum whose complement is connected in the plane, then C has the fixed point property. Partial solutions have been given of two types—one of these says that if the continuum has certain properties (such as being like a segment or a disk), it has the fixed point property; the other type says that if certain restrictions are placed on the mapping, some point is left fixed. In spite of the vast amount of time spent on this problem by some talented workers, the answer still eludes us. Perhaps some student coming along will succeed in verifying the conjecture that has baffled so many of his predecessors—or maybe someone will construct a counterexample showing that the conjecture is false.

If one were asked to name a plane continuum with more than one point which was topologically equivalent to each continuum in it with more than a point, one might think of a segment. (See Fig. 3.) For many years no one knew of any example topologically different from a segment, although the question was raised in mathematical journals as to whether or not there were any. In 1948, another example was found. The problem as to whether or not there are still other examples is unsolved.

SOME EXAMPLES OF TOPOLOGICAL SPACES

What is a point?

If one is studying the real number system or a real line one might answer that a point is a real number. For analysis, the answer might be—a complex number. In plane analytics, a point is an ordered pair of real numbers—in solid analytics it is an ordered triple of real numbers.

In synthetic geometry and in topology, a point is an undefined term. If we have a collection of elements (which we call points), and certain subcollections of these elements (which we call neighborhoods), we may regard the original collection as a topological space if the neighborhoods satisfy certain conditions called axioms.

In this section we mention some abstract spaces somewhat out of the ordinary. We describe them by telling what the points and neighborhoods of these spaces are.

1. **Euclidean 4-space.** Here each point has four coordinates instead of three. (No mention need be made of time. In fact, any student who has worked problems with four variables may consider himself as having worked with 4-space.) A point is an ordered quadruple of numbers and a neighborhood with center at (x_0, y_0, z_0, w_0) and radius ϵ is the set of all points (x, y, z, w) such that $((x_0 - x)^2 + (y_0 - y)^2 + (z_0 - z)^2 + (w_0 - w)^2)^{1/2} < \epsilon$.

2. **Hilbert space.** The points of this space are sequences $(a_1, a_2, \cdots)$ of numbers such that $\sum a_n$ exists. The distance between two points $(x_1, x_2, \cdots), (y_1, y_2, \cdots)$ is$(\sum (x_n - y_n)^2)^{1/2}$. Neighborhoods are the interiors of generalized spheres.

3. **A function space.** A point in this abstract space is a continuous function $y = f(x)$ $(0 \leq x \leq 1)$. A neighborhood with center at $f_0(x)$ and radius ϵ is the collection of all functions $f(x)$ such that

$$| f(x) - f_0(x) | < \epsilon$$

for each x between 0 and 1 inclusive.

Metric spaces. Many of the spaces we deal with in topology are metric spaces. *Metric spaces* are topological spaces with distance functions $\rho(p, q)$ (called the distance between p and q). Distance functions satisfy the following conditions.

1. $\rho(x, y) \geq 0$, *the equality holding if and only if* $x = y$.
2. $\rho(x, y) = \rho(y, x)$ *(symmetry)*
3. $\rho(x, y) + \rho(y, z) \geq \rho(x, z)$ *(triangle condition)*.
4. $\rho(x, y)$ *preserves limit points.*

This fourth condition means that a point x belongs to the closure of a set M if and only if for each positive number ϵ, there is a point m of M such that $\rho(x, m) < \epsilon$.

Exercise 25. Describe a distance function for the above function space.

We shall now describe a topological space that has no distance function.

4. **Another function space.** This is a function space used extensively in studying linear spaces. Again, a point in this abstract space is a function $y = f(x)$ $(0 \leq x \leq 1)$ but this time we do not suppose that it is continuous. To get a neighborhood in this function space, we select a finite collection of numbers $x_1, x_2, \cdots, x_n$ between 0 and 1 inclusive and a positive number ϵ. The neighborhood $N(f_0(x); x_1, x_2, \cdots, x_n ; \epsilon)$ is the set of all functions $f(x)$ defined for $0 \leq x \leq 1$ and such that

$$| f_0(x_i) - f(x_i) | < \epsilon$$

for $(i = 1, 2, \cdots, n)$. To get another neighborhood, we select another function, another finite set of points, and another positive number ϵ.

If p is a *limit point of a set* X in a metric space, there is a sequence of different points of X converging to p.

In the function space we have just described, let $f_0(x)$ be the function such that $f_0(x) = 0$ for $0 \leq x \leq 1$. Let M be the collection of all functions that take on the value 0 at a finite collection of numbers and on the value 1 everywhere else. Although $f_0(x)$ is a limit point of M, no sequence of points of M converges to $f_0(x)$. This shows that the function space we have described is not a metric space.

Exercise 26. If $f_1(x)$, $f_2(x)$, $\cdots$, is a sequence of points of M in the above function space, what is a neighborhood of $f_0(x)$ that contains no point of $f_1(x)$, $f_2(x)$, $\cdots$?

Exercise 27. If p is a limit point of a set M in a metric space, how might one select a sequence of points of M converging to p?

The topologist is acquainted with dozens of different spaces—he may invent new spaces as the needs arise. Some of these spaces are much simpler than the ones we have described and others are more complex.

Queer spaces play an important role in topology. A wealth of examples of spaces with certain properties helps our understanding of these properties. It suggests to us that theorems are likely to follow from these properties.

Also, queer spaces are useful as counterexamples. There is no more convincing way of showing that one set of properties do not imply a second set than by exhibiting a space with the first set of properties but without the second set. Of the problems solved during recent years which had been regarded as difficult unsolved problems for many years, a goodly number of them were solved by the use of counterexamples.

TRENDS TOWARD ABSTRACTNESS

One of the aspects of modern mathematics is the trend toward abstractness. This is not only true in topology but also in algebra, analysis, and other branches of mathematics. Let us examine how this trend has affected topology.

In proving a theorem in plane geometry, the student does not use all the properties of the plane but merely uses certain of the axioms. Similarly, in proving a topological theorem in the plane, one might note that his proof depends only on certain properties of the plane. The theorem would then hold true in any space with these essential properties. Let us illustrate this with an example.

THEOREM. *The sum of two closed plane sets is closed.*

This theorem can be proved without mentioning many properties of the plane. There is no need in mentioning the x-axis, y-axis, or straight lines. The theorem can be proved by using the definitions of closed sets, sum, limit point, along with the fact that if N_1, N_2 are two neighborhoods containing the same point, there is a neighborhood N_3 of the point in $N_1 \cdot N_2$.

Exercise 28. Prove the above theorem on the basis of the definitions and the fact that if a point of the plane lies in each of two neighborhoods N_1, N_2, there is a neighborhood N_3 of the point in $N_1 \cdot N_2$.

We can prove the theorem about the sum of two closed sets being closed in any of the spaces mentioned on pages 327–29. One method of studying topology is to consider what theorems can be proved in a space with certain properties. One may not even know what space is under consideration as long as it is known that the space has the properties used in proving the theorem. Naturally the mathematician is interested in getting the maximum in the way of results or conclusions from a minimum in the way of hypotheses or axioms.

If one is able to prove that certain theorems are true on the basis of certain properties of a space, at some time in the future a use may be found for a particular space with these properties, and the theorems proved before the space was invented will be known to hold.

An advantage (other than giving the imagination and mind ample working space) in dealing with abstract spaces is that our theorems tell us more than things about ordinary points and also reveal truths about collections of other sorts—such as collections of functions such as described in Examples 3 and 4 on pages 328 and 329.

There are other abstract spaces which have for points other sorts of unexpected things—closed sets, continua, matrices.

We shall not give here the axioms for topological spaces frequently used. They include the condition that if N_1, N_2 are neighborhoods of the same point, there is a neighborhood of the point in $N_1 \cdot N_2$. The interested reader may find such axioms on pages 1, 2 of Whyburn's *Analytic Topology* and elsewhere.

Types of unsolved problems. We finish this chapter by mentioning some types of unsolved problems.

As we have indicated, some theorems go about as follows: Each space with a certain set of properties has some other property. Quite a few of the many mathematical papers published each year are devoted to proving this kind of theorem. We shall not discuss here the properties of interest but they include such things as being compact, bicompact, and having distance functions.

A more concrete type of theorem tells us that certain theorems (such as the Jordan curve theorem) hold in the plane or in some other familiar space. During the past forty years we have made much headway in proving theorems for the plane. In more recent years some headway has been made at hacking away at some of the complicated theorems in 3-space.

Intermediate between the abstract and concrete types of theorems are the characterization theorems. Show that a well-known space (such as a line, plane, or 3-space) satisfies a certain minimum set of conditions —then show that any space that satisfies these same conditions must be topologically equivalent to the well-known space.

We have some good topological characterizations of the line. However, one of the interesting unsolved problems (Souslin problem) is in this area. It gives a certain set of conditions satisfied by the line and asks if any space which satisfies them must be homeomorphic with a line.

Topological characterizations of the plane have been discovered during the past 40 years. An unsolved problem for many years in this area may be stated as follows: Is the connected compact metric space S topologically equivalent to the surface of a sphere if it has more than one point and satisfies the following conditions?

1. The neighborhoods of S are connected.
2. The complement of no simple closed curve is connected.
3. The complement of each pair of points is connected.

The answer was given in the affirmative in 1946 (Kline sphere characterization).

Although topological characterizations of 3-space have been given, these are still complicated. In view of the recent spurt toward increasing our knowledge of the topological properties of 3-space, there is hope for improvements here.

It is difficult to specify the types of unsolved problems that are of most interest to research topologists. Researchers vary in their interests and the field is wide. Things in which one worker is intensely interested may not even concern someone else at the moment, because he may be too busy working on something else. Although there are many unsolved problems in topology, those of interest are not likely to have easy solutions. However, we have talented young people coming along each year who are taking their places as ingenious researchers and making contributions that help push the frontier of mathematics farther back.

ANSWERS TO THE EXERCISES

1. For one keyboard, the answer was B and X. Other letters were grouped as follows: (A, R), (C, I, J, L, M, N, S, U, V, W, Z), (D, O), (E, F, G, T, Y), (H, K), (P, Q). If one regards the letters as occupying area rather than being curves, each letter is equivalent to a disk or a disk with one or two holes in it.

2. We use functional notation to describe the homeomorphism—that is the point corresponding to x is called $f(x)$. Let c be a point of the interior of the triangle and $f(c)$ be the center of the circle. For each point p on the triangle, let $f(p)$ be the point on the circle such that the ray from c through p points in the same direction as the ray from $f(c)$ through $f(p)$. Each point q of the first ray corresponds to the point $f(q)$ of the second ray such that $f(q)$ divides $f(c)$, $f(p)$ in the same ratio (internally or externally) that q divides c, p. (See Fig. 8).

3. First we describe a homeomorphism g which takes the plane onto itself and the circle with equation $x^2 + y^2 = 1$ onto the given graph. If p is a point of the interior of the circle, $g(p)$ is on the same vertical line with p and divides the points where the vertical line hits the graph in the same ratio that p divides the points where the line hits the circle. If p is beneath the circle, $g(p)$ is on the same vertical line with p and the same distance below the graph as p is below the circle. If p is a point of the plane that is neither interior to, below, or on the lower semicircle of the circle, $g(p) = p$.

Suppose f is a homeomorphism such as given in Exercise 2 of the plane onto itself that takes the given triangle onto the circle with equation $x^2 + y^2 = 1$. Then gf is the required homeomorphism—to find the point that corresponds to p, find the point $f(p)$ that corresponds to p under the homeomorphism f and then find the point $gf(p)$ which corresponds to this point $f(p)$ under g.

4. Construct interval AGC and interval BD as shown in Figure 9. Also construct the "kite-shaped" figure shown in Figure 10. We describe the homeomorphism f of the polygon $ABCDEF$ onto the triangle $A'C'D'$ by steps.

(i) Let f be a homeomorphism that takes AB linearly onto $A'B'$—if p is a point of AB, $f(p)$ divides A', B' in the same ratio that p divides A, B. Extend f to take BC, CD, DE, EF, FA, AG, GC, BD linearly onto $B'C'$, $C'D'$, $D'E'$, $E'F'$, $F'A'$, $A'G'$, $G'C'$, $B'D'$.

(ii) We now extend f so that it takes the interior of $ABCG$ onto the interior of $A'B'C'G'$. Let p and $f(p)$ be points of the interiors respectively of these two triangles. If x is a point other than p of the interior of $ABCG$, extend the segment from p through x until it intersects triangle $ABCG$ at a point y. Now $f(y)$ has already been defined in the first step. The point $f(x)$ is the point that divides $f(p)$, $f(y)$ in the same ratio that x divides p, y.

(iii) Similarly extend f to take the interior of triangle BCD onto the interior of triangle $B'C'D'$.

(iv) To extend f to the interior of $ABDEF$, take a point p of this interior such that no ray from p intersects $ABDEF$ in two points (we might choose p on the line which bisects angle E) and let $f(p)$ be any point on the interior of $A'B'-D'E'F'$. For each point x other than p of the interior of $ABDEF$, let y be the point where the ray from p through x intersects $ABDEF$ and $f(p)$ be the point that divides $f(p)$, $f(y)$ in the same ratio that x divides p, y.

(v) To extend the homeomorphism f to the exterior of $AGCDEF$, consider the same points p, $f(p)$ used in Step iv. For each point x of the exterior of $AGCDEF$ let y be the point where the segment from p to x hits $AGCDEF$. Then

$f(x)$ is the point that divides $f(p)$, $f(y)$ exteriorly in the same ratio that x divides p, y.

5. Points of $X-r$, t. Limit points of $X-p$, q, t.

6. p, q, s, t.

7. (a) A triangle, (b) points between two parallel lines, (c) rational points (with rational abscissas and ordinates), (d) the plane.

8. Show that each limit point of the closure belongs to the closure.

9. $\bar{G} = G +$ segment from $(0, 1)$ to $(0, -1) + \{(1/\pi, 0)\}$.

10. (a) 1, 2, 7, 8. (b) 1, 5, 7. (c) 1, 2, 3, 4, 5, 8. (d) All.

11. (a) Projection of p on x-axis. (b) Projection of p on y-axis. (c) Distance from p to horizontal lines if they are equidistant from p. (d) Distance from p to vertical lines if they are equidistant from p.

12. If for each pair of horizontal planes, one above and one below p, there is a neighborhood of p such that each point of the graph in, above, or below the neighborhood is between the horizontal planes.

13. We show that each neighborhood U of $f(p)$ contains a point of $f(X)$. Let V be a neighborhood of p such that $f(V \cdot X) \subset U$. Since p is a limit point of X, V contains a point x of X and $f(x)$ is in U.

14. If $f(X)$ were the sum of two mutually separated (non-null) sets A, B, then X would be the sum of the sets $f^{-1}(A)$, $f^{-1}(B)$. These two sets are mutually exclusive because f is single valued, and each is an open subset of X by Exercise 15. Hence if $f(X)$ were the sum of two mutually separated (non-null) sets A, B, X would be the sum of two mutually separated (non-null) sets $f^{-1}(A)$, $f^{-1}(B)$.

15. Suppose p is a point of $f^{-1}(D)$ where D is an open subset of Y. If U is a neighborhood of $f(p)$ with $U \cdot Y \subset D$, the continuity of f implies that there is a neighborhood V of p such that $f(V \cdot X) \subset U$. Then $V \cdot X \subset f^{-1}(D)$. The sum of all such V's for all points p of $f^{-1}(D)$ is an open set W and $W \cdot X = f^{-1}(D)$ is an open subset of X.

16. If U is an open subset of Y containing $f(p)$, we have that $f^{-1}(U)$ is an open subset of $X-f^{-1}(U) = D \cdot X$ where D is open. We can use a neighborhood V of p in D to show that the definition of continuity is satisfied at p. Since f is continuous at each point p of X, it is continuous.

17. We show that each infinite collection y_1, y_2, $\cdots$ of points ($y_i \neq y_j$ if $i < j$) of $f(X)$ has a limit point in $f(X)$. Let x_i be a point of $f^{-1}(y_i)$. Then x_1, $x_2 \cdots$ has a limit point x_0 in X. Let K be infinite set formed by subtracting $f^{-1}(x_0)$ from x_1, x_2, $\cdots$ Since x_0 is a limit point of K, $f(x_0)$ is not a point of $f(K)$, then by Exercise 13, $f(x_0)$ is a limit point of $f(K)$ and hence of y_1, y_2, $\cdots$.

18. If G is a collection of domains covering $f(X)$, the inverse of each of these is an open subset of X by Exercise 15. If X is bicompact, a finite number of these inverses cover X and the collection of corresponding elements of G covers $f(X)$.

19. We show that if K is an infinite collection of points in a bicompact space X, then K has a limit point in X. For each point p of K let G_p be the sum of p and the complement of K. No finite subcollection of the G_p's covers X so either the G_p's do not cover X or some G_p is not open. If some point q of X does not lie in any G_p, this point q is a limit point of K because it belongs to $\bar{K}$ but not to K. If some G_p is not open the corresponding point p is a limit point of K because if it were not, G_p would be the sum of two open sets—the complement of $\bar{K}$ and a neighborhood containing p but no other point of K.

20. Let L be the set of points of the x-axis associated with numbers n that

have infinitely many numbers of the collection which are bigger than n while R is the rest of the x-axis.

21. Suppose pq is the segment with left end p and right end q, L_p is the part of the line to the left of p and L_q is the part of the line to the right of q. The segment pq is connected because the assumption that it is the sum of two mutually separated (non-null) sets A, B leads to the contradiction that the line is not connected since it is the sum of two mutually separated sets A', B'; A' is A, $A + L_p$, $A + L_q$, $A + L_p + L_q$ according as A contains neither p nor q, A contains p but not q, A contains q but not p, A contains both p and q; B' is defined in a similar fashion.

22. Suppose that S is compact (non-null) subset of the line. Denote a fixed point of S by s_0. If S were not bounded, for each integer i there would be a point s_i in S which would not belong to the neighborhood with center at s_0 and radius i. This is impossible since S is compact but s_1, s_2, $\cdots$ would have no limit point. If S were not closed there would be a point x_0 not of S which is a limit point of S. For each integer i there would be a point s_i of S in the neighborhood with center at x_0 and radius $1/i$. This is impossible since S is compact and s_1, s_2, $\cdots$ would not have a limit point in S.

23. Let L be the collection of all points p of the x-axis such that a finite subcollection of G covers each point of S to the left of p. Let R be the rest of the x-axis.

24. Suppose f is a mapping of a horizontal segment S into itself, L is the set of points of S that move to the right under f, and R is the set of points of S that move to the left. Unless f leaves some point fixed, the left end of S belongs to L and the right end belongs to R. If a point p of S moves to the right, it follows from the continuity of f that there is a neighborhood of p such that each point of S in the neighborhood moves to the right. Hence, L is an open subset of S. Also, R is an open subset of S. Since S is connected (Exercise 20), it is not the sum of two mutually separated sets L and R. Any point of $S - (L + R)$ is left fixed under f.

25. $\rho(f(x), g(x))$ is the maximum value of $| f(x) - g(x) |$.

26. $N(f_0(x); y; \frac{1}{2})$ where y is a value such that $1 = f_1(y) = f_2(y) = \cdots$.

27. For each integer i let m_i be a point of M whose distance from p is less than $1/i$. Then m_1, m_2, $\cdots$ converges to p.

28. To show that the sum $A + B$ of two closed sets A, B is closed, we need to show that each limit point of $A + B$ belongs to $A + B$.

Suppose p is not a point of $A + B$. Then it is not a point of A (from the definition of sum). Since A is given to be closed, p is not a limit point of A (definition of closed). Hence, there is a neighborhood N_1 of p which contains no point of A (definition of limit point). Likewise, p is not a point or limit point of B and there is a neighborhood N_2 of p which contains no point of B.

There is a neighborhood N_3 of p which lies in both N_1 and N_2. (We cannot justify this from the definitions.) No point of A is in N_3 since no point of A is in N_2—also, no point of B is in N_3. Hence N_3 contains no point of $A + B$ since it contains no point of A and no point of $A + B$. Hence, $A + B$ is closed.

BIBLIOGRAPHY

1. BING, R. H. "Examples and Counterexamples." *Pi Mu Epsilon Journal* 1: 311–17; 1953.

2. COURANT, RICHARD, and ROBBINS, HERBERT. *What Is Mathematics?* New York: Oxford University Press, 1941. Chapter 5.
3. HAHN, HANS. "Geometry and Intuition." *Scientific American* 190: 84–91; March 1954.
4. HALL, D. W., and SPENCER, D. E. *Elementary Topology.* New York: John Wiley & Sons, 1955.
5. HOCKING, J. G. "Topology." *Astounding Science Fiction* 53: 96–110; March 1954.
6. KELLEY, J. L. *General Topology.* New York: D. Van Nostrand Company, 1955. Chapter 1.
7. LEFSCHETZ, SOLOMON. *Introduction to Topology.* Princeton, N. J.: Princeton University Press, 1949. Introduction and Chapter 1.
8. MESERVE, BRUCE E. "Topology for Secondary Schools." *The Mathematics Teacher* 46: 465–74; 1953.
9. NEWMAN, H. M. A. *Elements of the Topology of Plane Sets of Points.* Cambridge, Mass.: Cambridge University Press, 1951.
10. OGLIVY, C. S. *Through the Mathescope.* New York: Oxford University Press, 1956. Chapter 11.
11. SIERPINSKI, WACLAW. *General Topology.* (Translated by C. C. Krieger.) Toronto: University of Toronto Press, 1952. Chapters 1 and 2.
12. TUCKER, A. W., and BAILEY, H. S., JR. "Topology." *Scientific American* 182: 18–24; January 1950.
13. WHYBURN, G. T. "Analytic Topology." *American Mathematical Society,* Colloquium Publications, Vol. 28, New York, 1942. Chapter 1.
14. WILDER, R. L. "Some Unsolved Problems of Topology." *American Mathematical Monthly* 44: 61–70; 1937.

XI

The Theory of Probability

HERBERT ROBBINS

IN this chapter we shall try to give some idea of probability theory as a branch of mathematics. The first three sections are of a general introductory nature, while those which follow present a variety of special problems with the idea of exhibiting some of the main techniques of the theory.

The theory of probability originated in Italy and France during the sixteenth and seventeenth centuries when mathematicians first subjected to rational analysis the physical phenomena of random variation as typified by various games of chance: dice, cards, lotteries, and so on. Since then the mathematical content of the theory has steadily deepened, and at the same time the domain of applicability of the theory has widened from its original concern with games of chance to embrace the phenomena of random variation in the physical, biological, and social sciences. Today the theory of probability is a fundamental part of scientific methodology.

THE NATURE OF PROBABILITY

Consider an experiment $\mathcal{E}$ which can be performed repeatedly, at least in theory, under essentially the same conditions. We suppose that each trial of $\mathcal{E}$ must result in one and only one element or "point" ω (*omega*) of a set[1] Ω (*omega*) of all possible *outcomes*. The set Ω is called the *sample space* of the experiment, and any subset A of Ω is called an *event*. (For example, if $\mathcal{E}$ consists of tossing a die, Ω contains six elements $\{1, 2, 3, 4, 5, 6\}$, and if we have bet that we will obtain an even number we will be interested in the event $A = \{2, 4, 6\}$.) Suppose $\mathcal{E}$ is performed a certain number of times n and that in exactly $n(A)$ of these the outcome ω is found to be an element of A. The ratio $n(A)/n$ is called the *relative frequency* of the event A in the given sequence of n trials. It has long been observed that for certain types of experiments $\mathcal{E}$ and certain asso-

---

[1] For a discussion of sets see Chapter III.

ciated events A the ratio $n(A)/n$ for large n will have approximately the same value from one set of trials to another, and accordingly we regard $n(A)/n$ as an experimental value of a constant $P(A)$ associated with the event A under the experimental conditions ε, and called the *probability* of A. In much the same way we regard certain measurements taken on a bar as experimental values of a constant called the "length" of the bar. If we imagine a sequence of measurements to be taken with greater and greater precision we can suppose that the sequence would converge to the length of the bar, and in the same way we can suppose that if n were allowed to become infinite in a random experiment ε the values $n(A)/n$ would tend to $P(A)$ as a limit:

$$(1) \qquad\qquad P(A) = \lim_{n \to \infty} \frac{n(A)}{n}.$$

(The limit relation (1) is meant in an empirical physical sense, since $n(A)$ is not a mathematically defined function of n to which the rigorous "ϵ, δ" definition of limit could be applied (see page 221).)

Motivated by (1) the formal mathematical apparatus for probability theory is set up in the following way. Let Ω be any abstract set of elements ω, and let $P(A)$ be any real-valued set function with values ≥ 0 defined for all A in some family $\mathfrak{F}$ of subsets of Ω. We shall assume that the family $\mathfrak{F}$ contains as an element Ω itself and that $P(\Omega) = 1$, and that $\mathfrak{F}$ is closed under the set-theoretical operations of complement and finite or denumerable union. Finally we assume *the addition rule*:

If A_1, A_2, $\cdots$ is any sequence, finite or denumerable,[2] of *disjoint* sets in $\mathfrak{F}$ then[3]

$$(2) \qquad\qquad P(\bigcup A_i) = \sum P(A_i),$$

the sum on the right being a finite sum or an infinite series according as the sequence A_1, A_2, $\cdots$ is finite or denumerable.

The family $\mathfrak{F}$ of subsets of Ω for which $P(A)$ is assumed to be defined will in some cases be the family of *all* subsets of Ω, but in general will be more restricted. However, in the more elementary parts of probability theory it is customary to consider only sample spaces Ω consisting of a *finite or denumerable* set of points ω, and in such cases $\mathfrak{F}$ can always be taken as the family of all subsets of Ω. Each point ω of Ω is a one-point subset of Ω for which the probability $P(\omega) \geq 0$ is defined, and from (2)

---

[2] An infinite set of elements which can be put into one-to-one correspondence with the positive integers is said to be *denumerable*, or *countable*.

[3] This formula is read: The P-function of the union of sets A_i is equal to the sum of the P-functions of the sets A_i.

we can obtain the probability $P(A)$ of any subset of Ω as the finite sum or infinite series

(3)
$$P(A) = \sum_{\omega \in A} P(\omega).$$

The only condition on the $P(\omega)$ is that

(4)
$$P(\Omega) = \sum_{\omega \in \Omega} P(\omega) = 1.$$

Once probabilities $P(\omega) \geq 0$ are given, subject to (4), the solution of any problem in probability theory where the sample space Ω is finite or denumerable reduces to the evaluation of a sum of the form (3).

Before considering any specific problems in probability theory we shall introduce the important concepts of *conditional probability* and *independence*. Consider any two events A and B, in general not disjoint subsets of Ω, and denote their intersection by $A \cap B$. Suppose we perform n trials of the experiment $\mathcal{E}$ and compute the *relative frequency of the occurrence of A within only the subset of trials in which event B occurs*; *i.e.* the ratio $\dfrac{n(A \cap B)}{n(B)}$.

This ratio can be rewritten as

$$\frac{\dfrac{n(A \cap B)}{n}}{\dfrac{n(B)}{n}}$$

and in accordance with (1) will tend, as $n \to \infty$, to the ratio

$$P(A \cap B) \mid P(B),$$

providing $P(B) \neq 0$. Accordingly we define the *conditional probability of the event A given the event B* as the ratio

(5)
$$P(A \mid B) = \frac{P(A \cap B)}{P(B)},$$

assuming $P(B) \neq 0$. (The number $P(A)$ itself may then be called the *unconditional probability* of the event A.) Equation (5) may be written in the form

(6)
$$P(A \cap B) = P(B) \cdot P(A \mid B),$$

and is then known as the *product rule* of probability. If we interchange the roles of A and B we obtain the analogous formula

(7)
$$P(A \cap B) = P(A) \cdot P(B \mid A).$$

In general the unconditional probability $P(A)$ will differ from the conditional probability $P(A \mid B)$. If, however, $P(A) = P(A \mid B)$, then we say that A is *independent of B*. From (6) we then have

(8) $$P(A \cap B) = P(A) \cdot P(B).$$

Substituting (8) in (7) we find that $P(B) = P(B \mid A)$, so that B is then also independent of A. When three events A, B, C are concerned we can write

(9)
$$P(A \cap B \cap C) = P(A \cap B) \cdot P(C \mid A \cap B)$$
$$= P(A) \cdot P(B \mid A) \cdot P(C \mid A \cap B).$$

If all conditional probabilities such as $P(B \mid A)$ and $P(C \mid A \cap B)$ are equal to the corresponding unconditional probabilities we say that A, B, and C are independent events, and in this case

(10) $$P(A \cap B \cap C) = P(A) \cdot P(B) \cdot P(C).$$

A similar definition of independence and formula for the probability of joint occurrence hold for more than three events.

Example 1. In the simplest case the sample space will consist of a *finite* number of elements ω_1, ω_2, $\cdots$, ω_r and each ω_i will have the same probability, which from (4) must be $1/r$. If A contains exactly k of these elements then by (3), $P(A) = k/r$; in other words, the *probability* of an event A is the ratio of the number k of outcomes ω of $\mathcal{E}$ which are "favorable" to A (*i.e.* belong to A) to the total number r of possible outcomes of $\mathcal{E}$. This is the classical definition of probability found in most of the older textbooks but is only applicable to a special class of problems, such as those involving cards or dice.

For example, we mention the following problem, one of the first in the history of mathematical probability. How many times must a pair of dice be tossed in order that the probability of obtaining at least one "double-six" be approximately $\frac{1}{2}$? Solution: A single toss of a pair of dice has $6 \cdot 6 = 36$ possible outcomes and we regard them all as equally probable. Similarly, n tosses of a pair of dice yield a sample space Ω consisting of $(36)^n$ elements of equal probability. Those with no double-six consist of $(35)^n$ elements. Hence those with at least one double-six consist of $(36)^n - (35)^n$ elements, and the event A of obtaining at least one double-six in n tosses of a pair of dice has probability

$$P(A) = \frac{(36)^n - (35)^n}{(36)^n} = 1 - \left(\frac{35}{36}\right)^n.$$

By logarithms we find that

$$P(A) = .4914 \text{ if } n = 24$$

$$P(A) = .5055 \text{ if } n = 25.$$

This is in accordance with the experience of the Chevalier de Méré who found that he lost money in the long run in betting on the occurrence of a double-six in $n = 24$ tosses.

Example 2. Let the experiment $\mathcal{E}$ consist in tossing a coin repeatedly until heads appears for the first time. The sample space Ω here consists of the denumerably infinite sequence of elements

$$\omega_1 = (H)$$

$$\omega_2 = (T, H)$$

$$\omega_3 = (T, T, H), \text{ etc.},$$

together with the outcome of an infinite sequence of tails,

$$\omega_0 = (T, T, T, \cdots).$$

If the coin is unbiased the probability of heads at any one toss is $\frac{1}{2}$, and by (10) it is reasonable to assign to the above outcomes the probabilities

$$P(\omega_1) = \frac{1}{2}$$

$$P(\omega_2) = \frac{1}{2} \cdot \frac{1}{2} = \frac{1}{2^2}$$

$$P(\omega_3) = \frac{1}{2} \cdot \frac{1}{2} \cdot \frac{1}{2} = \frac{1}{2^3}$$

$$\vdots$$

$$P(\omega_0) = 0,$$

the last expressing our intuitive idea that an infinite sequence of tails is, though not "impossible," highly unlikely to say the least. We then have, in comformity with (4)

$$\sum_{i=0}^{\infty} P(\omega_i) = \sum_{i=1}^{\infty} \frac{1}{2^i} = 1.$$

Let A denote the event that heads first occurs at an *odd* number of throws. Then

$$P(A) = P(\omega_1) + P(\omega_3) + P(\omega_5) + \cdots$$

$$= \frac{1}{2} + \frac{1}{2^3} + \frac{1}{2^5} + \cdots = \frac{1}{2}[1 + \frac{1}{2^2} + \frac{1}{2^4} + \cdots]$$

$$= \frac{1}{2}\left[1 + \frac{1}{4} + \left(\frac{1}{4}\right)^2 + \cdots\right] = \frac{1}{2}\left[\frac{1}{1 - \frac{1}{4}}\right] = \frac{2}{3},$$

while if B denotes the event of heads occurring first at an *even* number of throws we find similarly,

$$P(B) = P(\omega_2) + P(\omega_4) + P(\omega_6) + \cdots$$

$$= \tfrac{1}{2^2} + \tfrac{1}{2^4} + \tfrac{1}{2^6} + \cdots$$

$$= \tfrac{1}{2}[\tfrac{1}{2} + \tfrac{1}{2^3} + \tfrac{1}{2^5} + \cdots] = \tfrac{1}{2}\cdot\tfrac{2}{3} = \tfrac{1}{3}.$$

Or, more simply, since

$$1 = P(\omega_1) + P(\omega_2) + P(\omega_3) + P(\omega_4) + \cdots$$

$$= P(A) + P(B),$$

it follows that

$$P(B) = 1 - P(A) = 1 - \tfrac{2}{3} = \tfrac{1}{3}.$$

Example 3. In our first two examples the sample space of the experiment $\mathcal{E}$ contained either a finite or a denumerable number of points ω, and the computation of the probability of any event required only finite summation or the evaluation of an infinite series. We shall now consider an experiment $\mathcal{E}$ whose sample space Ω consists of a continuum of points[4] and where probabilities must be found by integration. Suppose a marksman shoots at a target whose center is at the origin of an x, y-plane, and for the moment suppose that we are only interested in the x-coordinate X of the point of impact of the bullet, which can theoretically be any real number. For any fixed value, say $X = 1$, the probability of the event $X = 1$ will be zero, and if we ask for the probability of the event A which consists of the x-coordinate of the point of impact lying, say, within the interval $-1 \leq x \leq 1$ then it is not possible to obtain $P(A)$ by summation. In cases like this we assume that the probability of any event A, in this case some region of the x-axis, is given by the integral of some non-negative function $f(x)$ over A,

$$P(A) = \int_A f(x)\, dx,$$

the function $f(x)$ being such that

$$\int_{-\infty}^{\infty} f(x)\, dx = 1.$$

$f(x)$ is called the *probability density* of the random variable X. Likewise, the probability of the y-coordinate of the point of impact of the bullet

---

[4] A continuum of points is a set of points such as all the points on the axis of reals.

lying in any region B of the y axis will be given by the integral

$$\int_B g(y)\ dy,$$

where $g(y)$ is the probability density of the random variable Y.

A priori the functions $f(x)$ and $g(y)$ are quite arbitrary. It is typical of the power of probability theory, however, that if we are willing to make certain plausible physical assumptions about the process of shooting at the target we can determine the mathematical form of these functions precisely. Let R denote any rectangle in the x, y-plane specified by the inequalities $a \leqq x \leqq b$, $\alpha \leqq y \leqq \beta$. We shall assume that the x and y deviations of the bullet from the center $(0, 0)$ of the target are *independent* of one another, which amounts to assuming that

$$P(R) = P(a \leqq X \leqq b)\cdot P(\alpha \leqq Y \leqq \beta) = \int_a^b f(x)\ dx \cdot \int_\alpha^\beta g(y)\ dy$$

$$= \int_R\!\!\int f(x)g(y)\ dx\ dy.$$

By subdividing the x, y-plane into small rectangles and passing to the limit we see that, in fact, this formula for $P(R)$ as a double integral holds for any region R, rectangular or not. Next, we shall assume that the problem possesses *circular symmetry* about $(0, 0)$, so that the probability $P(R)$ would be unchanged by a rigid rotation of the x, y-plane about $(0, 0)$. This amounts to assuming that the integrand $f(x)\cdot g(y)$ can be written in the form

(11) $$f(x)\cdot g(y) = \phi(r^2)$$

where $r^2 = x^2 + y^2$ is the square of the distance of (x, y) from $(0, 0)$. Differentiating both sides of (11) partially with respect to x and y and dividing by the original equation (11) we obtain

(12) $$\frac{f'(x)}{2xf(x)} = \phi_1(r^2)\frac{g'(y)}{2yg(y)} .$$

Since this is an identity in x and y it follows (why?) that the left-hand side of (12) is a constant c, and hence

$$[\log f(x)]' = 2cx$$

$$\log f(x) = cx^2 + d,$$

$$f(x) = e^d \cdot e^{cx^2} = De^{cx^2}.$$

Since necessarily

(13) $$\int_{-\infty}^{\infty} f(x)\, dx = 1$$

it follows that c must be negative, so we can write

$$f(x) = De^{-x^2/2\sigma^2}$$

where we have set $c = -(1/2\sigma^2)$, σ being an arbitrary positive constant. From the formula

$$\int_{-\infty}^{\infty} e^{-x^2/2}\, dx = \sqrt{2\pi},$$

proved in advanced calculus, it follows easily that D must have the value $1/\sqrt{2\pi}\sigma$ in order for (13) to hold, so that finally

(14) $$f(x) = \frac{1}{\sqrt{2\pi}\,\sigma}\, e^{-x^2/2\sigma^2},$$

the so-called *normal probability density function* (with mean 0 and standard deviation σ). A similar formula holds for $g(y)$.

From (14) it follows that in shooting at the target

(15) $$P(a \leq X \leq b) = \frac{1}{\sqrt{2\pi}\sigma} \int_a^b e^{-x^2/2\sigma^2}\, dx.$$

The value of σ characterizes the accuracy of the marksman in the x direction. To see this, let us determine the value of a such that

$$P(-a \leq X \leq a) = \tfrac{1}{2}.$$

We have from (15), setting $t = x/\sigma$,

(16) $$P(-a \leq X \leq a) = \frac{1}{\sqrt{2\pi}\sigma} \int_{-a}^{a} e^{-x^2/2\sigma^2}\, dx = \frac{1}{\sqrt{2\pi}} \int_{-a/\sigma}^{a/\sigma} e^{-t^2/2}\, dt,$$

and from tables of the integral

$$\Phi(x) = \frac{1}{\sqrt{2\pi}} \int_{-\infty}^{x} e^{-t^2/2}\, dt$$

we find that the right-hand side of (16) has the value $\tfrac{1}{2}$ for $a/\sigma \cong \tfrac{2}{3}$. Thus approximately

(17) $$P(-\tfrac{2}{3}\sigma \leq X \leq \tfrac{2}{3}\sigma) = \tfrac{1}{2}.$$

The larger the value of σ the wider an interval about the origin must be in order to contain about one half of all shots.

RANDOM VARIABLES

Let Ω with points ω be the sample space of a random experiment $\mathcal{E}$. Any real-valued function $X = X(\omega)$ defined for the points of Ω is called a *random variable*. A trial of the experiment will result in a point ω and hence in a numerical value of X. (As an example, the experiment might consist of a set of n tosses of a coin. Each point ω would then consist of a sequence of n letters H or T, and the random variable of interest might be the total number X of H's in the sequence. Or, as in the preceding section, the experiment might consist in shooting at a target and the random variable X might be the x-deviation of the point of impact.) The important thing about a random variable is its *probability distribution*: For any values $a \leq b$ we want to know the probability $P(a \leq X \leq b)$ that X will have a value between a and b. When Ω consists of a finite or denumerably infinite set of points ω, the possible values of any random variable X will also be finite or denumerably infinite. We denote them by x_1, x_2, $\cdots$, and say that X has a *discrete distribution*. Then $P(X = x_i) = P(x_i) =$ sum of all probabilities attached to points ω of Ω for which $X(\omega) = x_i$, and by the addition rule,

$$(18) \qquad P(a \leq X \leq b) = \sum_{a \leq x_i \leq b} P(x_i).$$

On the other hand when Ω contains more than a denumerable set of elements ω, $X = X(\omega)$ will in general have more than a denumerable set of values, and the probability $P(a \leq X \leq b)$ will usually be assumed to be given by the integral

$$\int_a^b f(x) \, dx$$

where $f(x)$ is a non-negative function with total integral 1 from $-\infty$ to $+\infty$. In this case X is said to have a *continuous distribution*.

Two important numerical characteristics of the probability distribution of a random variable X are the *mean* (or *expectation*) and *variance*, defined and denoted as follows:

$$(19) \quad \mu \, (mu) = E(X) = \text{mean of } X = \sum_i x_i P(x_i) \quad \text{or} \quad \int_{-\infty}^{\infty} x f(x) \, dx,$$

$$\sigma^2 \, (sigma \text{ squared}) = E(X - \mu)^2 = \text{variance of } X$$

$$(20) \qquad \qquad = Var \, X = \sum_i (x_i - \mu)^2 P(x_i) = \int_{-\infty}^{\infty} (x - \mu)^2 f(x) \, dx.$$

In the subsequent discussion we will confine ourselves to the discrete case to save space. It will be seen that μ is a measure of central tendency of the distribution, being a weighted average of the values of X, each value weighted by its probability. If $P(x_i)$ is regarded as a mass concentrated at x_i then the center of gravity of the resulting mass distribution on the x-axis would be

$$(21) \qquad \frac{\sum_i x_i P(x_i)}{\sum_i P(x_i)} = \sum_i x_i P(x_i) = \mu.$$

Likewise, σ^2 is a weighted average of the square of the distances of the values x_i from the mean value μ, and measures the extent to which the distribution is concentrated about its mean value μ. In terms of mass distribution σ^2 would represent the moment of inertia about the center of gravity. A probability distribution for which σ^2 is small has most of its probability concentrated close to its mean value μ. In the extreme case in which $\sigma^2 = 0$ all the probability would be concentrated at one point, the value μ.

Mean and variance are special cases of the *expectation of a function* $f(X)$ *of a random variable* X:

$$(22) \qquad E(f(X)) = \sum_i f(x_i)P(x_i).$$

For $f(x) = x$ we have

$$E(X) = \sum_i x_i P(x_i) = \mu,$$

and for $f(x) = (x - \mu)^2$ we have

$$E(X - \mu)^2 = \sum_i (x_i - \mu)^2 P(x_i) = \sigma^2.$$

A useful relation is

$$\sigma^2 = E(X - \mu)^2 = \sum_i (x_i - \mu)^2 P(x_i) = \sum_i (x_i^2 - 2\mu x_i + \mu^2) P(x_i)$$

$$= \sum_i x_i^2 P(x_i) - 2\mu \sum_i x_i P(x_i) + \mu^2 \sum_i P(x_i)$$

$$= \sum_i x_i^2 P(x_i) - 2\mu^2 + \mu^2 = \sum_i x_i^2 P(x_i) - \mu^2$$

or

$$(23) \qquad \sigma^2 = E(X^2) - [E(X)]^2.$$

An intuitive interpretation of the concept of the expectation $E(f(X))$ of a function of a random variable can be given as follows. Let the random experiment be performed n times and let each value x_i occur n_i times, $\sum_i n_i = n$. Then the *average value* of the function $f(X)$ taken over this set of n trials will be

(24)
$$\frac{\sum_i n_i f(x_i)}{n}.$$

But our interpretation of probability in terms of relative frequency implies that for large n we will have

$$\frac{n_i}{n} \cong P(x_i).$$

Hence the value of (24) should be nearly

(25)
$$\sum_i P(x_i)f(x_i) = E(f(X)).$$

Thus $E(f(X))$ represents our expectation of the *average value of $f(X)$ in a long series of trials.*

Thus far we have considered only one random variable $X(\omega)$ defined on the sample space Ω of a random experiment. Often there will be more than one interesting numerical function to be considered. Suppose for example that two such functions $X(\omega)$ and $Y(\omega)$ are defined on the same sample space Ω, and that X has values x_i and probability distribution $P(x_i)$, while Y has values y_j and probability distribution $P(y_j)$. A single trial of the experiment will then produce a pair of numbers (X, Y), say the pair (x_i, y_j). We define

$$P(x_i, y_j) = P\,[(X, Y) = (x_i, y_j)]$$
$$= \text{sum of all probabilities attached to points } \omega \text{ of } \Omega$$
$$\text{for which simultaneously}$$

$$X(\omega) = x_i \text{ and } Y(\omega) = y_j.$$

The numbers $P(x_i, y_j)$ define the *joint probability distribution* of X and Y. We now have

$$\sum_i \sum_j P(x_i, y_j) = 1.$$

Moreover, by the addition rule,

$$\sum_j P(x_i, y_j) = \text{probability that } X = x_i = P(x_i),$$

$$\sum_i P(x_i, y_j) = \text{probability that } Y = y_j = P(y_j).$$

The numbers $P(x_i)$ and $P(y_j)$ are sometimes called the *marginal distributions* of X and Y.

A very important case is that in which the events $X(\omega) = x_i$ and $Y(\omega) = y_j$ are *independent* for every i and j. This is equivalent to requiring that for all i and j,

$$P(x_i, y_j) = P(x_i) \cdot P(y_j).$$

Let $X = X(\omega)$ and $Y = Y(\omega)$ be any two random variables defined on the same sample space Ω, independent or not. We can then form the random variable $X + Y$ whose value for any outcome ω of $\mathcal{E}$ is $X(\omega) + Y(\omega)$. We shall show that

(26) $$E(X + Y) = E(X) + E(Y).$$

To prove (26) let

$$P(x_i, y_j) = P[X = x_i \text{ and } Y = y_j]$$
$$P(x_i) = P[X = x_i], \ P(y_j) = P[Y = y_j],$$

with the relations

$$P(x_i) = \sum_j P(x_i, y_j), \quad P(y_j) = \sum_i P(x_i, y_j).$$

Then

$$E(X + Y) = \sum_i \sum_j (x_i + y_j) P(x_i, y_j)$$
$$= \sum_i \sum_j x_i P(x_i, y_j) + \sum_i \sum_j y_j P(x_i, y_j)$$
$$= \sum_i \left(\sum_j P(x_i, y_j)\right) x_i + \sum_j \left(\sum_i P(x_i, y_j)\right) y_j$$
$$= \sum_i P(x_i) x_i + \sum_j P(y_j) \cdot y_j = E(X) + E(Y).$$

Now consider the random variable $X \cdot Y$, *under the assumption that X and Y are independent, i.e. that*

$$P(x_i, y_j) = P(x_i) \cdot P(y_j).$$

Then

$$E(X \cdot Y) = \sum_i \sum_j x_i y_j P(x_i y_j) = \sum_i \sum_j x_i y_j P(x_i) P(y_j)$$
$$= \left(\sum_i x_i P(x_i)\right)\left(\sum_j y_j P(y_j)\right) = E(X) \cdot E(Y).$$

Thus we have proved that

(27) $E(XY) = E(X) \cdot E(Y)$ *if X and Y are independent.*

In the same way we can show that if X_1, X_2, $\cdots$, X_n are *any* random variables defined on the same sample space then

(28) $E(X_1 + X_2 + \cdots + X_n) = E(X_1) + E(X_2) + \cdots + E(X_n),$

while if X_1, X_2, $\cdots$, X_n are *independent* then also

(29) $E(X_1 \cdot X_2 \cdots X_n) = E(X_1) \cdot E(X_2) \cdots E(X_n).$

Suppose now that X_1, X_2, $\cdots$, X_n are *independent* random variables defined on the same sample space, with individual means and variances given by

$$E(X_i) = \mu_i, \quad Var(X_i) = \sigma_i^2 .$$

From (28) we have that

$$E(X_1 + \cdots + X_n) = \sum_{i=1}^{n} \mu_i .$$

Likewise

$$Var(X_1 + \cdots + X_n) = E[(X_1 + \cdots + X_n) - (\mu_1 + \cdots + \mu_n)]^2$$

$$= E\left[\sum_{i=1}^{n} (X_i - \mu_i) \right]^2$$

$$= E\left[\sum_{i=1}^{n} (X_i - \mu_i)^2 + \sum\sum_{i \neq j} (X_i - \mu_i) \cdot (X_j - \mu_j) \right]$$

$$= \sum_{i=1}^{n} E(X_i - \mu_i)^2 + \sum\sum_{i \neq j} E[(X_i - \mu_i)(X_j - \mu_j)].$$

But

(30) $E[(X_i - \mu_i)(X_j - \mu_j)] = E(X_i X_j - \mu_i X_j - \mu_j X_i + \mu_i \mu_j)$

$$= E(X_i X_j) - \mu_i \mu_j - \mu_j \mu_i + \mu_i \mu_j$$

$$= E(X_i X_j) - \mu_i \mu_j$$

and since we have assumed that the random variables X_1, X_2, $\cdots$, X_n are independent it follows from (27) that the value of (30) is 0 whenever $i \neq j$. Hence

(31) $Var(X_1 + \cdots + X_n) = \sum_{i=1}^{n} \sigma_i^2.$

Thus *for a sum of independent random variables the mean of the sum is the sum of the means and the variance of the sum is the sum of the variances.* More generally, for any linear combination with constant coefficients

$$Y = a_1X_1 + a_2X_2 + \cdots + a_nX_n$$

of *independent* random variables,

(32)
$$\begin{cases} E(Y) = \sum_{i=1}^{n} E(a_iX_i) = \sum_{i=1}^{n} a_i\mu_i, \\[2mm] Var\,(Y) = \sum_{i=1}^{n} Var\,(a_iX_i) = \sum_{i=1}^{n} a_i^2\sigma_i^2. \end{cases}$$

In particular, if all the independent random variables $X_1, X_2, \cdots, X_n$ have the same mean μ and variance σ^2, then setting $a_i = 1/n$ in (32) we obtain

(33)
$$\begin{cases} E\left(\dfrac{X_1 + \cdots + X_n}{n}\right) = \mu, \\[4mm] Var\left(\dfrac{X_1 + \cdots + X_n}{n}\right) = \dfrac{\sigma^2}{n}. \end{cases}$$

Equations (33) are fundamental in statistics, for they show that the arithmetic mean of a series of n independent observations on a random variable X has a distribution with the same mean value μ as the original distribution but a variance σ^2/n which tends to zero as $n \to \infty$, so that the probability distribution of the arithmetic mean tends to concentrate around the constant μ for large n, thus providing a reliable estimate of μ when μ is unknown.

To make more precise the relation between the degree of concentration of a distribution and the size of its variance we reason as follows. Let X be any random variable with values x_i and probabilities $P(x_i)$, with mean μ and variance σ^2. Mark off on the x-axis an interval with center at μ and extending a length $t\sigma$ on either side of μ where t is an arbitrary positive number. Call this interval A,

$$A : \mu - t\sigma < x < \mu + t\sigma$$

By definition

(34)
$$\sigma^2 = \sum_i (x_i - \mu)^2 P(x_i).$$

If we denote by $\sum_i'$ the sum over only those values of i for which x_i lies *outside* the interval A then clearly

(35)
$$\sigma^2 \geq \sum_i' (x_i - \mu)^2 P(x_i).$$

But every term $(x_i - \mu)^2$ in this sum is greater than or equal to $t^2\sigma^2$, and hence

$$(36) \qquad\qquad \sigma^2 \geq t^2\sigma^2 \sum_i' P(x_i)$$

The sum of probabilities in (36) is simply the probability of the event $|X - \mu| \geq t\sigma$. Hence we can write

$$\sigma^2 \geq t^2\sigma^2 P[\,|X - \mu| \geq t\sigma\,]$$

or

$$(37) \qquad\qquad P[\,|X - \mu| \geq t\sigma\,] \leq 1/t^2.$$

This is *Chebyshev's inequality*. It is valid for any $t > 0$, although for $t \leq 1$ the statement is trivially true, and states that for any random variable X with mean μ and standard deviation σ the total amount of probability lying outside an interval extending a length $t\sigma$ on either side of μ cannot exceed $1/t^2$, and for large t this is near 0. For example, setting $t = 3$,

$$(38) \qquad\qquad P[\,|X - \mu| \geq 3\sigma\,] \leq \tfrac{1}{9}.$$

It is important to note that Chebyshev's inequality gives only an upper bound to a probability; for most distributions the probability of (38), for example, will be considerably less than $\tfrac{1}{9}$. For example, if it is known that X is *normally* distributed with mean μ and variance σ^2 then

$$P[\,|X - \mu| \geq 3\sigma\,] = \frac{1}{\sqrt{2\pi}} \int_{-\infty}^{-3} e^{-t^2/2}\, dt + \frac{1}{\sqrt{2\pi}} \int_{3}^{\infty} e^{-t^2/2}\, dt$$

$$= 2\left(1 - \frac{1}{\sqrt{2\pi}} \int_{-\infty}^{3} e^{-t^2/2}\, dt\right)$$

$$\cong 2(.00135) = .0027,$$

much less than $\tfrac{1}{9}$. It is the complete generality of Chebyshev's inequality, applying as it does to any random variable possessing a finite variance σ^2, that gives it its great importance in the theory of probability.

THE BINOMIAL AND ASSOCIATED DISTRIBUTIONS, LIMIT THEOREMS

Consider an experiment $\mathcal{E}$ in which there are only *two* possible outcomes, "success" and "failure," with respective probabilities p and $q = 1 - p$. (For example, $\mathcal{E}$ might consist of tossing a coin, with heads defined as success and tails as failure.) If $\mathcal{E}$ is performed n times independently, the total number of successes X will be a random variable with possible values $x = 0, 1, \cdots, n$. Let $P_n(x)$ denote the probability that $X = x$, *i.e.* the probability of the event consisting in obtaining exactly x successes out of the n trials. One way in which this event can occur is for the results of the n trials to be successively

(39)
$$\underbrace{S, S, \cdots, S,}_{x} \quad \underbrace{F, F, \cdots, F}_{n-x}$$

the probability of which is, by the assumption of independence of the n trials,

(40)
$$p \cdot p \cdots p \cdot q \cdot q \cdots q = p^x q^{n-x}.$$

But there are in all

$$\binom{n}{x} = \frac{n(n-1)(n-2)\cdots(n-x+1)}{x!} = \frac{n!}{x!(n-x)!}$$

different orders in which the xS's and the $(n-x)F$'s could appear in (39) (the number of different permutations of n letters of which x are S's and $(n-x)$ are F's), each such order of outcomes having the same probability (40) of occurrence, so that the total probability corresponding to the event $X = x$ is

(41)
$$P_n(x) = \binom{n}{x} p^x q^{n-x}.$$

A random variable X with values $x = 0, 1, 2, \cdots, n$ and with $P(X = x)$ given by (41) is said to have a *binomial distribution* with parameters n, p, the name arising from the fact that the terms (41) are the successive terms in the expansion of $(q + p)^n$ by the binomial theorem:

(42)
$$(q + p)^n = q^n + \frac{n}{1!} q^{n-1} p + \frac{n(n-1)}{2!} q^{n-2} p^2 + \cdots + p^n$$
$$= \sum_{x=0}^{n} \binom{n}{x} q^{n-x} p^x = \sum_{x=0}^{n} P_n(x).$$

Since $q + p = 1$ the left-hand side of (42) has the value 1, so that the sum of all the $P_n(x)$ for $x = 0, 1, 2, \cdots, n$ is 1, as it should be.

To compute the numerical values of $P_n(x)$ for $x = 0, 1, 2, \cdots, n$ it is convenient to start with $P_n(0) = q^n$ and to use the recurrence relation (easily verified from (41))

(43)
$$P_n(x) = \frac{p}{q} \cdot \frac{(n-x+1)}{x} \cdot P_n(x-1)$$

for $x = 1, 2, \cdots, n$.

To see qualitatively how the values $P_n(x)$ vary as x increases from 0 to n we observe from (43) that the ratio

$$(44) \qquad \frac{P_n(x)}{P_n(x-1)} = \frac{p}{q} \cdot \frac{(n-x+1)}{x}$$

of each pair of successive terms will be greater than 1 as long as

$$(45) \qquad \frac{p}{q} \cdot \frac{(n-x+1)}{x} > 1,$$

i.e. as long as

$$(46) \qquad x < (n+1)p.$$

Thus the successive terms $P_n(0)$, $P_n(1)$, $P_n(2)$, $\cdots$ steadily increase as x increases as long as $x < (n+1)p$. As soon as $x > (n+1)p$ the terms begin to decrease steadily. *It follows that the maximum value of $P_n(x)$ is reached when x is the greatest positive integer which is less than $np + p$.* Henceforth we shall denote this integer by ν; ν is always within 1 of np since it is less than $np + p$ and greater than $np + p - 1 = np - q$. Hence we can write with an error less than 1, $\nu \cong np$. (If $np + p$ is itself an integer then both $\nu = np - q$ and $\nu + 1 = np + p$ afford equal maxima to $P_n(x)$; if $np + p$ is not an integer then the maximum of $P_n(x)$ is unique.) It will be instructive for the reader to compute by (43) the value of $P_n(x)$ for $p = \frac{1}{2}$ and $n = 1, 5, 10$ and also for $p = \frac{1}{3}$ and the same values of n, in each case drawing a diagram in which the abscissa is x and the ordinate $P_n(x)$.

The probability $P_n(\nu)$ corresponding to the most probable value $\nu \cong np$ of X can be found for large n, when the direct evaluation of the factorials in (41) is difficult, by using *Stirling's formula*, proved in advanced calculus, which states that for any integer n,

$$(47) \qquad n! = \{\sqrt{2\pi}\, n^{n+1/2} e^{-n}\} e^{r_n},$$

where the exponent r_n satisfies the inequalities

$$(48) \qquad \frac{1}{12n + 1} < r_n < \frac{1}{12n},$$

so that e^{r_n} is close to $e^{1/12n} \cong 1$ for large n. Thus

$$(49) \qquad n! \sim \sqrt{2\pi}\, n^{n+1/2} e^{-n}$$

for large n, in the sense that the ratio

$$\frac{n!}{\sqrt{2\pi}\, n^{n+1/2} e^{-n}} = e^{r_n}$$

tends to 1 as $n \to \infty$. Thus

$$P_n(\nu) = \frac{n!}{\nu!(n-\nu)!} \, p^\nu q^{n-\nu}$$

$$\sim \frac{\sqrt{2\pi} \, n^{n+1/2} \, e^{-n} p^\nu q^{n-\nu}}{\sqrt{2\pi} \, \nu^{\nu+1/2} e^{-\nu} \cdot \sqrt{2\pi} \, (n-\nu)^{n-\nu+1/2} e^{-(n-\nu)}}.$$

Replacing ν by its approximate value np this simplifies to

$$\frac{n^{n+1/2} p^{np} q^{nq}}{\sqrt{2\pi} \, (np)^{np+1/2} (nq)^{nq+1/2}} = \frac{1}{\sqrt{2\pi npq}},$$

so that finally

(50) $$P_n(\nu) \sim \frac{1}{\sqrt{2\pi npq}} \qquad \text{as} \quad n \to \infty,$$

in the sense that

(51) $$\lim_{n \to \infty} \frac{P_n(\nu)}{\dfrac{1}{\sqrt{2\pi npq}}} = 1.$$

Thus for large n even the maximum probability $P_n(\nu)$ tends to 0 like $1/\sqrt{n}$. For example, if a coin with $p = q = \frac{1}{2}$ is tossed $n = 100$ times the most probable number of heads will be $\nu = 50$ and the probability $P_{100}(50)$ will be approximately

$$\frac{1}{\sqrt{2\pi npq}} = \sqrt{\frac{2}{\pi}} \cdot \frac{1}{\sqrt{100}} = \sqrt{\frac{2}{\pi}} \cdot \frac{1}{10} \cong .08$$

To find the *mean* and *variance* of the binomial distribution of X we observe that we can write

(52) $$X = X_1 + X_2 + \cdots + X_n$$

where $X_i = 1$ if the ith trial is a success and $X_i = 0$ if the ith trial is a failure. The auxiliary random variables X_i are independent. It follows that

$$EX_i = q \cdot 0 + p \cdot 1 = p,$$
$$EX_i^2 = q \cdot 0^2 + p \cdot 1^2 = p,$$
$$\sigma_{X_i}^2 = EX_i^2 - (EX_i)^2 = p - p^2 = p(1-p) = pq,$$

and hence by (52) and (32)

$$EX = p + p + \cdots + p = np,$$
(53) $$\sigma_X^2 = pq + pq + \cdots + pq = npq,$$
$$\sigma_X = \sqrt{npq}.$$

The values (53) could of course be computed directly from the general definition of mean and variance:

$$EX = \sum_{x=0}^{n} P(x) \cdot x = \sum_{x=0}^{n} \frac{n!}{x!(n-x)!} p^x q^{n-x} \cdot x,$$

etc., but the method used above is much simpler and more instructive.

Applied to the binomial distribution, Chebyshev's inequality (37) states that for any positive number t,

$$P[|\ X - np\ | \geq t\sqrt{npq}] = P[|\ X/n - p\ | \geq t\sqrt{pq/n}] \leq 1/t^2,$$

or, replacing t by $\epsilon\sqrt{n/pq}$, where ϵ is an arbitrary positive number,

(54) $P[|\ X/n - p\ | \geq \epsilon] \leq pq/n\epsilon^2.$

If we hold ϵ fixed in (54) and let $n \to \infty$ it follows that

$$\lim_{n\to\infty} P[\ |\ X/n - p\ | \geq \epsilon] = 0$$

so that

(55) $$\lim_{n\to\infty} P[\ |\ X/n - p\ | < \epsilon] = 1.$$

This relation expresses one of the oldest and most important limit theorems of probability theory, due to Jacob Bernoulli: *For any positive ϵ, no matter how small, the probability that in n independent trials the success ratio X/n will deviate from the true probability p of success by less than ϵ tends to 1 as n becomes infinite.* Thus although the probability distribution $P_n(x)$ of X itself becomes more and more spread out from 0 to n as $n \to \infty$, the probability distribution of X/n becomes more and more concentrated in an arbitrarily small neighborhood of the point p. It is for this reason that we can estimate the unknown probability p of any event E by taking as our estimate of p the proportion of times, X/n, that E occurs in n independent trials of the experiment in question. Inequality (54) shows, for example, that for $\epsilon = .1$

$$P\left[\left|\frac{X}{n} - p\right| \geq .1\right] \leq \frac{100\ pq}{n} \leq \frac{100 \cdot \frac{1}{4}}{n} = \frac{25}{n},$$

since the product $pq = p(1 - p)$ takes its maximum $\frac{1}{4}$ when $p = 1 - p = \frac{1}{2}$. Thus

(56) $$P\left[\left|\frac{X}{n} - p\right| \geq .1\right] \leq \begin{cases} .25 & \text{if } n \geq 100 \\ .1 & \text{if } n \geq 250 \\ .01 & \text{if } n \geq 2500, \text{ etc.} \end{cases}$$

Actually these inequalities are very crude and can be sharpened considerably by more refined computations, as we shall see later.

It is often necessary to compute the value of $P_n(x)$ in the binomial distribution (41) for large values of n and small values of p. Suppose that for fixed x we let $n \to \infty$ and $p \to 0$ in such a way that the product $\mu = np = EX$ is held constant. Then

$$P_n(x) = \frac{n(n-1)(n-2) \cdots (n-x+1)}{x!} p^x q^{n-x}$$

$$= \frac{1\left(1 - \frac{1}{n}\right)\left(1 - \frac{2}{n}\right) \cdots \left(1 - \frac{x-1}{n}\right)}{x!} (np)^x \cdot q^{-x} \cdot q^n$$

$$= \frac{1\left(1 - \frac{1}{n}\right)\left(1 - \frac{2}{n}\right) \cdots \left(1 - \frac{x-1}{n}\right)}{x!} \mu^x \left(1 - \frac{\mu}{n}\right)^{-x} \cdot \left(1 - \frac{\mu}{n}\right)^n.$$

Since for any constant a,

$$\lim_{n \to \infty} \left(1 + \frac{a}{n}\right)^n = e^a,$$

it follows that for fixed x,

$$(57) \qquad \lim_{\substack{n \to \infty \\ p = \mu/n}} P_n(x) = \frac{1 \cdot 1 \cdot 1 \cdots 1}{x!} \mu^x \cdot 1^{-x} \cdot e^{-\mu} = e^{-\mu} \cdot \frac{\mu^x}{x!}.$$

A random variable X with values $x = 0, 1, 2, \cdots$ is said to have a *Poisson distribution* with parameter μ if $P[X = x]$ is given by

$$(58) \qquad P(x) = e^{-\mu} \cdot \frac{\mu^x}{x!}.$$

The relation (57) asserts that for large n and small p the binomial distribution $P_n(x)$ is approximated by the Poisson distribution $P(x)$ with parameter $\mu = np$.

Considered as a distribution in its own right the Poisson distribution (58) is important in many fields of application. We observe that it is in fact a legitimate probability distribution on the set of integers 0, 1, 2, $\cdots$ since the series expansion

$$e^a = \sum_{x=0}^{\infty} \frac{a^x}{x!}$$

assures us that

$$\sum_{x=0}^{\infty} P(x) = \sum_{x=0}^{\infty} e^{-\mu} \frac{\mu^x}{x!} = e^{-\mu} \cdot e^{\mu} = e^0 = 1.$$

If X is a random variable with the Poisson distribution (58) its mean and variance are easily found to be

$$(59) \qquad\qquad EX = \mu, \qquad \sigma_X^2 = \mu.$$

These values are what we would expect from the derivation of the Poisson distribution as the limit of the binomial with $np = \mu = \text{constant}, n \to \infty$, $p \to 0$, since under these circumstances

$$\text{binomial mean} = np = \mu$$

$$\text{binomial variance} = npq \to \mu \cdot 1 = \mu.$$

An amusing example of the use of the Poisson distribution is provided by the following problem. Given that there are "on the average" μ raisins per cookie in a certain batch of cookies, what is the probability that a cookie chosen at random will contain exactly x raisins?

We reason as follows: Let the total volume of dough from which the batch of cookies is made be V and the volume of a single cookie be v. Into the volume V we assume that n raisins are placed independently and at random, so that for any single cookie the probability of each raisin landing in it is $p = v/V$. The probability of a single cookie containing exactly x of the n raisins is then given by the binomial formula

$$P_n(x) = \binom{n}{x} p^x q^{n-x}.$$

The average number of raisins per cookie is

$$\frac{\text{total no. of raisins}}{\text{total no. of cookies}} = \frac{n}{\left(\dfrac{V}{v}\right)} = n \cdot \frac{v}{V} = np = \mu.$$

If now both n and V increase while v and μ are fixed then $n \to \infty$, $p = v/V \to 0$, $np = \mu$ and hence by (57)

$$(60) \qquad P_n(x) \to P(x) = e^{-\mu} \cdot \frac{\mu^x}{x!}; \qquad x = 0, 1, \cdots.$$

Thus, for example if there are on the average 3 raisins per cookie the probability of a single cookie containing *less than* 3 raisins will be approximately

$$(61) \qquad e^{-3}\left(1 + \frac{3}{1!} + \frac{3^2}{2!}\right) = e^{-3}\left(\frac{25}{4}\right) \cong .31.$$

From the point of view of statistical inference, if it is found empirically that the proportion of cookies in a large sample which contain less than

3 raisins is appreciably larger than the value (61) it would be reasonable to infer that $\mu < 3$.

It is of great interest to consider the binomial distribution *for fixed* p as $n \to \infty$, so that $\mu = np \to \infty$ and the Poisson approximation is useless. We have already seen that if $\nu \cong np$ denotes the most probable value of a random variable X with a binomial distribution then

$$(62) \qquad P_n(\nu) \sim \frac{1}{\sqrt{2\pi npq}} \qquad \text{as} \qquad n \to \infty,$$

and that

$$\mu = EX = np, \qquad \sigma_X = \sqrt{npq}.$$

If we introduce the *reduced random variable*

$$Y = \frac{X - \mu}{\sigma_X} = \frac{X - np}{\sqrt{npq}},$$

then for all values $n = 1, 2, \cdots$ we will have

$$EY = 0, \qquad \sigma_Y = 1.$$

To get an idea of the distribution of Y as n increases it is useful to plot in a y,z-coordinate plane the value of y as abscissa and the value of $\sqrt{npq} \cdot P[Y = y]$ as ordinate. Since X runs from 0 to n in increments of 1, Y will run from $-np/\sqrt{npq}$ to $nq/\sqrt{npq}$ in increments of $1/\sqrt{npq}$, and for $x = 0, 1, \cdots, n$

$$P\left[Y = \frac{x - np}{\sqrt{npq}}\right] = P[X = x] = P_n(x) = \frac{n!}{x!(n - x)!}p^x q^{n-x}.$$

If the endpoints of the ordinates, whose coordinates are

$$(63) \qquad \left(\frac{x - np}{\sqrt{npq}}, \sqrt{npq} \cdot P_n(x)\right); \qquad x = 0, 1, \cdots, n,$$

are plotted carefully in the y, z-plane it will be found that as $n \to \infty$ they tend more and more closely to lie on a smooth curve whose equation is

$$(64) \qquad z = \frac{1}{\sqrt{2\pi}} e^{-y^2/2},$$

already encountered in (14) (Fig. 1). We shall not give an analytical proof of this fact here, except to observe that for the most probable value $x = \nu$ the point (63) becomes

$$(65) \qquad \left(\frac{\nu - np}{\sqrt{npq}}, \sqrt{npq} \cdot P_n(\nu)\right),$$

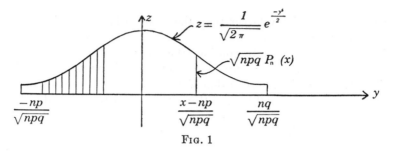

FIG. 1

and from (62) and the fact that ν is within 1 of np it follows that the point (65) tends to the limiting position

$$\left(0, \frac{1}{\sqrt{2\pi}}\right),$$

which is the point where the curve (64) intersects the z-axis.

The relation between the points (63) and the curve (64) may be expressed by the statement that if $y = \dfrac{x - np}{\sqrt{npq}}$ then

$$(66) \qquad\qquad \sqrt{npq}\, P_n(x) \sim \frac{1}{\sqrt{2\pi}}\, e^{-y^2/2},$$

the approximation being better on the whole the larger the value of n.

It is remarkable that although the probability distribution of X is unsymmetric when $p \neq \frac{1}{2}$, it nevertheless tends to become symmetric about the value np as $n \to \infty$.

The value of the approximation (66) goes far beyond that of facilitating the computation of a single value of $P_n(x)$. Suppose in fact that we require the probability that the binomially distributed random variable X shall lie within the two limits

$$(\mu + a\sigma_x,\ \mu + b\sigma_x) = (np + a\sqrt{npq},\ np + b\sqrt{npq})$$

where a and b are any two constants such that $a < b$. This will be equal to $1/\sqrt{npq}$ times the sum of the ordinates corresponding to points on the y-axis in the interval (a, b) or, equivalently, to the sum of the areas of rectangles erected on each of these ordinates and of width $1/\sqrt{npq}$. This will be approximately equal to the area under the curve

$$z = \frac{1}{\sqrt{2\pi}}\, e^{-y^2/2}$$ between a and b, and the approximation becomes better

and better for fixed a, b as $n \to \infty$. That is, for any fixed numbers $a < b$, we have

(67) $\quad \lim_{n \to \infty} P\left[a < \dfrac{X - np}{\sqrt{npq}} < b\right] = \dfrac{1}{\sqrt{2\pi}} \displaystyle\int_a^b e^{-v^2/2}\, dy = \Phi(b) - \Phi(a),$

where by definition

(68) $\qquad\qquad\qquad \Phi(x) = \dfrac{1}{\sqrt{2\pi}} \displaystyle\int_{-\infty}^x e^{-v^2/2}\, dy$

is the function giving the area under the "normal" curve $1/\sqrt{2\pi}\, e^{-v^2/2}$ from $y = -\infty$ to $y = b$. The total area under this curve is shown in advanced calculus to be 1, and by symmetry

$$\Phi(x) = 1 - \Phi(-x).$$

Tables of Φ are readily available. A short one is given below:

x	$\Phi(x)$
0	.50000
1	.84134
2	.97725
3	.99865
4	.99997

For example, in 10,000 tosses of an unbiased coin: $n = 10,000$; $p = q = \frac{1}{2}$; $np = 5,000$; $\sqrt{npq} = 50$. Choosing $b = 2$, $a = -2$ in (67) we see that the number of heads obtained will lie between $np - 2\sqrt{npq} = 4,900$ and $np + 2\sqrt{npq} = 5,100$ with probability $\Phi(2) - \Phi(-2) = 2\,\Phi(2) - 1 \cong .95$.

A random variable Y continuously distributed from $-\infty$ to ∞ is said to be *normal* (with mean 0 and variance 1) if for any two numbers $a < b$,

$$P[a < Y < b] = \Phi(b) - \Phi(a).$$

Thus we can say: *If X is a binomially distributed random variable with parameters n, p and if p is fixed while $n \to \infty$, then the distribution of the reduced random variable*

$$Y = \dfrac{X - np}{\sqrt{npq}}$$

tends to the normal. This theorem, due to Abraham de Moivre, ranks with Bernoulli's theorem in importance.

We have already observed that a random variable X with a binomial distribution can be regarded as the sum

$$X = X_1 + \cdots + X_n$$

of n independent random variables, each with values 0 or 1 and with distributions

$$P[X_i = 0] = q, \qquad P[X_i = 1] = p,$$

so that

$$EX_i = p, \qquad \sigma_{X_i} = \sqrt{pq}.$$

Bernoulli's law of large numbers states that for any $\epsilon > 0$, no matter how small,

$$\lim_{n \to \infty} P\left[\left|\frac{X}{n} - p\right| < \epsilon\right] = 1$$

while de Moivre's theorem states that for any two numbers $a < b$,

$$\lim_{n \to \infty} P\left[a < \frac{X - np}{\sqrt{npq}} < b\right] = \frac{1}{\sqrt{2\pi}} \int_a^b e^{-y^2/2}\, dy.$$

Both of these theorems admit of the following far-reaching generalizations. Let

$$X = X_1 + \cdots + X_n$$

be a sum of n independent random variables, each with the same probability distribution, completely arbitrary except that

$$\mu = EX_i \qquad \text{and} \qquad \sigma = \sqrt{E(X_i - \mu)^2}$$

exist as finite numbers. Then for any $\epsilon > 0$,

$$(69) \qquad \lim_{n \to \infty} P\left[\left|\frac{X}{n} - \mu\right| < \epsilon\right] = 1,$$

and for any two numbers $a < b$,

$$(70) \qquad \lim_{n \to \infty} P\left[a < \frac{X - n\mu}{\sigma\sqrt{n}} < b\right] = \frac{1}{\sqrt{2\pi}} \int_a^b e^{-y^2/2}\, dy.$$

Thus the average value $(X_1 + X_2 + \cdots + X_n)/n$ becomes more and more concentrated in probability about the constant μ, while the distribution of the reduced random variable

$$\frac{(X_1 + \cdots + X_n) - n\mu}{\sigma\sqrt{n}}$$

becomes more and more nearly normal as $n \rightarrow \infty$. The relation (69) is called the *Law of Large Numbers* while (70) is called the *Central Limit Theorem*. The proof of (69) can be carried out exactly as in the binomial case, while the proof of (70) is considerably more difficult and will not be given here. Both theorems are of the greatest importance in applications of probability theory to statistical problems.

A PROBLEM IN COMBINATORIAL PROBABILITY

Suppose the cards in one deck are laid out individually face down and then the cards of another deck are well shuffled and laid down one by one on the cards of the first deck. We say there is a *match* whenever the two cards in a single pile are identical. The total number of matches between the two decks will vary from trial to trial and therefore forms a random variable X. What is the probability P_k that X will have the value $k = 0, 1, 2, \cdots$?

We shall suppose that there are n different cards in each deck, and for simplicity we shall denote them by the integers $1, 2, \cdots, n$. The result of a single trial of the experiment will then be represented by the two sequences of integers

$$(71) \qquad \begin{cases} 1, 2, 3, \cdots, n-1, n \\ r_1, r_2, r_3, \cdots, r_{n-1}, r_n, \end{cases}$$

each number r_i of the second row representing the card of the second deck which is paired with card i of the first deck. The total number of possible outcomes of the experiment is equal to $n!$, the total number of different permutations $r_1, r_2, \cdots, r_n$ of the integers $1, 2, \cdots, n$, and we shall suppose all these permutations to be equally likely. Let A_k denote the number of these permutations in which there are exactly k matches, *i.e.* in which $r_i = i$ exactly k times in (71) as i ranges from 1 to n. Then $P_k = A_k/n!$, and it remains to find a formula for the numbers A_k.

As a first step we shall find the number A_0 of permutations (71) with *no* matches. Since this number depends on the size n of the decks we shall for the moment denote it by b_n to indicate this dependence. We can then find a recursion formula for b_n by considering what the situation would be if there were $n + 1$ cards in each deck and therefore b_{n+1} permutations of the integers $1, 2, \cdots, n + 1$ with no matches. Consider first the class of all such permutations with no matches in which card 1 of the second deck is paired with card $n + 1$ of the first deck, *i.e.* in which $r_{n+1} = 1$. Card $n + 1$ of the second deck must then either be paired with card 1 of the first deck or with some other card. If it is paired with card 1 of the first deck then cards 2, 3, $\cdots$, n of the second deck are paired in some order with the same cards of the first deck, and the number of ways this can be done with no matches is b_{n-1} by definition. On the other hand, if card $n + 1$ of the second deck is not paired with card 1 of the first deck we can for the moment relabel it as "1" and consider the number of ways in which the cards "1", 2, $\cdots$, n of the second deck can be paired with cards 1, 2, $\cdots$, n of the first deck without any matches, this number being b_n. Thus in all there are $b_{n-1} + b_n$ ways of pairing the two decks of $n + 1$ cards each without any matches, in such a way that card 1 of the second deck is paired with card $n + 1$ of the first deck. But any of the cards 1, 2, $\cdots$, n of the second deck could be paired with card $n + 1$ of the first deck and for each of these n cases there would be $b_{n-1} + b_n$ permutations of the remaining cards of the second deck without any matches with the first deck. Thus we have the recursion formula

$$(72) \qquad\qquad b_{n+1} = n(b_{n-1} + b_n) \qquad\qquad (n = 2, 3, \cdots).$$

If we now denote the *probability* of no matches in pairing two decks of n cards each by $a_n = b_n/n!$ then from (72) we have the recursion formula

$$(n + 1)!\, a_{n+1} = n[(n - 1)!\, a_{n-1} + n!\, a_n]$$

or after simplification,

$$a_{n+1} = \frac{n}{n + 1} \cdot a_n + \frac{1}{n + 1} \cdot a_{n-1},$$

which can be written in the form

$$(73) \qquad\qquad a_{n+1} - a_n = -\frac{1}{n + 1}(a_n - a_{n-1}).$$

This suggests introducing the differences

$$d_{n+1} = a_{n+1} - a_n,$$

in terms of which (73) becomes

$$(74) \qquad\qquad d_{n+1} = -\frac{1}{n + 1} \cdot d_n.$$

It is obvious that $b_1 = 0$ and $b_2 = 1$, so that $a_1 = 0/1! = 0$ and $a_2 = \frac{1}{2}! = \frac{1}{2}$ and hence

$$d_2 = \frac{1}{2} - 0 = \frac{1}{2}.$$

Now by (74) we have successively

$$d_3 = -\frac{1}{3} \cdot d_2 = -\frac{1}{2.3},$$

$$d_4 = -\frac{1}{4} \cdot d_3 = \frac{1}{2.3.4},$$

(75)

.

.

.

$$d_n = \frac{(-1)^n}{n!}.$$

It follows that for $n \geq 2$

$$a_n = a_1 + (a_2 - a_1) + (a_3 - a_2) + \cdots + (a_n - a_{n-1})$$

$$= d_2 + d_3 + \cdots + d_n = \frac{1}{2!} - \frac{1}{3!} + \frac{1}{4!} - \cdots + (-1)^n \frac{1}{n!},$$

and hence that for two decks of n cards each the probability of no matches is

$$(76) \quad P_0 = P[X = 0] = a_n = \frac{1}{2!} - \frac{1}{3!} + \frac{1}{4!} - \cdots + (-1)^n \cdot \frac{1}{n!}.$$

It remains to find $P_k = P[X = k]$ for two decks of n cards each. To do this we must determine the numbers A_k defined above, and this is now quite simple. Consider first the number of permutations of the integers $1, 2, \cdots, n$ in which the integers $1, 2, \cdots, k$ match and the rest do not. There are b_{n-k} of these. But there are $\binom{n}{k}$ ways of choosing k pairs to match out of the n pairs, to each of which correspond b_{n-k} permutations of the remaining integers without matches. Hence

$$A_k = \binom{n}{k} b_{n-k},$$

and thus finally by (76),

$$P_k = \frac{A_k}{n!} = \frac{1}{n!} \cdot \frac{n!}{k!(n-k)!} \cdot (n-k)! \, a_{n-k}$$

(77)

$$= \frac{1}{k!} \left[\frac{1}{2!} - \frac{1}{3!} + \frac{1}{4!} - \cdots + (-1)^{n-k} \cdot \frac{1}{(n-k)!} \right].$$

This formula is valid for $k = 0, 1, 2, \cdots, n - 2$. We also have

(78)
$$P_{n-1} = \frac{A_{n-1}}{(n-1)!} = \frac{\binom{n}{n-1}}{(n-1)!} b_1 = 0$$

(which is obvious anyway as there *cannot* be exactly $n - 1$ matches in two decks of n cards), and

(79)
$$P_n = \frac{A_n}{n!} = \frac{1}{n!},$$

since A_n is obviously 1. Our problem is now completely solved.

As a numerical example we have for $n = 5$ the following probability distribution for the number of matches:

Number of matches	0	1	2	3	4	5
Probability..............	.367	.375	.167	.083	0	.008

It is noteworthy that the probability of *no* matches in two decks of n cards is by (76)

(80)
$$P_0 = \frac{1}{2!} - \frac{1}{3!} + \frac{1}{4!} - \cdots + (-1)^n \cdot \frac{1}{n!}.$$

The series expansion, valid for all x,

$$e^x = 1 + \frac{x}{1!} + \frac{x^2}{2!} + \frac{x^3}{3!} + \frac{x^4}{4!} + \cdots,$$

shows that

(81)
$$e^{-1} = 1 - 1 + \frac{1}{2!} - \frac{1}{3!} + \frac{1}{4!} - \cdots + (-1)^n \frac{1}{n!} + \cdots.$$

Since this is an alternating series with decreasing terms it follows that the error involved in stopping with the term $1/n!$ in (81) is less than the value of the next term, $1/(n + 1)!$. Hence with this approximation we can write

(82)
$$P_0 = e^{-1} \cong .368$$

for any large n. The probability of obtaining *at least one* match is then

(83)
$$1 - P_0 \cong .632.$$

GAMBLER'S RUIN

The early concern of probability theory with games of chance is well illustrated in the problem of "gambler's ruin." Peter and Paul play a succession of games, in each of which the stake is one dollar. The probabilities of winning each game for Peter and Paul are respectively p and $q = 1 - p$. Initially Peter has a dollars and Paul b dollars, and they agree to play until one or the other has lost all his money. What is the probability P that it is Peter who will be ruined?

At any stage of play Peter will have a certain capital, say x dollars, $x = 0, 1, \cdots, a + b$, and Paul will have $a + b - x$ dollars. Let $f(x)$ denote the *conditional probability* that it is Peter who will ultimately be ruined during the subsequent play, given that he has x dollars at the moment and that Paul has $a + b - x$ dollars. The probability that Peter will ultimately be ruined when his initial capital is a dollars is then $P = f(a)$. In order to find $f(a)$ we shall in fact find the values of the function $f(x)$ for all $x = 0, 1, \cdots, a + b$. It is clear that

$$(84) \qquad\qquad f(0) = 1, \qquad f(a + b) = 0,$$

for if at any moment Peter has 0 dollars he is already ruined and the play stops, while if he has $a + b$ dollars Paul is already ruined. Now we assert that $f(x)$ must satisfy the following *difference equation*:

$$(85) \quad f(x) = pf(x + 1) + qf(x - 1); \qquad x = 1, 2, \cdots, a + b - 1.$$

For, given that Peter's capital is x at the moment, for him to be ruined during the subsequent play he must *either*

(A) win the next game *and* then, with a capital of $x + 1$ dollars, go on to be ruined: $P(A) = p \cdot f(x + 1)$, *or*

(B) lose the next game *and* then, with a capital of $x - 1$ dollars, go on to be ruined: $P(B) = q \cdot f(x - 1)$, so that by the addition rule,

$$f(x) = P(A) + P(B) = p \cdot f(x + 1) + q \cdot f(x - 1),$$

as was to be shown.

We now have the mathematical problem of solving the difference equation (85) subject to the boundary conditions (84). As in the analogous problem of solving a linear differential equation with constant coefficients, we first try to find solutions of (85) of the form $f(x) = A^x$, where A is some constant. For this to be a solution it is necessary and sufficient that

$$A^x = pA^{x+1} + qA^{x-1}$$

or

$$pA^2 - A + q = 0,$$

the two solutions of which are found to be $A = 1$ and $A = \lambda = q/p$. For the moment we shall assume $p \neq \frac{1}{2}$ so that 1 and λ are different. We can then form the "general" solution of (85) containing two arbitrary constants

$$f(x) = c_1 \cdot 1^x + c_2 \cdot \lambda^x = c_1 + c_2 \lambda^x.$$

But c_1 and c_2 must be chosen so as to satisfy the boundary conditions (84), so that

$$c_1 + c_2 = 1, \qquad c_1 + c_2 \lambda^{a+b} = 0,$$

the solution of which is

$$c_1 = \frac{-\lambda^{a+b}}{1 - \lambda^{a+b}}, \qquad c_2 = \frac{1}{1 - \lambda^{a+b}}.$$

Hence the function $f(x)$ that we seek is

$$f(x) = \frac{-\lambda^{a+b}}{1 - \lambda^{a+b}} + \frac{\lambda^x}{1 - \lambda^{a+b}},$$

and the probability P of Peter's ultimate ruin is

(86) $$P = f(a) = \frac{\lambda^a(1 - \lambda^b)}{1 - \lambda^{a+b}} \qquad (\lambda = q/p, p \neq \frac{1}{2}).$$

If we denote the probability of Paul's ultimate ruin by Q, then by symmetry it follows that Q will be obtained from (86) by replacing λ by $1/\lambda$ and interchanging a and b. In this way we find that

(87) $$Q = \frac{1 - \lambda^a}{1 - \lambda^{a+b}}.$$

From (86) and (87) it follows that $P + Q = 1$ so that either Peter or Paul is "certain" to be ruined eventually, *i.e.* the probability $1 - (P + Q)$ that the sequence of play will last forever under the given conditions is zero.

To cover the excluded case $p = q = \frac{1}{2}$ we allow λ to tend to 1 in (86); by l'Hôpital's rule, or directly, we find in this case that

(88) $$P = \frac{b}{a + b}, \qquad Q = \frac{a}{a + b} \qquad (p = q = \frac{1}{2}),$$

so that the probabilities of ultimate ruin for each player are inversely proportional to their initial stakes.

It is of interest to consider the case in which $p > q$, so that $\lambda < 1$, and $b \rightarrow \infty$, a remaining constant. This will be the case if Peter represents a gambling house which has an advantage in the odds but which must play against all comers, *i.e.* against an adversary who is for all practical purposes infinitely rich. We see from (86) that in this case

$$(89) \qquad\qquad\qquad P \rightarrow \lambda^a \qquad\qquad (\lambda = q/p < 1, b \rightarrow \infty).$$

(Of course, in this case Paul can never be ruined, so that $1 - P = 1 - \lambda^a$ represents the probability of an infinite duration of play.) By choosing a large enough, λ^a can be made arbitrarily small. Thus a gambler with an advantage in the odds can reduce his probability of ruin to as small a number as desired by starting with a large but finite initial capital, even if his adversary is infinitely rich. For example, if $p = .55$, $q = .45$ so that $\lambda = 9/11$, and if $a = \log_{9/11} .001 \cong 35$ then as $b \rightarrow \infty$

$$P \rightarrow \lambda^a = .001.$$

The problem of gambler's ruin has another interpretation in terms of "random walk." Starting from the origin on a number axis, a point moves one unit to the right or left according as a coin falls heads or tails. The coin is tossed repeatedly and the moving point observed until it reaches either the point $-a$ or $+b$, where a and b are fixed positive numbers. The probability P computed above is then the probability that the point will reach $-a$ before it reaches $+b$, and similarly for Q. This interpretation of the problem has recently been used in statistics to obtain *sequential sampling plans* in acceptance sampling of industrial products. In this application it is important to find the probability distribution, and in particular the expected value, of the random variable representing the "duration of play," *i.e.* the number of steps taken by the moving point before it reaches either $-a$ or $+b$. We shall not go into this problem here.

A STOCHASTIC PROCESS

The disintegration of a radioactive substance is an example of a *stochastic process*—a sequence of events occurring in time and fluctuating according to the laws of chance. Suppose we could observe the process of disintegration with a Geiger counter and could mark a dot on a time axis at each instant that a single atom disintegrates. As time goes on we would find an irregular pattern in the appearance of the dots. During successive intervals of some fixed length t we might observe for example

3 dots, 1 dot, 0 dots, *etc.* After a long series of such intervals we could tabulate the relative frequencies with which an interval of length t contains 0, 1, 2, $\cdots$ dots and in this way form some idea of the values of the *probabilities* $p_n(t)$ that a time interval of length t chosen at random will contain exactly n dots, $n = 0, 1, 2, \cdots$. A priori all that we know about the functions $p_n(t)$ is that

(90) $$p_n(t) \geqq 0, \qquad \sum_{n=0}^{\infty} p_n(t) = 1,$$

and that

(91) $$p_0(0) = 1, \qquad p_n(0) = 0 \qquad \text{for } n = 1, 2, \cdots .$$

We shall now show how on the basis of certain reasonable probability assumptions we can find the explicit form of the functions $p_n(t)$. Our assumptions are as follows:

(A) The probability that a time interval will contain n dots depends only on the *length* of the interval and not on its position on the time axis.

(B) The numbers of dots in any two non-overlapping time intervals are two *independent* random variables.

(C) The probability of finding *more than* 1 *dot* in a small time interval of length h is small compared to h, *i.e.* it is a function $o(h)$ such that $o(h)/h \to 0$ as $h \to 0$.

With these assumptions we are ready to begin the argument.

The function $p_0(t)$ gives the probability of finding no dots in an interval of length t. We can determine the form of this function as follows. Consider two adjacent time intervals

$$I = (0, t), \qquad J = (t, t + h),$$

which together form the interval

$$K = (0, t + h).$$

There will be no dots in K if and only if there are no dots in I *and* none in J. Hence by assumptions (A) and (B),

(92) $$p_0(t + h) = p_0(t) \cdot p_0(h).$$

It is not hard to show that the only solutions of this functional equation such that $0 \leqq p_0(t) \leqq 1$ are of the form

(93) $$p_0(t) = e^{-\mu t}$$

where μ is some positive constant. We shall assume this fact here.

On the basis of (93) we have for small h

$$p_0(h) = e^{-\mu h} = 1 - \frac{\mu h}{1!} + \frac{(\mu h)^2}{2!} - \frac{(\mu h)^3}{3!} + \cdots$$

(94)
$$= 1 - \mu h + \frac{\mu^2 h^2}{2}\left[1 - \frac{\mu h}{3} + \cdots\right]$$

$$= 1 - \mu h + o(h),$$

where as before we denote by $o(h)$ any function of h such that $o(h) \, / \, h \to 0$ as $h \to 0$. From (90), assumption (C), and (94) it follows that

(95) $\quad p_1(h) = 1 - p_0(h) - \displaystyle\sum_{n=2}^{\infty} p_n(h)$

$$= 1 - [1 - \mu h + o(h)] - \sum_{n=2}^{\infty} p_n(h)$$

$$p_1(h) = \mu h + o(h),$$

since the sum or difference of two functions of the form $o(h)$ is again of this form.

Now returning to the intervals I, J, K introduced above we observe that for any $n = 1, 2, \cdots$ the same sort of argument gives the relation

$p_n(t + h) = p_n(t) \cdot p_0(h) + p_{n-1}(t) \cdot p_1(h)$

$$+ \sum_{j=2}^{n} p_{n-j}(t)p_j(h) = p_n(t)[1 - \mu h + o(h)]$$

$$+ p_{n-1}(t)[\mu h + o(h)] + o(h),$$

whence

$$\frac{p_n(t + h) - p_n(t)}{h} = - \mu p_n(t) + \mu p_{n-1}(t) + \frac{o(h)}{h}.$$

Holding t fixed and letting $h \to 0$ we obtain the differential equation

(96) $\qquad\qquad p'_n(t) = -\mu p_n(t) + \mu p_{n-1}(t) \qquad (n = 1, 2, \cdots).$

As we already know that $p_0(t) = e^{-\mu t}$ we have a sequence (96) of differential equations which can be solved successively for $p_1(t), p_2(t), \cdots$. To do this we set

$$f_n(t) = e^{\mu t}p_n(t),$$

for which

$$f'_n(t) = e^{\mu t}[p'_n(t) + \mu p_n(t)],$$

so that equations (96) become simply

(97) $\qquad\qquad f'_n(t) = \mu f_{n-1}(t) \qquad (n = 1, 2, \cdots),$

with $f_0(t) = 1$ and the boundary conditions

(98) $f_n(0) = e^{0\mu}p_n(0) = p_n(0) = 0.$

Successive integrations give

$f_1'(t) = \mu f_0(t) = \mu, \qquad f_1(t) = \mu t + C, \qquad f_1(0) = C = 0, \qquad f_1(t) = \mu t,$

$$f_2'(t) = \mu f_1(t) = \mu^2 t, \qquad f_2(t) = \frac{\mu^2 t^2}{2} + C = \frac{\mu^2 t^2}{2},$$

$$f_3'(t) = \frac{\mu^3 t^2}{2}, \qquad f_3(t) = \frac{\mu^3 t^3}{3!},$$

$$\cdot \quad \cdot \quad \cdot \quad \cdot \quad \cdot$$

$$f_n'(t) = \frac{\mu^n t^{n-1}}{(n-1)!}, \qquad f_n(t) = \frac{\mu^n t^n}{n!},$$

and hence finally

(99) $p_n(t) = e^{-\mu t} \cdot \frac{(\mu t)^n}{n!} \qquad (n = 0, 1, 2, \cdots).$

Thus if $X(t)$ denotes the random variable giving the number of dots in an interval of length t, then $X(t)$ has a *Poisson probability distribution* (58) with parameter μt, both mean and variance therefore being equal to μt.

To form an estimate of the value of μ we may take a large time interval $(0, t)$ and count the number of dots observed per unit time, *i.e.*

$$\hat{\mu} = \frac{X(t)}{t},$$

for the random variable $\hat{\mu}$ will have

$$E(\hat{\mu}) = \frac{\mu t}{t} = \mu,$$

$$E(\hat{\mu} - \mu)^2 = Var(\hat{\mu}) = \frac{\mu t}{t^2} = \frac{\mu}{t} \cong 0,$$

and therefore the probability distribution of $\hat{\mu}$ will be concentrated with probability near 1 in a small neighborhood of the unknown parameter μ. For example, if 3000 dots are observed in 1000 seconds the estimate $\hat{\mu}$ will be 3 which should not deviate much from the true value of $\hat{\mu}$.

For some purposes the preceding mathematical model of radioactive disintegration is also capable of describing other stochastic phenomena, for example the incidence of telephone calls at a central exchange during the period of greatest density of calls. Knowing the average number of calls per second during this period we can then compute from (99) the probability of getting during a short time interval a number of calls exceeding the capacity of the system. For example, if the average number of calls per second is 2, and if the system can handle, say, 5 calls per second, then the probability of the system being overloaded during a random interval of one second will be

$$\sum_{n=6}^{\infty} p_n(1) = 1 - \sum_{n=0}^{5} p_n(1) = 1 - e^{-2} \sum_{n=0}^{5} \frac{2^n}{n!} \cong .005.$$

BIBLIOGRAPHY

1. CRAMÉR, H. *The Elements of Probability and Some of Its Applications.* New York: John Wiley & Sons, 1955. An excellent book on the elementary level. Also gives an introduction to some of the basic ideas of mathematical statistics.
2. FELLER, W. *An Introduction to Probability Theory and Its Applications.* Vol. 1. New York: John Wiley & Sons, second edition, 1957. A modern classic on the mathematical theory of probability.
3. LOÈVE, M. *Probability Theory; Foundations, Random Sequences.* New York: D. Van Nostrand Company, 1955. A rigorous account of modern methods in probability theory. For the mathematically advanced reader.

XII

Computing Machines and Automatic Decisions[1]

CHARLES B. TOMPKINS

DURING the last few years a radical change has developed in the field of mathematics formerly known to the erudite as numerical analysis and to others as arithmetic. This change has been caused by the radical development of computing machines which use electronic components. At the same time there has been a remarkable development in the social (and antisocial) sciences—econometrics, organization theory, military theory, and so on. This development has been based to a considerable extent on developments in the field of operations research, initiated on a large scale by the military scientists during the war, and in the field of automation. Automation is an old field which is best described as a natural development of the mass production practices which became popular in this country half a century or more ago; this development has recently been influenced in a most essential way by the study of operations research and by the development of automatic computing equipment.

Automation demands that processes be set up for automatic decision by computing machines. The gain is not only the freeing of human beings for more important, more lucrative, or more enjoyable occupations, but also the smoothing of operations through the increased speed with which the simplest decisions required in the operations can be reached and through the electrical nature of the decision output, which can be communicated conveniently, accurately, and speedily to the operating machinery.

[1] The preparation of this paper was sponsored by the Office of Naval Research and the Office of Ordnance Research, U. S. Army. Reproduction in whole or in part is permitted for any purpose of the United States Government. The author acknowledges valued advice from Professor Clifford Bell of the University of California at Los Angeles, and Robert C. Crawford and John B. Kennedy of the Santa Monica High School in Santa Monica, California.

These developments have led to or been accompanied by radical studies in many classical subjects. For example, when it is realized that a machine can make a decision, a psychologist is likely to ask whether the machine can learn. This question has given rise to a series of arguments whose principal results have included reexamination of definitions of learning. The designers of computing machines have long talked of their machines in phraseology normally applied to the human brain. The machines have memories, for example, and they use words (coded expressions which may be either numbers or directions concerning the behavior of the machine). On the one hand, the memory of a machine may be considered to be faster but not more exotic than the scratch pad used by a human computer; on the other hand, the brain of the human computer also serves partially as a remarkably versatile but not always precise scratch pad with a phenomenal indexing system which permits immediate recovery of the most amazing material—ranging from limericks to telephone numbers. The brain also learns from experience, and it seems to make more complicated decisions than our present machines can make. These decisions are more complicated in method, but seemingly not more complicated in final outcome; one of the objects of the present paper is to show that the most complicated decisions can be reduced to a series of essentially trivial ones if explicit rules of decision are stated. However, the ability of the human brain to reason or to learn by "feel" or intuition seems not to have been imitated by our machines yet.

We have made and designed machines which seem to learn in an elementary way. We can even make machines which forget, and men have at least described how machines can become neurotic. In case a Society for the Prevention of Cruelty to Machines is formed to prevent us from driving machines crazy, I must quickly add that mysterious failures in machines have themselves many times driven anyone with as much computing machine experience as the author to the verge of insanity (or further), and one might even imagine that the human tendency to experiment or to deviate is a more or less self-healing type of transitory failure of routine reasoning similar to the transitory failures we have observed in machines.

As an example of rudimentary learning, we shall outline a description of a machine which learns to avoid sunburn (or more accurately light burn). We need do only a few simple things. First we must put into the machine some substance which is "hurt" by high light intensity so that the light burn actually exists. This could be a slowly photoactive chemical of some sort (and if a reversible reaction is chosen, the light burn can heal). Next, a steerable and mobile unit must be developed with two competing controls. One of these controls must be one (say, photoelectric in nature) which turns the unit away from light. Finally, means must be included to enhance the power of this control relative to the other after exposure to intense light; this means might be the photochemical reaction used to simulate the light burn or (say) the darkening of a photosensitive film through which the competing control signals must pass in the form of light beams. The result will be a device which learns through bitter experience to try to avoid high intensity light.

Qualitative descriptions, such as the one above, will generally be avoided here. A system of strictly accurate numerical description of some elementary decision processes and some computing techniques which are immediately susceptible to application in mechanical and particularly electronic equipment will be described. Some of the more remote and more romantic important applications of this development were described above; however, there is already a large field of important but not so romantic (except in their productive output) applications of these simple techniques, and this field of application is growing at an impressive rate now. Thus, although to the psychologist or to the romanticist the question as to whether a machine can learn is important, the present applications of computing machinery and machinery for making decisions are largely along more utilitarian lines.

Because of the present wide application of electronic equipment to these utilitarian tasks the descriptions here will largely stick to the elementary processes used in the tasks. However, there are other reasons for developing the elementary aspects of this subject and omitting the romantic "abstractions." The development of a science to a quantitative state is essentially the birth of the science—its passage from an embryonic to a developed state. At this time mathematics becomes important in the life of the science—as crucially important as proteins are in the life of a human. This exciting transition is currently taking place in the development of many of the social sciences, and the mathematics being introduced is at least greatly influenced by the mathematics which is important in machinery which computes and which makes decisions. Other mathematics enters into the development of these fields also, of course, but basically the machines are the only fully arithmetized deciders, computers and actors available; and the ones which are digital in operation are all based on a remarkably simple arithmetic system.

The author has always been something of a radical among teachers, and he has continuously urged revision of our high-school and college curricula in mathematics. In particular, in unnecessarily strong terms (and somewhat ineffective terms) he has urged greater attention to details in the axioms and proofs of Euclidean geometry and in algebra; he has taken the view that correct statements must eventually be easier to teach than partially correct statements. In addition he has taken the view that immediately applicable or fully illustratory statements probably attract the interest of more students than do remotely applicable or only partially illustratory statements. Thus the reader is invited to consider here, but not to regard as obviously socially acceptable, a view that in the future the curricula of our high schools might well include studies of some of the subjects to be expounded here—both because they are simple and complete and because they are important in the operation of much of the nation's more essential research, development, and production.

BINARY ANALYSIS OF DECISIONS

In the All-England tennis championships contested at Wimbledon, England, the draw for the 1955 play in the men's singles developed into the following pattern for the last four rounds:

```
Rosewall   ⎫
Merlo      ⎭  Rosewall ⎫
                       ⎬ Rosewall ⎫
Davidson   ⎫  Davidson ⎭          ⎪
Ayala      ⎭                      ⎬ Nielsen ⎫
                                  ⎪         ⎪
Shea       ⎫  Pietrangeli⎫        ⎪         ⎪
Pietrangeli⎭           ⎬ Nielsen ⎭          ⎪
                       ⎪                    ⎪
Segal      ⎫  Nielsen  ⎭                    ⎬ Trabert
Nielsen    ⎭                                ⎪
                                            ⎪
Flam       ⎫  Patty    ⎫                     ⎪
Patty      ⎭           ⎬ Patty ⎫            ⎪
                       ⎪       ⎪            ⎪
Larsen     ⎫  Hoad     ⎭       ⎬ Trabert ⎫ ⎪
Hoad       ⎭                   ⎪          ⎬⎭
                               ⎪          
Perry      ⎫  Drobney  ⎫       ⎪
Drobney    ⎭           ⎬ Trabert⎭
                       ⎪
Kumar      ⎫  Trabert  ⎭
Trabert    ⎭
```

This information, which was extracted from the periodical *World Tennis* for August 1955, is typical of all such tournament pairing.

The process illustrated is a decision process. The decision to be made was the identity of the recipient of a trophy. This decision is compatible with an assumption that superiority in tennis is a transitive relation: that is, if a player x is better than y and if y is better than z, then x is better than z. It is not a good idea to examine this assumption here, and the initial hunches of the seeding committee (who try to separate strong players so that they meet each other in the later rounds in order to be able to examine their play more fully for a later decision as to relative ability) are also best set aside for the moment. Under the idealization mentioned, the chart above means that two players to the left of a bracket competed and that the player whose name appears to the right won the match. Play continued from left to right until finally Mr. Trabert defeated the only other player left in the tournament. Since only the identity of the best player was wanted (not the worst and not a complete ranking list ordering all players—an embarrassing thing at best) the players who had lost were not required to compete further.

In this way the problem of determining the best player among the sixteen remaining at this point of the tournament (the earlier rounds were carried out in a similar way, but space is precious here) was factored into fifteen problems of finding who was the better player of a pair. The problem of determining the better player of a pair was solved in the elementary way of requesting them to compete until one had won three sets.

The ease with which this decision was arithmetized is at least partially due to the simplifying assumption of transitivity in the prowess function of players; indeed the ranking committees may have a hard time with the 1955 world rankings because Mr. Hoad, whose performance here was relatively inglorious, defeated Mr. Trabert in the Davis Cup Challenge Round a few weeks later. It has been shown by example (see item 1 in the Bibliography following this chapter) that the ranking committees must define the meaning of their rankings considerably more precisely than they now do if an automatic decision process is to lead to a determination of a linear ranking of the players.

The decisions of modern computing machines are similar to those of the Wimbledon tournament committee. They are based on comparisons between pairs of quantities. The operation of the system is likely to be precisely that illustrated in the Wimbledon draw sheet.

One slight difficulty should be noted. It is clear that in each round at Wimbledon there were twice as many players as there were in the succeeding round. This requires that initially the number of players is a power of 2. This is difficult to arrange in a free country at times; the difficulty is solved by inventing fictitious players (named Bye) who always lose. Thus, if 104 players were entered in the tournament, 24 players all named Bye and all fictitious would be entered by the tournament committee, and all would lose in the first round (no two are allowed to play each other) without their opponents having to hit a ball. This would bring the abstract entry list to 128, which is the seventh power of 2; the tournament would be completed in 7 rounds.

Modern automatic decision methods are precisely those of the tennis tournament committee which were outlined above. They suffer exactly the same weaknesses in that only the amount of judgment built into the process can be exercised automatically. Thus, one might say that a Davis Cup win is more important than a Wimbledon win to both Hoad and Trabert; if this is correct, there can be little doubt but that Hoad should have a higher final ranking on the evidence shown above than Trabert (at the time of writing the results of other important tournaments were not available, since these tournaments had not yet been played); if, on the other hand, we gave these results to a machine and added that Trabert wanted most to win at Wimbledon and that Hoad had always wanted most to win in the Davis Cup round we could expect nothing better than that the Society for the Prevention of Cruelty to Machines would be after our scalp for driving our machine to neurotic state if we demand a definite ranking.

Before getting into details, it might be well to note that most machines would give a definite ranking under these conditions—but only because of the way in which the decision would be demanded. The assumption is made above that every match ends in a definite win. Baseball is a somewhat more violent game than tennis—at least so far as conversation between players and the umpire is concerned. Possibly because of this there is included in the umpire's code a means of resolving ties. The tie goes to the runner. Thus, in baseball the binary decision is used to decide whether a batter is still a batter or not, whether a base runner is out or not, and so on. However, in baseball there is a realistic appraisal of the probability that an umpire will believe that a ball reached a first baseman and a runner reached the base at precisely the same time (a completely unlikely concurrence), and under these conditions (with minor exceptions) the umpire is instructed to rule as though the base runner had reached the base before the ball was caught by the fielder. Normally decision machines are instructed in the same way, even though there may be no logic behind the instruction; thus, it is still likely that the man, not the machine, will end up neurotic, and that the machine will be completely (although not necessarily wisely) instructed on how to reach a decision.

Precisely, then, our automatic decisions are based on comparisons. We are to decide between n courses of action. If n is not a power of 2, we augment the field of competition by adding (formally) several fictitious courses of action which we surely will not take. We have a means of comparison between alternative courses of action, and finally we conduct a tournament just like a tennis tournament with the exception of a rule that (say) ties are decided in favor of the competitor listed first. Clearly any system which is transitive can be handled in this way; any system which requires real judgment (in that suitable competitive criteria cannot be set explicitly and transitively) is not well suited to current automatic decision. Systems in which the binary decision process is not applicable (because of absence of transitivity) can be set up on paper, and if they can be described explicitly and arithmetically an automatic decision process can be set up. They are susceptible to a binary formulation with the complexity assigned to the scoring system.

THE BINARY NUMBERING SYSTEM

For machines (which do not receive wordy or graphic information avidly) it is convenient to number the positions in the tournament draw illustrated above. Furthermore, for reasons which are partly obvious and which will be partly introduced later, it is convenient to have the digits in the numbers restricted to two values; these values are normally 0 and 1. As we read against the stream from right to left in the tournament draw diagram, we (1) add a digit 0 at the right end of the number whenever we go *up* and to the left, (2) add a digit 1 at the right end of the number whenever we go *down* and to the left. Thus, the position of Nielsen in the final round of the tournament above would be numbered 0, and that of Trabert would be numbered 1—a single digit being used to denote the relative position of two objects. Similarly, the pair of digits 00 would be used to number the position occupied by Rosewall in the next earlier (the semifinal) round, 01 would denote Nielsen's position, 10 Patty's, and 11 Trabert's. A complete numbering of the positions would look like this:

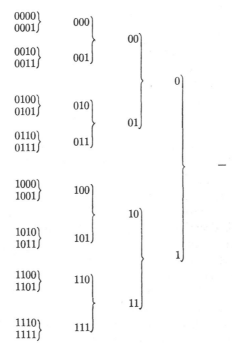

Since most digital computers have a fixed number of digits (the term *binary digit*, frequently shortened to *bit*, is used to describe a mark which is either 1 or 0), the whole design above could not easily be shown on an automatic machine. However, it offers examples, in each of its vertical columns, of how binary computing machines count, how many different situations or numbers can be represented, and so on.

In particular, since each round of the tournament contains twice as many players as the preceding round, and since each step to the left increases the number of bits by one, it is clear that a machine whose numbers are exactly n bits in length can have these numbers assume 2^n different values, or can use a number to designate uniquely one of 2^n different situations or actions. It is clear (through mathematical induction) that no two numbers in any column can be the same, for to get from the right end of the "draw sheet" to the numbers in question it must have been necessary at least once to move up and to the left to get to one of the numbers and down and to the left to get to the other; the two numbers will differ at least in the bit corresponding to this fork.

Furthermore, we can detect an orderly counting system as we look down any column. Suppose that the numbers in the left column are assigned increasing values (so that, for example, the players in the left-hand column are numbered, decimally, from 0 through 15); the system used to get from one number to the next higher is generally the same as that used in the decimal system. The smallest number which can be represented is one whose bits are all zeros. The largest is one whose bits are all ones. If x is a number whose bits are not all ones, the next larger number, $x + 1$, is obtained by changing the least significant 0 (the 0 furthest to the right in the number) to 1 and changing all the 1's to the right of it to 0. In any counting system of the general type we use, the process of proceeding from one number to the next higher is defined in a similar way: The least significant digit whose value is not that of the highest digit (the least significant digit not a 9 in the decimal system) is raised in value by 1 and all digits to its right are set to the value 0.

Once this rule is described, it is possible to build up the whole arithmetic based on numbers in binary representation by mathematical induction. Thus, the processes of addition of long numbers by adding digits and carrying, subtraction digit by digit, multiplication, and division can be described in just the terms used in the early grades of our elementary schools. However, the tables which must be remembered are much simpler and smaller, for there are only two digits instead of ten.

One can easily work out the whole arithmetic scheme for himself, and little detail will be included here. The subject of different bases (the base is sometimes called the radix of the arithmetic system) has fascinated many people, and numerous bases have been used. Thus, the binary system was used long before vacuum tubes were invented, and other systems have used bases as high as twenty (used by the Mayans) and as high as sixty (the reader will find a vestige of this if he consults his watch and reads time in hours, minutes, and seconds). It is not appropriate here to go into details of the development of numbering systems and means of converting a number expressed in one system to one expressed in another; however, the reader may refer to item 2 of the Bibliography for this chapter for a well-referenced description of this subject. (See also Chapter II.) A summary of some of the operations in binary arithmetic will be needed, however, and it will be given without proof below.

Just as the sum of two n-digit decimal numbers may contain $n + 1$ digits, so the sum of two n-bit numbers may contain $n + 1$ bits. This is inconvenient in many machine applications, but nothing can be done about it. Thus, it might be best to think of adding two binary numbers (computing machines normally limit their addition to two numbers at a time and do not add up a long column of numbers like those in the work books in our elementary schools) in terms of adding a 0 to the left of each (they are assumed to have n bits each before this addition). The sum will be an $n + 1$ bit number. The addition can be done digit by digit. There will be three inputs to each digital sum and two outputs: The three inputs will be the two corresponding digits from the numbers to be added and the carry from the preceding addition; the outputs will be the new digit in the position being considered and the carry to the next later digit. The addition table is the following:

Digit from first number	Digit from second number	Carry digit from earlier addition	New digit	Carry digit to next addition
0	0	0	0	0
0	0	1	1	0
0	1	0	1	0
0	1	1	0	1
1	0	0	1	0
1	0	1	0	1
1	1	0	0	1
1	1	1	1	1

The carry introduced to the addition of the least significant digit is 0, and the carry digit from the $(n + 1)$st digit must necessarily be 0, for the two digits from the addends are both 0.

A similar table for subtraction can be prepared; it involves taking values from the more significant digits rather than carrying them to the more significant digits.

Once addition and subtraction are understood, there is no difficulty in passing on to long multiplication and long division. Here the binary system is most attractive, for no real multiplication table is required. The only possible digits are 0 and 1; if the multiplier digit considered is 0 no addition is required and the required shifts are carried out preparatory to considering the next multiplier digit; if the multiplier digit is 1 the multiplicand (properly shifted) is added to the number being built up as the product. The necessity of multiplying the multiplicand by a digit greater than one arises in all systems of arithmetic except the binary system, but here no such requirement develops. Division is similarly simplified.

Actually, electronic devices frequently are built to add all n bits of numbers simultaneously. This process is one in which it is convenient to distinguish between the *augend* and the *addend*: The addend is added to the augend. The addition process is in two steps—the digital addition and the carry. The digital addition is governed by a simple addition table used first without carry; the two corresponding digits are added according to the entries in the table above with the carry digits being taken as zero. (Thus there are only four entries in the table.) Next the carry is generated, propagated, and received. A carry is generated by a digit which in the augend was 1 and which in the sum before the carry is 0. All such digits develop carry digits simultaneously, and they are propagated to the left in the sum. Any digit receiving a carry digit is complemented; that is, it is changed from whichever value it has to the other possible value—1's are changed to 0's and 0's are changed to 1's. The carry is stopped by any digit which had a 0 value before receiving the carry, and it is transmitted by any digit whose value before the carry was 1.

Subtraction is carried out in the same way (although something must be said about the possibility of generating negative numbers). Explicitly, subtraction starts by the same addition without carry as addition, and it is followed by carries generated simultaneously by digits which were 0 in the minuend and which are 1 in the difference before carry. These carries are propagated to the left, complementing all digits they reach, and being stopped when they reach a digit whose value in the sum before complementation is 1.

Negative numbers are represented in several different ways in the various computing machines which have been built. In every case, one of the binary digits serves as a sign digit. In some machines the number is represented by a sign and its absolute value, just as decimal numbers are written. In some, positive numbers are represented with the most significant digit zero, and negative numbers are represented as complements; that is, the negative number (whose absolute value is restricted to be less than 2^{n-1}, where n is the number of bits in the representation of the numbers) is added algebraically to the number 2^n, and the result is a number whose most significant digit is 1. Negative numbers are then recognized by the value of this most significant digit and they are treated accordingly. This system is used in most desk calculating machines (where the negative number is added to a power of 10). In electronic machines, the complementation is usually accomplished by complementing each digit individually (which is equivalent to adding $2^n - 1$ to the negative number) and then one is added to the resulting number.

It is easy to complement binary digits in the electronic machines, and in many cases the adjusting addition is omitted. Thus, the complemented number will consist of the digits individually complemented. The reason for this is an irreversibility of electronic components, which differs from the reversible operation of the wheels in a desk calculator. The electronic computer almost invariably subtracts by complementing and adding or adds by complementing and subtracting and it seldom both adds and subtracts; to build a machine which would add and subtract in elementary ways (such as the ways mentioned above) has seemed unattractive because of the demand for two carry propagation lines, or for one line with two sets of controls.

The possibility of increasing the number of digits when two numbers are added leads to a real danger that the sum of two large positive numbers will be interpreted by a complementing machine to be a negative number, for the most significant digit which is reserved to indicate the sign may fill with a carry from the next digit. This difficulty is inherent in complementing systems, and it must be overcome by careful planning. It exists in desk calculators, but a being with judgment operates these machines and gets a chance to exercise that judgment after each addition or subtraction.

If the digits are complemented individually to convert a number from one sign to another, there are two representations for the number zero. One of these has all bits 0 and the other has all bits 1; the second of these numbers corresponds to -0. If a zero is generated by adding a positive and a negative number, the resulting representation will invariably have all bits 1; on the other hand, if a zero is generated by subtracting a number from itself, the resulting representation will invariably have all bits 0. Since the all zero representation is less shocking to most users than the all one representation, at least one series of major machines uses digital complementation and subtraction instead of addition.

If the individual bits are to be complemented, the rules of addition must be changed slightly; addition or subtraction is to be carried out as before except that the least significant digit must receive the carry from the most significant digit. Again, this rule is given here without proof; the reader should be able to convince himself that it is sound.

The question of representing fractions must also be solved in any useful number system. Machines generally use a binary fraction representation similar to the decimal fraction representation. A number expressed decimally extends in two directions from the decimal point. The value of the number is the sum of the products obtained by multiplying each digit by a power of 10 corresponding to its position in the number; thus the digit to the left of the decimal point is multiplied by $1 = 10^0$, and the power of ten is raised one for each step to the left and lowered one for each step to the right. The same system may be used for any base; the base is raised to the zero-th power for the digit to the left of the radical point and to a power one higher for each step to the left and one lower for each step to the right. Binary fractions behave just as conveniently as decimal fractions, and there is no problem after this definition has been accepted.

A final question which must be settled is what to do about numbers whose representation for some reason must be in a base other than two. The most frequently met numbers of this type are decimal numbers. These numbers can all be converted to binary numbers, but the conversion process is burdensome and if each of many numbers is to be subjected to a small number of calculations a better approach is desirable.

It has been indicated above (and it will be stated more explicitly below) that electronic computers prefer to make binary decisions and not to try to distinguish between more than two choices. This is because a vacuum tube is able to tell easily and with a high degree of reliability whether or not a substantial current is flowing in a particular circuit, but they tell how much current is flowing with a greatly reduced reliability and only at the cost of greatly enhanced circuitry. Thus, it is common for these machines to carry out decimal arithmetic, when it is required, in terms of binary decisions.

The ten decimal digits can be assigned arbitrary marks from the left-hand column of the numbered display of positions in the tennis tournament above; no later column has enough entries. That is, four binary digits may be used to code a single decimal digit.[2] After this is done, the decimal arithmetic may be programmed with the type of electronic decision elements which will be described below—the configurations simply become a little more complicated than they are for binary arithmetic. Several different codings have been suggested and used; they are chosen for simplicity in the resulting circuitry, and this may depend sensitively on which arithmetic operations must be carried out. One obvious coding is to code each decimal digit by assigning its binary equivalent, so that the decimal digits from 0 through 9 correspond, in order, with the first ten groups of four bits each in the left column of our original binary array. Another scheme is to move everything down three steps in the column; if this is done, complementing a decimal digit with respect to 9 is equivalent to complementing each bit in its coded representation. Other schemes for accomplishing this easy complementation have been proposed; some involve assigning values, 1, 2, 2, and 4 to the individual

---

[2] Using *two* as a base and the concept of place value, the decimal digits, 1 through 9, are represented by the respective symbols: 1, 10, 11, 100, 101, 110, 111 1000. 1001.

bits in the representation of the decimal digits and of extracting easily complemented combinations from the resulting possibilities.

Most readers will find that the particular representation chosen is of no great importance until equipment is being built; however, the fact that decimal arithmetic can be reduced to complicated binary decisions is of importance. The addition and multiplication tables for decimal numbers can be translated to the coded binary equivalents, and the resulting operations can be scheduled through the use of the elementary circuits which will be described below.

ELECTRONIC ELEMENTS CAPABLE OF BINARY DECISION AND COMPUTATION

High-speed electronic computing depends upon the existence of electronic circuits each of which can take one of several different stable conditions and which can influence other circuits or be influenced by other electrical stimuli in a way which depends upon the state it is in. In particular several circuits are known in which the number of stable states is two; these are the so-called *flip-flop circuits* or *Eccles-Jordan circuits*. Such circuits can be made to behave in many different ways, but fundamental in each is a pair of electronic elements connected to influence each other so that either can conduct electrical current but not both simultaneously. These pairs can be set by a variety of means, and they can be connected to give a variety of reactions. The commonest way of setting them is by a voltage or current pulse. The reactions may be any of several: A pulse applied to one point in the circuit may turn the "off" conductor on and the other off no matter which is conducting, so that the effect is roughly that of pulling the chain on the pull chain light switch; pulses to another point may assure that following the pulse a particular half of the circuit is conducting whether or not it was previously conducting, so that the effect is roughly that of pushing on one or the other button of a wall push button switch in a lighting circuit; the circuit may be arranged to stay in a particular state except for a fixed time after being disturbed, so that being set to its extraordinary state (by either of the types of stimuli mentioned above) it will return to its ordinary state after this time has passed; it may be arranged so that it will remain in either state for a fixed time and then shift to the other and so oscillate until interrupted. The active elements may be vacuum tubes, magnetic elements, transistors, or any of a variety of other elements. The reader can find a fairly complete but old discussion of some of the engineering aspects of these elements in reference 3 following this chapter.

In particular, these flip-flop elements can be used to represent binary digits, and an array of n such elements can be used to represent an n-bit number. They may be interconnected with a variety of electrical and electronic components and augmented by others. The components include permanent magnets which may be set electrically to either of two magnetic states for the storage of bits, gates by means of which the ability of a source to transmit a signal along a conductor can be controlled electrically, delay units which will accept an electrical signal at one place and time and deliver it considerably later at another place, valves which will permit signals to go in one direction but not in another, and so on.

The machines usually operate with more or less regularly timed pulses transmitted over various lines. The presence of a pulse is frequently denoted by a 1 and absence of a pulse at a time considered is denoted by a 0. These are the marks, of course, which appear in the binary arithmetic described earlier. Similarly, the two states of a flip-flop unit are denoted 0 and 1 depending on which bit is represented by the individual states.

All these component patterns are usually arranged into circuits capable of carrying out a few simple algebraic operations on the marks 0 and 1. A function of a single binary digit which may be desired either for the flip-flop or for the dynamic circuits is the complementation function already described. For flip-flops it is carried out in one way, and for pulses on a line it is carried out in another way, but algebraically the function is the same. Its values are given below:

COMPLEMENTATION

Argument	0	1
Functional Value	1	0.

It is not feasible to consider the electronic production of this function here, but it should be noted that this function is perfectly easily produced at any point of a circuit.

Several functions of a pair of bits are required in computing equipment. One of these is a function of two bits occurring in a fixed flip-flop unit during successive time periods; it is used to generate the carry pulse. It might be called *development*. Its functional values are given below:

DEVELOPMENT

	Second Argument	
First Argument	0	1
0	0	0
1	1	0.

The *functional value* is the value of a carry digit originating with the given digit after the first step in the parallel addition described above. The carry for a subtraction is generated by a similar function in which the argument bits are complemented—a process accomplished electrically by connecting the leads to the other element of the flip-flop unit.

The other most common binary functions occurring in computing machines are the "or" circuit and the "and" circuit. The first is designed to transmit a pulse if one is received from one or more of two inputs, (or, in practice, more than two), and the "and" circuit is designed to transmit a pulse if and only if pulses are received on both of two input lines. Algebraically, these functions are:

OR

Arguments	0	1
0	0	1
1	1	1

AND

Arguments	0	1
0	0	0
1	0	1.

Since the functions are symmetric in their arguments it makes no difference in the tables which is the first and which is the second argument.

An algebra can be built up on these functions, and this algebra is constantly used in the design of switching systems, such as the complicated systems which are used in digital computers. The algebra is frequently referred to as Boolean, and much effort has been given to its development recently. An extensive report of the use of algebra of this kind in the design of computing circuits is contained in reference 4 of the Bibliography (see end of chapter).

It might be noted that there is likely to be a slight difference between circuits which are expected to receive simultaneous pulses and circuits which are expected to receive one continuous signal and a pulse on its two input lines; the latter is frequently referred to as a gate.

This algebra is used to reduce the complicated circuit diagrams to manageable form for analysis. Its importance may be realized from a description, which will be written below largely without the use of the Boolean algebra, of one stage of the simplest parallel adding circuit outlined earlier.

This description will be largely symbolic. Lines will indicate conductors, boxes with words in them will indicate units capable of carrying out the functions described below. In addition, the following symbol will be used to indicate a flip-flop circuit (Fig 1):

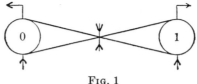

Fɪɢ. 1

The lines to the center of the cross will cause complementation if a pulse is introduced through either of them; if a pulse is introduced through the line into the lower part of one of the small circles, it will cause the flip-flop to assume the state corresponding to the number written in that circle, and the lines from the top of the circle corresponding to the state of the flip-flop will contain an output signal, the other line will contain no output signal.

In the illustration below, an "add pulse" is gated from the clock to all the digits of the addend simultaneously at a time t_{add} chosen to carry out the addition. At a time Δt seconds later a carry pulse initiate signal is introduced from the clock over a line in the lower right corner; it goes into an "and" circuit. The carry pulse line is connected with similar digital units, the next less significant digit on the right, the next more significant digit on the left. A little study will convince the reader that this circuit will carry out the addition process described above, leaving the sum in the augend register.

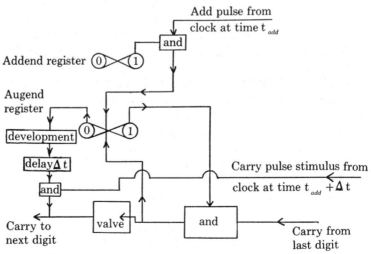

Fɪɢᴜʀᴇ 2. One digit of a parallel binary adder

INTEGRATION OF PRECEDING SECTIONS

The path followed up to this point started with a tennis tournament, wandered through a numbering system naturally arising from the tennis tournament, went into the arithmetic based on this numbering system, and then examined superficially electronic components which may help in carrying out this arithmetic. The path was cyclic in a way which the reader might not have noticed: It started with a discussion of decisions, and particularly binary decisions, and it ended with a heroic diagram showing how electronic elements which in the long run do nothing but make binary decisions (to transmit a pulse or not, to change state or not, to assume the state corresponding to digit 0 or the state corresponding to digit 1) can carry out the arithmetic which somehow presented itself in the intervening sections.

(It is probable that the confusion described above is an example of a research problem in mathematical logic, and it might be cited to the inquiring minds who feel that they should be able to understand why mathematics is not a static thing, fixed since the time of Euclid. The general description of binary decision is easily absorbed, but its exposition and description are difficult. The arithmetization of logical functions is the business of the mathematical logician, and major advances in computing techniques have been contributed by mathematical logicians, but there is still no clear arithmetization of the processes which are described here. This remains a research problem, as does the problem of effective coding of the large computing instruments already built and those larger machines now planned.)

Some of the binary decision processes introduced at the beginning of the paper are required in the operation of the large machines to be described later. Therefore it might be well to try to summarize the remarks made so far and the connections between them.

The tennis tournament introduced the idea of deciding between two alternatives on the basis of comparative scores. At the upper left corner of the tennis tournament there was a decision as to whether Mr. Rosewall or Mr. Merlo would play in the second round; this decision turned out to be that Mr. Rosewall would play because he was the first to win three sets.

In many applications, the computation of the scores involves some arithmetic function. Thus, in an inventory problem a decision as to whether to order an item or not to order it arises. It is to be ordered if the stock on hand is sufficiently low and the demand is sufficiently high to justify an order. Some formula is provided for this, and eventually the stock on hand may be compared with the level at which reorder is justified. This comparison is a binary decision. It may be carried out electronically by the circuit patterns we have described. In particular, the maximum level at which reorder is justified may be subtracted from the observed level; if the answer is positive there is to be no order, but if the aswer is non-positive (that is, negative or zero) there is to be a reorder. The decision can be made to depend, then, on the most significant or sign digit of a number in binary representation, and the decision may be carried out electrically (as indeed is happening in many places) by use of a property of the electronic unit representing this digit—it is able to furnish electrical signals whose nature depend upon which of the two possible stable states it occupies at the time the signal is demanded.

At present, the decisions of automation seem all to be of this type. An automatic decision is made to sponsor or not to sponsor some activity depending upon the value of one binary digit in one number which is computed in connection with the operation being carried out automatically.

The calculations leading to the result also seem to require some decisions, and these decisions are made in the same way.

Finally, it should be noted that the binary nature of the decision must be viewed a little carefully. Thus, in connection with reorder the activity might be whether to reorder or not; if there is to be a reorder the number ordered might well vary—it might be the number appearing at a particular place in the computing machine, for example. Thus, the decision is not whether or not to reorder 1,000 units, but rather whether to order the number of units specified by a number in a given place or not to order at all.

Even in one match of a tennis tournament, the scoring involves a tremendously tedious number of binary decisions. After a point is decided a decision must be made as to whom to credit it—if x won the point add one point to his score; if x did not win the point add one point to y's score. Then the scorer must ask whether the player receiving the point has scored as many as four points in the game, if not, the game continues, if so, a continuing analysis must be made. If a continuing analysis is required, the scorer asks whether the player who just won the point has as many as two more points than his opponent; if he does, he has won the game and the analysis continues, if not, the game continues with no further analysis.

If the analysis is continued, the question of whether a set has been completed must be attacked in much the same way as the question of whether the game was won. The game score is set back to zero, the set score is adjusted by adding a game to the credit of the person who just won the game, and then the tedious questioning resumes; has the player who just won the game won as many as six games, and has he won as many as two more games than his opponent.

If a set has been won, the scorer adds one set to the match score of the player winning it and resets the set score to zero. He then asks whether the winning player has won as many as three sets; if so, the match is over and all that remains is for the players to shake hands. If the winning player has not won as many as three sets the scorer asks whether the total number of sets played is exactly three; if it is the scorer must determine whether either player wants a rest, and if so, the rest is called for ten minutes. If the total number of sets played is not exactly three or if neither player wants a rest, play is continued. There are still a few minor decisions concerning which court is used by each player and who serves.

These decisions are made by the human brain in a time which used to be considered to be incomparably shorter than the time required to play the tennis match. Similar decisions were made in seemingly similarly inconsequential time in connection with many production processes. However, as the number of such decisions began to grow into the millions per day or per hour, this notion of inconsequential time came in for a revision, and automation was born. An automatic computer can make such decisions in a few millionths of a second and, what is more important, it can transmit them in electrical form to whatever machine is to be governed by them in comparably short times.

The same effect turns up in long arithmetic calculations. Here, in the old days, an operator introduced numbers to a desk calculator, and this calculator did all the work. The operator was fairly intelligent, and he could make decisions in times incomparable with the times required for the machine to carry out the numerical calculations (and for him to write them down). The electronic machinery partially described above, however, can carry out an addition in a few millionths of a second, and it suddenly becomes clear that the decision function, the function of operating the machine, and the function of keeping track of the intermediate results become the slow functions; they were originally incomparably faster than the machine or at least comparable in speed with the machine.

The next few sections of this report will deal with the problems encountered in expediting the decision, programming, and recording functions of a computation. They will describe (with the paucity of detail used above) some of the engineering developments made to expedite these operations, but their main concern will be logical and mathematical developments.

Then some attention will be paid to the strange new phenomena which develop because of radical changes in the relative rates (for example) with which a number can be computed and found in a table, and the new numerical analysis being developed as a result of the new computing equipment will be discussed.

STRUCTURE OF THE COMMON MACHINE

The description of the common general purpose digital computer which will be attempted here will be out of date almost immediately because of the development of more versatile machines, on the one hand, and the development of more specialized machines (or at least of machines whose versatility lies in other directions) on the other hand; these last machines are being developed particularly for business applications and for automation. The purpose of this section is only to illustrate means which have been used to overcome some of the problems which arise with the new high-speed electronic computing components, and it by no means describes any existing or proposed machine.

These problems are largely the result of the change in relative times required for the operations involved in computing which were discussed above.

The arithmetic unit of a modern electronic digital computer is likely to be analogous to the arithmetic unit of a modern desk calculating machine; the electronic digital computer is likely (but not certain) to be binary and the desk calculator is likely to be decimal, but these are not important logical differences from the point of view to be taken here.

The electronic machine has an accumulating register which is capable of addition and which can shift numbers to the right, one bit at a time. This unit can be reset to contain nothing but 0's automatically. In many machines this register is called the A-register, and this notation will be used here.

The A-register may be n bits in length or $2n$ bits in length, where n is the basic number length of the machine. For the standard desk calculator it is $2n$ digits in length, but most electronic calculators have n-bit A-registers in order to economize on components, particularly in the carry propagation lines.

The nature of the construction of the A-register is not important to the considerations here; it is a parallel unit in many machines (that is, all digits are added simultaneously, as described above), but others have serial adders.

If the A-register is only n bits in length, provision is usually made for it to shift its least significant digit into another register (which will be described later under the designation of M-register) which shifts its digits at the same time the digits of the A-register shift.

The purpose of the shifting is to permit multiplication. If two n-bit numbers are multiplied, the product may have as many as $2n$ bits, and the purpose of providing space, either in the A-register or in the A-register and the M-register, for $2n$ bits is to permit retention of the whole product. If the A-register itself is $2n$ bits long, accumulative products may be built up (until the register overflows with a number too large); that is, the product of two numbers may be added to a number already in the A-register. If the A-register is only n bits long (and if the M-register does not have provision for carry initiation and propagation) this accumulative multiplication cannot be carried out in a single operation. However, a more detailed description of multiplication should await a description of addition.

For the addition, the augend is placed in the A-register, and the addend is placed in another register, which will be called the S-register here. (The designations are chosen to indicate that the A-register is the accumulating—or adding—register, the M-register holds a multiplier, and the S-register receives numbers from the storage unit, which will be described later.) The connections between the S-register and the A-register might include those indicated in the figure on page 388. (In addition, the A-register has components which permit the right shift.) Minor modifications are required if the machine is a subtracting machine rather than an adding machine, and for other similar deviations.

Upon the receipt of the add pulse and the later carry initiate pulse, the addition is carried out, and the answer is left in the A-register.

Additional variations are possible between machines when multiplication is described. Multiplication will be described here generally in terms of a machine whose A-register is n bits long with a shift into the M-register. There will be required, in addition to the other registers, a counter which can count up to n shifts and stop the multiplication at that point.

The A-register is cleared before the multiplication, the counter is cleared, the multiplier is placed in the M-register, and the multiplicand is placed in the S-register. For each bit of the multiplier, the following steps are carried out, the bit controlling the operation being the one in the least significant position of the M-register:

(a) The machine makes a binary decision based on the value of bit examined (that is, the bit in the least significant position of the R-register), carrying out step (b) below if this bit has value 1, skipping to step (c) if this bit has value 0;

(b) If this step is to be carried out, the machine adds the number in the S-register to the number in the A-register;

(c) The machine shifts the number in the R-register one bit to the right, discarding the bit which was least significant, and the number in the A-register one bit to the right with its least significant bit shifting into the most significant position in the R-register;

(d) The machine adds one to the number in the product counter;

(e) The machine makes a binary decision, returning to step (a) above if the number in the product counter is less than n and signalling that the multiplication is complete if the number in the product counter has reached the value n.

At this stage the least significant n digits of the product are contained in the M-register and the most significant digits are in the A-register. If a rounded product of n bits is desired, a final binary decision is made based on the value of the most significant bit in the R-register; if this bit has value 1 the number in the A-register is increased by 1, if it has value 0 the number in the A-register is left alone.

As was noted above, subtraction may be carried out by complementing and adding. Some machines make no provision for division. Those which have division built in as an automatic function do it by successive subtraction using binary decisions in the usual way. The quotient is built up in the M-register, the dividend is introduced through the A-register, and the divisor is introduced through the S-register. Since the most significant part of the quotient is computed first (as opposed to the least significant part of the product) provision for left shift must be made one way or another; in the A-register this is easy, for the number in this register is simply added to itself to shift one bit to the left (shifting a bit to the left is equivalent to multiplication by two), but some special provision for this must be made in the M-register.

With the equipment described above, an addition is normally performed in a few millionths of a second, and it seems that laboratory and possibly commercial equipment can be developed without radical change of techniques to perform the addition in something less than 5×10^{-7} seconds. The times depend upon the size of the numbers (which determines the longest carry possible); the times quoted are for numbers about forty bits in length—which correspond roughly in size to numbers twelve decimal digits in length.

It becomes clearly imperative to develop some method faster than human intervention to get the operands to the various registers and to get the answers out from the registers and into a position to be used. For the desk machine the input is by human operation of a manual keyboard, and output is either by reading a register (and hand copying the number shown) or by a printer attached to the machine.

The automatic digital computer uses an automatically consulted memory or store (the word store will be used here to go with the designation used for the S-register) for this purpose. This store usually contains room for enough numbers to carry out the calculation; the number of numbers which may be stored is usually a power of two, for binary machines, and it varies from an inconvenient low value of $256 = 2^8$ in at least one machine through average convenient values between $1024 = 2^{10}$ and $4096 = 2^{12}$ to an inconveniently large value (at least equipment-wise) of $16,384 = 2^{16}$. This last value is not inconvenient if it is divided into a high access speed and a low access speed part, which is done on some machines. If access times sufficiently long are allowed, the machine has essentially infinite storage capacity, but this interferes with the effective utilization of the high-speed arithmetic units. Great ingenuity is shown by the manufacturers of computing machines to increase the effective size of their stores, and the numbers above are given as modest examples and not as the best attainable values.

The physical nature of the store may be magnetic, electronic, or electrosonic. The bits may be stored as electrically created permanent magnets (either in individual cores or on a rotating drum), as charged spots on the face of an ordinary cathode ray tube similar to a television tube, or as sound pulses traveling through a length of mercury or some other substance. Other storage mediums have been suggested.

It might be pointed out here that these possible components are more fully described in reference 3 of the Bibliography, following this chapter, and that reference 5 outlines the development of computing equipment at least in terms of the equipment available at the time the book was written several years ago.

Any number in the store is specified by the location of the cell in which it is stored. These storage cells are numbered serially (in the binary system for a binary machine—hence the usual practice of providing a number of cells which is a power of two). A number desired for computation is called for by specifying its "address"; that is, the serial number of the storage cell containing the desired number.

A high speed communications network is provided in the machine to expedite the loading and unloading of the various registers involved in the computation. There is a communications channel between the S-register and every storage cell; there is a channel between the S-register and the A-register; and there is a channel between the S-register and the M-register. All these channels are two way channels permitting transfer of numbers in either direction.

A complete multiplication operation would involve the following steps:

(1) The A-register is cleared and the product counter is cleared;

(2) The number in the cell designated to contain the multiplier is transferred to the S-register;

(3) The number in the S-register is transferred to the M-register;

(4) The number in the cell designated to contain the multiplicand is transferred to the S-register;

(5) The multiplication proceeds as outlined in the steps (a) through (e) listed above;

(6) The number in the A-register is transferred to the S-register;

(7) The number in the S-register is transferred to the storage cell designated to receive the most significant half of the product;

(8) The number in the M-register is transferred to the S-register;

(9) The number in the S-register is transferred to the cell of the store designated to receive the least significant half of the product.

After this last transfer, the machine signals that it is ready for more.

The speed of this operation is still higher than the reaction time of the human operator. The whole operation described in (1) through (9) above is carried out on many currently operated machines in much less than one thousandth of a second, and in some machines in less than one ten thousandth of a second. Thus, it seems desirable to introduce instructions to the machine automatically.

This automatic introduction of instructions is accomplished by means of coded commands. In most machines the coded commands are the same length as the numbers, and they are likely to be stored in the store used for numbers. However, they are routed to a control register, and this register decodes the commands and interprets them in terms of the elementary machine operations described above.

The coded commands are highly stylized, and the problem of coding commands is another problem in mathematical logic which has developed during the last few years.

The whole multiplication operation described above may be coded into a single command in some machines; other machines would require two or three commands or even four for the operation. In the one command version, the command would contain in specified positions the following information: the operation to be performed (multiplication of two numbers and storage of both halves of the product); the address in the store of the first operand; the address in the store of the second operand; the address in the store for the first half of the result; the address in the store for the second half of the result.

The number of different arithmetic operations which a machine can carry out varies from machine to machine; it is as low as ten on some machines, as high as 70 or 80 on others. However, every time the operation is to be performed, it must be introduced to the machine in the highly stylized code developed for the machine, with precise specifications of the addresses from which the operands are to be extracted and the addresses to which the results are to be routed and a precise coded description of the operation to be performed. The machine will exercise no judgment in removing misprints, and it will not tolerate sloppy encoding of commands. Nor will the machine be perturbed in the slightest if the numbers occurring during the arithmetic operation become too large to be accommodated (it might casually discard the most significant parts) or become too small to be significant. This type of worry is pretty well delegated to the analyst who presents the problem to the machine (although many machines do have overload warnings of various kinds).

It is clear that once the commands to the machine have been coded they may be introduced through automatic equipment at rates of speed high enough to keep the arithmetic unit of the machine operating at reasonable speeds. There remains only one difficulty; if the machine can carry out three multiplications or 16 additions each thousandth of a second (which is a low estimate), then it must receive somewhere between 10^8 and 5×10^8 commands each hour it computes. Even if a fast coder could write a command each second, it would take 3,000 or more such coders working with the machine to keep the machine busy. Since it is frequently easier to carry out a command on paper than to direct the machine to carry it out, some additional automatic help is clearly required here. This help is provided by having the machine write its own code. Explicitly, the coder codes the machine to write and to execute a code to solve a problem; he cannot write every command which is to be carried out in solving that problem.

This process, which is the most intricate of the processes involved in operating a digital computer, is brought about by admitting the coded commands to the arithmetic units of the machine, where they can be modified. Usually only the addresses are modified. Another crucially important item is the inclusion in the design of the machine a mechanism whereby the sequence of commands executed by the machine is to be determined in a fairly versatile way. This will become clearer if the description of the command is extended.

The commands have been described above in terms of their arithmetic effects, and the implicit assumption has been made that the commands are introduced in some order by some outside operator. Actually, in a modern general purpose machine the coded command specifies two effects:

(1) It specifies the arithmetic effect, as described above;

(2) It specifies the location of the next command to be carried out by the machine, and this location may be influenced by the result of the calculation specified.

Thus, it is essential that the commands themselves be in some serially numbered store. In most machines the *ordinary progress* through the command sequence is the sequence of positions in this store. However, at least some commands may vary this sequence either absolutely or conditionally. Thus some commands may direct an arithmetical operation and also direct the control unit to look in an extraordinary place for the next command; this permits cyclic computations (but gives no means for escaping from the cycle). For example, the commands in positions 0 through k might be followed in an ordinary manner by the commands in the following cells (1 through $k + 1$, respectively) but the command in cell $k + 1$ might require unconditionally that it be followed by the command in cell 0. This would bring about a cyclic sequence of commands with no escape possible. However, if the command in position $k + 1$ were to be followed extraordinarily by the command in position 0 in case the number in the A-register after the arithmetic operation of the command (if any) is non-negative and ordinarily by the command in position $k + 2$ in case this number is negative (for example), the command sequence will depend upon the outcome of the calculation. Then, the cycle can be repeated a fixed number of times (possibly changing the addresses of commands in the process) and abandoned under the same type of binary decision which is built into the produce counter in the multiplication command. The tally might initially be set to a value one less than the number of cycles desired and decreased by one at the end of each cycle just before the conditional jump command; it would remain non-negative until the proper number of cycles had been completed, and then it would be negative at the end of the next cycle.

Similarly, of course, the binary decision can be made to depend on the relative size of any two numbers (or in general on the value of any one bit selected in accordance with the command logic formulated for the machine).

The general structure of the machine, then, is likely to contain the following provisions. There will be a store suitable for storing numbers or commands encoded to look like numbers. The store will communicate directly with an S-register, the cell of the store which at a particular time communicates with the S-register being designated by an address register. The S-register has two-way communication with the A-register (including the complicated communications necessary for addition or subtraction) and two-way communication with the M-register; in addition, a channel is provided permitting the S-register to transmit to (but it need not receive information from) a control register (C-register). There is also a control counter, which keeps track of the address from which the next command is to be taken. Facilities for the transfer of numbers from this counter to the address register of the store must be provided. The C-register accepts encoded commands from the S-register; it decodes these into addresses and operations, and it must have facilities to transmit the addresses to the appropriate register—its own counter if there is to be an extraordinary change of command sequence, and to the store's address register to control the selection of operands for the arithmetic process and to control the storage of results of the arithmetic operations. It should be noted explicitly that the results of the arithmetic operations may be encoded commands to be stored for future use. The control counter must be able to count—to add 1 to the number it contains in order to provide passage to the command which ordinarily follows in a sequence.

The particular configurations of these units, and the commands which the machine's control units will carry out under the influence of the C-register make up what is known as the logic of the machine, and they determine the exact performance of the machine under any command sequence. Again, this is an extensive field partly mathematical (to determine the most suitable arithmetic operations), partly engineering (to determine the easily feasible arithmetic operations), and largely logical (to determine the best means of encoding for the control of these selected operations). Here there is only room to scratch the surface of this important field, to remark its growing importance as new electronic computing components are developed, and to outline an example of coding.

The example to be developed here is misleading in one important respect, already emphasized: A machine accepts only carefully encoded instructions exactly conforming with the precepts laid down in its logical structure. Here no specific structure will be assumed, and a general verbal description of part of an encoded computation will be written. This avoids completely the question of logical check of the finished encoding of the problem—another difficult and important topic of mathematical logic.

The assumptions of the problem must be stated explicitly, however. The problem will involve two streams of numbers already stored in the machine. Each stream will be in a set of cells in serial order of addresses, so that successive numbers of either stream are addressed by adding one to the preceding address. The streams are equal in length, and all numbers are non-negative (to avoid confusion below about sign). The numbers are to be considered as proper binary fractions, with the binary point at the left end of the numbers. Pairs of numbers will be multiplied, and only the most significant half of the products will be retained. The problem will be to accumulate the sum of the products of the corresponding numbers in the two streams; it is assumed that advance assurance is at hand that this sum will be a number less than one, and no precautions will be taken to avoid overflow. Various constants, including one command stored in a safe place where it will not be modified (so that the sequence of commands can be reused) are assumed to be available. It is assumed that a number remains intact in a cell or register when it is transferred *from* that place to some other place, but that the number in any cell or register is destroyed when another number is transferred *to* that place. A verbal description of the results to be obtained in various elementary sequences of commands is given below; the reader may perceive that these results may be obtained by elementary processes already described.

SEQUENCE I—RESULTS TO BE OBTAINED

Create a number whose value is zero and store it in the cell designated to receive the accumulated product. In a convenient cell (using a constant stored) set a tally to a value one less than the number of multiplications to be performed. Using a constant whose value is the coded value of the command to be performed (or constants and commands if the command structure of the machine requires more than one command for the multiplication) set encoded instructions (one or more commands) at the proper place in the command sequence (viz., the cells following the one containing the last command of the set now being performed) to cause the first multiplication to be carried out and its result to be stored (if this is necessary) in some cell not being used for other useful purpose.

Before continuing with a description of Sequence II (which will complete the operation) it might be well to note that each sequence will be a set of commands encoded to be like numbers in the machine. Some ambiguity has been admitted above because various readers will be familiar with different machines; for example, some machines demand that an answer be stored somewhere, others permit it to remain in the A-register. The general result of Sequence I is to get ready to carry out accumulative multiplication. Sequence II will be used once for each number pair, and it will accumulate the required product; the reuse of Sequence II will depend upon a binary decision as described above.

SEQUENCE II—RESULTS TO BE OBTAINED

Carry out the multiplication (of the first pair of numbers, initially, but this instruction will be modified as the sequence continues), putting the product (if it may not be held in the A-register) into some convenient storage cell not needed for other purposes. Add this product to the number stored in the cell reserved for the accumulated product and put the answer back into the cell reserved for the accumulated product. Add one to every operand address in the multiply command (or commands) and put the modified commands back in their place in the command sequence. Subtract one from the tally (created in Sequence I), put the decreased tally back in its storage cell, and perform a binary decision: If the modified tally is non-negative start at the beginning of Sequence II again, if it is negative carry through the ordinary sequence of commands to another part of the calculation.

The product will have been accumulated by the time the tally becomes negative.

Much has been written (here and elsewhere) about the general philosophy of coding. The reader will understand it only after he codes. Further words here will do no good, and the reader is encouraged to work over the above example himself.

SOME EARLY EFFECTS TO BE EXPECTED FROM THE DEVELOPMENT OF COMPUTERS

The full impact of the availability of automatic computers has not yet been realized by the mathematical, physical, or social sciences, but it might be safe to make a few predictions concerning what some of the effects will be.

In the first place, there can be no question of the development of a new branch of mathematical logic concerned with the design and control of these machines. This logic, at the moment, seems to be starting with Boolean algebra so far as application to switching circuits and the design of machines is concerned. However, radical developments must be expected. So far as coding is concerned, the logic has no real structure as yet, but one must be developing. For the first time in our life time the subject of mathematical logic, which is in many ways the most simple and the most elementary of all the mathematical sciences, is taking on a real applied importance; its role in decision processes certainly merits close attention by our educators to its development in the near future.

In the second place, there can be no question but that the subject of numerical analysis will develop. This is an old subject formerly avoided by most mathematicians as dull. The subject is how to get adequate numerical solutions to problems without undue expenditure of work. The new machines suggest new methods, which were not applicable to desk computers, and methods which were applicable to desk computers are not always applicable to the new machines. Because of the expense of providing large storage which can be consulted by an automatic machine, tables are stored in such a machine only as a last resort (presently, but this may change!), and the most complicated interpolation schemes are used. Conversely, old extrapolation schemes of solving differential equations (which schemes were designed to permit large steps in the solution process) may become less important, since they demand space which can be saved by using simpler schemes and much shorter increments. In short, both because so many new jobs demand knowledge of numerical analysis and because it is suddenly developing in a radical way, the study of numerical analysis in our colleges and universities is taking on a new importance.

The applications of this new equipment will be the most noticeable result of its development, of course. Already in mathematics and the physical and engineering sciences, solutions to many old problems which were formerly beyond the power of human computers have been obtained. These results have largely been obtained without the introduction of new thoughts or new methods of attack so far, but again new methods are developing.

It seems probable now, however, that more people will be directly affected by the new machinery in its applications to business and to ontrol than by any other applications. In these applications there is a new possibility in learning about complicated processes. This possibility is the enhancement of detail and realism with which complicated situations can be simulated. Again, this subject is of the social sciences and not a part of mathematics, so it should not be expounded at great length here. Still one should note that it is no longer necessary to achieve experience at a rate no faster than nature and your competitors furnish the experience, but the machines are able to simulate complicated social situations (inventories involving hundreds of items, including people, for example) with due attention to random effects. These simulations are now being set up by the social scientists, and they show every sign of becoming major instructional and research tools for the development of more leadership and judgment and for the relegation of routine decision to the decision-making machines.

BIBLIOGRAPHY

1. ARROW, K. J. *Social Choice and Individual Values*. New York: John Wiley and Sons, 1951.
2. ORE, OYSTEIN. *Number Theory and Its History*, especially p. 37–39 for binary arithmetic. New York: McGraw-Hill Book Co., 1948.
3. ENGINEERING RESEARCH ASSOCIATES. *High Speed Computing Devices*. New York: McGraw-Hill Book Co., 1950.
4. RICHARDS, R. K. *Arithmetic Operations in Digital Computers*. New York: D. Van Nostrand Co., 1955.
5. BERKELEY, E. C. *Giant Brains or Machines That Think*. New York: John Wiley and Sons, 1949.
6. BUSH, ROBERT R., and MOSTELLER, FREDERICK. *Stochastic Models for Learning*. New York: John Wiley and Sons, 1955.
7. THE ASSOCIATION FOR COMPUTING MACHINERY, *Journal*. New York: the Association, 1954+.
8. NATIONAL RESEARCH COUNCIL. *Mathematical Tables and Other Aids to Computation*. Washington: the Council, 1943+.

XIII

Implications for the Mathematics Curriculum[1]

TWENTIETH-CENTURY mathematics has a new look. Its form and structure are in many respects quite different from that of previous centuries. Teachers must understand this form, must know what 20th Century mathematics is about, if they are to prepare their pupils for living in the 20th Century.

This yearbook contains presentations of several important aspects of modern mathematics. It reflects the points of view of a number of prominent mathematicians and, in this respect, presents mathematics as what mathematicians do with it and say that it is. There has been no attempt to obtain a formal sequential development throughout this yearbook. There has been a modest effort to introduce fundamental concepts in the early chapters that would be of use in the study of later chapters. This concluding chapter aims to emphasize the general point of view of the authors and editors, to collect some of the implications stated in the previous chapters, and to assist teachers in their search for the practical significance of the yearbook relative to their daily work. The author of this chapter accepts full responsibility for any discrepancies that may arise between this and other chapters since other authors have not had an opportunity to make such comparisons.

New processes and new points of view in mathematics have provoked a revolution in many college courses during the past decade which has resulted in emphasis now being given to concepts and techniques that could be found only in graduate courses prior to that time. Accordingly, most teachers should expect the point of view of modern mathematics to be an extension of the one developed in their undergraduate training. In

---

[1] The author wishes to acknowledge and express his appreciation for the assistance of Howard F. Fehr who read the preliminary draft of this manuscript and made many excellent suggestions regarding both the organization and the detailed exposition.

many cases the new concepts and techniques are easier than the old. However, it is quite essential that one be willing to reorient his thinking if he wishes to appreciate fully the new point of view which so thoroughly permeates modern mathematics.

The significance of modern mathematics as a part of our scientific culture has been established. It represents a way of thinking that should be understood by high-school teachers and should underlie the mathematical instruction of junior and senior high-school students. While we shall not be teaching much modern mathematics at the high school level, we can not avoid providing our students with an introduction to mathematics. Will it be the mathematics of 300 B.C.? the mathematics of the 18th Century? or the mathematics of the 20th Century? The following sections are concerned with some of the aspects of 20th Century mathematics which are closely related to the present content of courses in the curriculums of secondary schools and which may be used effectively to introduce the point of view of modern mathematics while at the same time increasing the student's understanding of the topics that we now teach.

Since the grade levels of various topics differ according to the maturity and ability of the students as well as the attitudes of teachers and members of the school board, it is not possible to specify the precise grade level of the implications that we consider. However, certainly the implications suggested in arithmetic are applicable in the upper elementary grades and thereafter; those suggested in algebra, geometry, and trigonometry, whenever these subjects are taught either as separate courses or as parts of integrated programs. Several of the implications of modern mathematics apply at all levels as, for example, those concerning the concept of a number. The general point of view applies at all levels; the specific form of the implication varies with the maturity of the students.

SETS

Students make daily use of sets of books, pen and pencil sets, sets of table silver, etc. People of all ages and in all walks of life are constantly preoccupied with sets of elements—sets of blocks, sets of picture cards, sets of golf clubs, sets of teeth (possibly false), or simply the sets of fingers on their right hands. McShane has emphasized in Chapter III that the "habit of thinking in terms of sets has become routine for most mathematicians and has also become part of the everyday thinking of many other scientists too." In Chapter III there are applications of the theory of sets to such diverse and elementary sets of elements as points, purchasers of life insurance, atoms, members of a woman's club, Venn diagrams (logic), citizens of countries, baseball leagues, the solution of equations, switching circuits, counting, and mathematical induction. Applications of the theory of sets may also be found in our elementary schools.

In the early grades children often consider bundles of sticks and discover facts such as those that we now state as

$$2 + 2 = 4, \qquad 2 + 1 = 1 + 2, \qquad 2 \cdot 3 = 3 \cdot 2.$$

Even though these discoveries involve relationships among sets of sticks rather than among numbers, they are very important in the development of an understanding of numbers. Such discoveries of relationships among sets of objects may be used to provide a basis for all of the mathematical concepts considered in the elementary schools. As the pupils advance, they may use sets of elements in the development of concepts of number, variables, and function. Later they may consider an algebra of sets as developed in Chapter III.

At present it appears that the elementary schools are leading the secondary schools in the use of sets of elements. However, the interpretatations are slightly less obvious at the secondary level. We shall consider some of these interpretations in the following sections. Sets of elements may be used effectively with both slow and fast learners, in both general and traditional courses. The theory of sets is of fundamental importance in all branches of mathematics. It should permeate the thinking of teachers at all levels—kindergarten through graduate school.

NUMBERS AND OPERATIONS

Primitive man obtained his concept of a number by abstracting common properties of sets of elements. For example, all sets which may be placed in one-to-one correspondence with the set of elements x, x, x (or if you prefer with the set of elements 1, 2, 3) are associated with the number represented by "3." This dependence of numbers upon sets of elements is emphasized in Chapters II and III. Also sets of elements were used in the proofs of properties of numbers. Even though you may be reluctant to adopt some of the symbolism for ninth-grade classes, it should be clear that the basic properties of numbers ought to be presented to the students either as arising from properties of sets of elements or as arising from definitions which have been formulated so that the properties in the mathematical system will be consistent with those of the sets of elements. Furthermore, it should be emphasized that the properties of numbers depend upon the particular set of numbers under consideration. For example, addition is always possible in the set of positive integers; subtraction is always possible in the set of positive and negative integers and zero. At least by the ninth grade it seems reasonable to expect that most students will recognize the basic properties of the positive integers and also the inadequacy of the set of positive integers when differences and quotients are involved.

As we expand the student's concept of number, we should emphasize the importance of definitions of equality, sums, products, and order relations for the new numbers. These definitions are not only made in terms of the previous numbers but also so that as many as possible of the properties of the previous set of numbers will remain as properties of the new set. This emphasis upon definitions seems particularly important in the ninth grade when fractions are presented as rational numbers and when signed numbers are introduced. We should not keep the student mystified by saying "it works this way." Rather we should admit and emphasize to him that operations and relations among new numbers (or other objects) must be defined. Then we should make sure that the student understands the definitions that we are using and recognizes that the definitions have been selected so that as many as possible of the familiar properties of the numbers that he has been considering will be properties of the new set of numbers.

It would be helpful to secondary-school students if mathematicians could agree on their selection of definitions. However, such agreement is not yet evident. The important idea to convey to the students is the role of definitions rather than the relative merits of various sets of definitions. We have too many futile arguments regarding the reason for certain statements in elementary mathematics. Nearly all of these arguments are based upon a failure to recognize that we are working in a mathematical system based upon symbols and definitions. Something must be defined and accepted without proof in order to have an axiomatic system. Our students should recognize this if we present all new mathematical ideas and symbols properly. Even in the elementary school, mathematics should be recognized, at least by the teacher, as a formal discipline based upon defined relations using sets of symbols which may be interpreted for many situations.

The above emphasis upon mathematical operations implies that we consider primarily and thoroughly the fundamental operations of addition, subtraction, multiplication, and division at the high-school level. Some of the students entering high school have never discovered that there is a relationship between addition and subtraction and a relationship between multiplication and division. (See Chapter V.) They have had separate rules for each as well as rules for several cases of each. We have the responsibility of emphasizing to them that these operations have certain properties and are interrelated in certain ways. Think over some of the special cases (*e.g.*, the three cases in percentage) that are frequently covered in textbooks and you will find several situations in which an emphasis upon mathematical operations will simplify the teaching and improve the student's understanding of mathematics. The approach to our work through mathematical operations is typical of many of the changes in *mode of presentation rather than content* that are implied by modern mathematics.

The above ideas should be applied appropriately at all grade levels to the sets of numbers under consideration. At each level we are concerned with statements about numbers. In the elementary as well as the upper grades these statements often involve question marks which stand in place of number symbols. For example, the statement $1 + 1 = ?$ is undoubtedly considered in some form by many children even before they enter school. Symbols such as "?" which stand in place of number symbols are forerunners of the variables used in algebra.

The students should recognize that the existence of solutions, *i.e.*, the possibility of finding numbers which satisfy given statements, often depends upon the set of numbers under consideration. Recognition of this dependence involves a consideration of the possibility of performing mathematical operations and provides a basis for expanding the student's concept of number. The emphasis upon mathematical operations is desirable even when the basic properties of the operations are not formally listed. At least by the ninth grade we should be able to emphasize the dependence of mathematical operations upon the set of numbers under consideration. Certainly before any student leaves high school special attention should be given to the operations that are permissible in each of the sets of numbers that he has considered. For college-preparatory students this might include the definition of a field (see Chapter V) with a corresponding systematic consideration of addition and multiplication.

VARIABLES

The use of a letter such as x, y, and t as a variable or more precisely as a symbol representing any element of a specified set of numbers seems too often to be reserved for the academic intelligentsia. Cannot any child who is able to understand and answer the question implied by the statement

$$1 + 1 = ?$$

also specify the particular element of the set of integers for which the statement

$$1 + 1 = x$$

holds? Granted that our phraseology in speaking to a child would require simplification, we should still recognize that the statements

$$1 + 1 = ? \quad \text{and} \quad 1 + 1 = x$$

each involve variables. Also in each case the pupil is expected to identify a subset of the set of integers (in this case a subset with only one member) such that the statement holds whenever the symbol "?" or "x" represents an element of the subset. The two problems are precisely the same even though they may arise at very different levels in our usual sequence of courses.

A recognition of the similarity of the above statements does not imply that algebra should be taught in kindergarten. It does imply that, at least by the time a child reaches the junior high school, the discussion of numbers should emphasize the role of numbers as symbols and the use of symbols such as "?" as variables. The change of notation from "?" to "x" is simply a matter of convenience and efficiency. The grade level at which symbols such as "x" are first used appears to require further consideration. Indeed, a recognition of the proper role of symbols should noticeably facilitate the teaching of formulas in the junior high school.

The desirability of specifying a precise concept of variable is obvious. Unfortunately, such a precise concept does not exist in mathematics today. Different groups of mathematicians use the word in different ways. The logicians distinguish between free variables and bound variables. This distinction removes many difficulties but does not appear usable when algebra is first introduced. In elementary algebra the following is a useful interpretation of variable: *A variable is a symbol that may represent any element from a specified set of elements.* Many of us have a preconceived notion that a variable should represent any one of several different elements. This notion is not shared by logicians who think of a variable as a symbol that appears as a "dummy" to hold a place for a number symbol.

FUNCTIONS

The modern concept of function replaces an earlier concept of single-valued function and is well-described in Chapter III. It is a special case of a relation. McShane's illustrative examples based upon "descendant," collections of cards, and sets of ordered pairs may be modified and extended for use in secondary schools. "The central idea of function is that

with a given first element there are never two different second elements."
The word "function" is commonly used in two ways:

(i) A *function* is a single-valued relation between the elements of a
set and the elements of a second set (not necessarily distinct from the
first); a mapping of one set into another. A precise way of saying this is
that a function is a set of ordered pairs of elements such that the first
elements are from one set and the second elements are from the other,
and there are no two pairs with equal first elements and unequal second
elements.

(ii) A *function* is a rule or expression which specifies the relationship
between the elements of the two sets just mentioned (*i.e.*, gives instructions for constructing the ordered pairs just mentioned).

We shall consider two comparisons of these interpretations of the
function concept:

(1) The relationship between x and $x + 5$ may be represented graphically for every real number x as a set of ordered pairs $(x, y) = (x, x + 5)$.
It is also commonly expressed by the equation $y = x + 5$ which provides
a rule for associating with each number x a corresponding number $x + 5$
which is also denoted by y.

(2) A relationship between x and $\sin x$ may be represented as in a
graph or a table by sets of ordered pairs of the form $(x, \sin x)$. It is also
commonly expressed by $y = \sin x$ or simply by $\sin x$.

As in the case of "variable" it is not possible to say that any particular
interpretation of "function" is *the* correct interpretation. Rather we
must recognize that each interpretation has useful applications and each
interpretation connotes many of the intuitive aspects of the other. Our
ultimate aim is understanding. Teachers should know what is meant by
these usages of "function" and be able to make effective use of the function concept.

Under each interpretation of "function" a relation or rule is implied.
Furthermore, as Langer points out in Chapter VIII, an analytic expression of this rule is not necessary. Under the interpretation of a function as a set of ordered pairs, the association of a particular second element with each first element indicates the relation; the first element is a
value of the *independent variable*; and the second element is a value of
the *dependent variable*. Corresponding definitions of these terms may be
obtained for other interpretations of "function." In each case the set of
all possible values of the independent variable is the *domain* of the function; the set of all possible values of the dependent variable is the *range*
of the function. The important aspects of function are based upon the
rule, the domain (the numbers which may be used in the rule), and the

range (the numbers which may be obtained by following the rule). These aspects may be stated very precisely in terms of ordered pairs. They may also be represented as in a particular table of values for sin x. They may be represented by an analytic expression such as $x^2 + 1$ with the set of real numbers as the domain of x and the set of real numbers greater than or equal to 1 as the range of values for the function. They may also be represented in many other ways. (See Chapters VII and VIII by Randolph and Langer respectively.)

In the secondary school the concepts of set and of a relation between sets deserve considerable emphasis. The terminology of domain and range is helpful as soon as the maturity of the students warrants its usage.

COORDINATES

Numbers are used as coordinates throughout our society. As an individual, each member of our society has a coordinate in several frames of reference—date of birth, age, height, weight, home address, telephone number, social security number, driver's license number, and so forth. Often a single coordinate on a number scale will suffice to identify or describe an individual element of a set. At other times the frame of reference involves several number scales. For example, the size of a rectangular solid may be described using three numbers. However, if the temperature and the time of the measurement are significant, five numbers are required. Many other examples of the use of sets of numbers may be found including the statistics used to describe certain features of contestants in beauty contests and the sets of numbers used to designate games won, lost, and tied by athletic teams. Our students often use numbers as coordinates with frames of reference involving one, two, three, four, and even more scales. Our task as teachers is to develop an understanding of coordinate systems and to use them effectively in our development of mathematical systems.

At the ninth-grade level coordinates may be used in the introduction of variables, the introduction of the solution of equations, and in the graphical representation of relations and functions. Throughout the curriculum graphical representations of relationships and graphical solutions of problems deserve a greater emphasis. In graphing relations on the xy-plane it is often advantageous to use different scales on the two axes. This is permissible whenever the distance relation

$$d = \sqrt{(x_1 - x_2)^2 + (y_1 - y_2)^2}$$

is not used as a measure of the distance on the graph.

Tenth-grade students can make effective use of the distance relationship, midpoint of a line segment, slope, and the equation of a line. The concept of a vector, the location of a point by means of its distance and direction from a fixed point, and the use of polar coordinates are all related in a manner that can be understood by tenth-grade students. Line values of sines and cosines are now considered in most high schools. Equally advantageous are the definitions of all trigonometric functions by means of coordinates, the solution of oblique triangles with reference to coordinate systems, and the derivation of trigonometric formulas using coordinates and the distance relationship. For example, any triangle ABC may be considered on a coordinate plane as having vertices at $A(0, 0)$, $B(c, 0)$, and $C(b \cos A, b \sin A)$ as in Figure 1. The distance

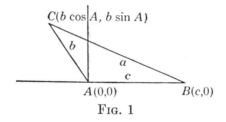

FIG. 1

relationship for BC then gives

$$a^2 = (b \cos A - c)^2 + b^2 \sin^2 A$$
$$= b^2 (\cos^2 A + \sin^2 A) - 2bc \cos A + c^2$$
$$= b^2 + c^2 - 2bc \cos A.$$

The formula for $\cos (A - B)$ may be derived in a similar manner without any restriction upon the quadrants in which the terminal sides of the angles may be situated.

MATHEMATICAL SYSTEMS

References to mathematical systems and mathematical models have been made throughout the preceding chapters. In the introduction (Chapter I) Newsom discusses mathematical systems and the fact that "the same system may occur frequently as the underlying pattern in many diverse real and ideal situations." Niven (Chapter II) discusses number systems and the use of number systems as mathematical models "to describe the real world." McShane (Chapter III) discusses sets of elements and states that: "The fundamental device of applied mathematics is to make up a sort of dictionary by which certain material objects correspond to certain pure mathematical structures and certain physical procedures correspond to certain mathematical operations in those structures." In other words, applied mathematicians are interested in mathematical models, in relationships between physical and mathematical systems.

Allendoerfer (Chapter IV) mentions the "growing group of scholars who are attempting to apply the axiomatic method to a variety of fields of knowledge" and think of this task as the construction of " 'mathematical models' of portions of their subjects." He described a mathematical model as "nothing more than a deductive system in which the undefined words and axioms represent certain aspects of the observed realities." He recognized model building as necessarily involving the selection of the details which are most important for the purposes of the mathematician just as a painting involves the selection of the details which are most important for the purposes of the artist. "Our abstract mathematical theories are abstractions from the real world; they are models of the relationships which we have observed experimentally. The conclusions that we draw from these models are conjectures about new relationships in the real world . . . . The value of our mathematics is that without it we seldom would have been able to guess the relationships which our mathematics has suggested to us." (See Allendoerfer's section entitled "Axioms," p. 70–76.)

MacLane (Chapter V) emphasizes the axiomatic development of algebra. He thinks of a model as a relation between an axiomatic system and another mathematical or physical system. The algebra of vectors, developed by Prenowitz (Chapter VI), had its origin in the 19th century search for a mathematical representation of physical quantities such as forces and velocities.

References could be made to other chapters. However, it should be clear from the above that modern mathematics is concerned with mathematical systems and the interpretations of these systems as models of other systems and of the various aspects of the physical universe. At the secondary-school level one of the most important implications of modern mathematics is that algebraic and geometric systems may be interpreted to represent the same situations. We often associate algebra with sets of numbers and unspecified representatives of these sets. We associate geometry with sets of points. We need to recognize that associated with the geometry of points on a line there is an algebra of the coordinates x of the points on the line; associated with plane geometry there is an algebra of the coordinates of the points considered as ordered pairs of numbers (x, y); associated with solid geometry there is an algebra of ordered triples of numbers (x, y, z). Algebra and geometry may both be used to study either sets of numbers or sets of points. The properties of the sets of numbers (or points) remain the same, but the point of view and the techniques used vary. Frequently we may discover properties using one point of view that we had previously missed when considering the set from a different point of view. This emphasis upon using both algebraic and geometric techniques is reflected in many new programs as well as in the classrooms of many fine teachers of traditional programs.

When mathematical systems are interpreted as representations of various aspects of the physical universe, both the relationship and the distinction between the sets of physical objects and the set of mathematical symbols should be emphasized. At the elementary school level this may be done by emphasing that numbers have properties of their own that do not depend upon the particular set of physical objects under consideration by the pupils. No formal treatment is implied or suggested —simply careful distinctions such as that between "3" and "3 apples." Such distinctions should also be made at all levels. Difficulties with units of measure, confusions regarding proofs of such statements as $2 + 2 = 4$, and many other troublesome problems may be removed by recognizing number symbols as elements of a mathematical system, a model that may be used or applied in several different situations.

Some of the most recent mathematical models may be found in the theory of games which includes models for several economic systems and even for the voting power of various coalitions in the United Nations. Details regarding these and other picturesque models should be forthcoming in our professional journals.

Mathematical systems may be abstract axiomatic systems created without reference to the physical universe or they may be systems developed as mathematical models of certain aspects of the physical universe. Our students first encounter mathematical systems as models. Few high-school students appreciate abstract mathematical systems devoid of references to the physical universe. However, all students work with mathematical systems and all mathematical systems are axiomatic systems involving abstractions. Thus our task is to teach students, commensurate with their maturity, to recognize the abstractions inherent in the mathematical models and the necessary axiomatic basis for the mathematical systems with which they are working.

PROOF

The importance of proofs in geometry is well recognized (see Chapter IX). The concept of mathematics as composed of mathematical systems which are at the same time axiomatic systems implies that proofs are an important aspect of all areas of mathematics. In algebra the treatment of objects other than numbers leads to new algebraic operations as well as to a better understanding of the old ones. As mentioned by MacLane in Chapter V, this requires a formal basis for algebraic operations—axioms, definitions, proofs, and so forth. Students need to rediscover axiomatic systems all the way through their study of mathematics. In these systems we, as teachers, should emphasize the need for unproved axioms (or postulates) and undefined terms.

Think of any one of the proofs that $2 + 2 = 4$. Many people feel that proofs of this sort are superfluous since any child knows that when a bundle is formed of two sticks and then two more sticks are added, there are a total of four sticks. The child knows this. He can count the sticks. But this does not prove that $2 + 2 = 4$. It simply demonstrates that a set or bundle consisting of two sticks and two sticks contains four sticks. A formal proof that $2 + 2 = 4$ shows that the mathematical system is consistent in this respect with the sets of sticks. The proof concerns a property of the mathematical system—the set of integers; it does not have any influence on the physical world. A recognition of this significance of the proof removes the circularity that some seem to have suspected as being present in mathematical proofs. Admittedly we often use apparent properties of physical objects to obtain postulates for our mathematical systems. For example, we may use properties of sets of discrete objects to obtain Peano's Postulates and then develop the set of positive integers from these postulates. However, the proofs of properties of positive integers must be regarded as just that—proofs of properties of positive integers such as $2 + 2 = 4$. They are not proofs for sets of objects. We work in the mathematical system instead of in the phyiscal world. Only by working formally in the mathematical system can the students gain the insight that is necessary for an appreciation of the mathematical system as an entity in its own right with a multitude of applications.

We lose many excellent opportunities for emphasizing deductive thinking when we restrict that topic to geometry. Often it is easier in algebra. We should make deductive thinking a part of all formal work in mathematics and let the students know when definitions are involved in our reasoning rather than appearances in an application of the mathematical system. Informal work with applications has a very important role, but it should be kept in its subsidiary position.

We often think of proofs as direct or indirect. An increasing number of teachers are recognizing that a proof in a mathematical system is a proof in an axiomatic system and that some knowledge of elementary symbolic logic is very useful. Allendoerfer introduces several aspects of logic in Chapter IV. Among these in the section entitled "Derived Implications" (p. 83) he points out that an implication and its contrapositive have the same truth table: If either one is true, they are both true; if either one is false, they are both false. This relationship between an implication and its contrapositive provides the logical basis for the method of indirect proof. An indirect proof of a statement is a direct proof of the contrapositive of the statement.

STATISTICAL THINKING AND MACHINE COMPUTATION

Somewhat in contrast with the postulational thinking of formal mathematics, we should recognize two other aspects of modern mathematics—statistical thinking and machine computation. These are relative newcomers that have not yet been completely accepted by many mathematicians. However, their influence upon the thinking of people concerned with quantitative concepts is very great and appears to be increasing rapidly. Their influence upon the high-school curriculum is also becoming evident.

The statisticians usually consider groups of people or, in general, populations of elements. For example, suppose that a clothing manufacturer was interested in the heights of children in grades 7 and 8 in the United States. Considerable information could be obtained by considering only such children in the public schools of Detroit, that is, by taking the children in grades 7 and 8 in the public schools of Detroit as a sample of all such children in the United States. The clothing manufacturer would find considerable variation in the heights of the children, but these variations would be clustered about a so-called average. He would be interested not only in the average but also in the distribution of the heights about that average. The use of samples of a set of elements to gain information about a large population and the study of the manner in which the elements of a given set of data are distributed about their average are important aspects of statistical thinking.

Statistical thinking is based upon probability, a study of chance or random variation (see Chapter XI). Traditionally, we think of probability in terms of the number of heads obtained when a coin is flipped fifty times and similar examples involving drawing cards from a deck, or rolling dice. As a trivial example, consider the case of a coin which has been flipped ten times and shown heads each time. Since by the theory of probability this sequence of events should be expected for an unbiased coin only about once in a thousand times, we might be suspicious that the coin was more likely to show heads than tails and we might then flip it some more to test our suspicions. If it showed heads thirty consecutive times, we should be highly suspicious that the coin was biased, but we could not be sure. If a billion unbiased coins were each flipped thirty times, the theory of probability indicates that one of them probably would show heads every time.

The importance of probability and statistics is based upon the fact that we of necessity live in an uncertain world and guide our lives according to the probabilities that we place on the occurrence of various events. In some cases we buy insurance in order to share the uncertainties with others and not be handicapped by rare incidents such as having our house struck by lightning and burned to the ground. In most cases we take our own risks and assume that the events that affect us will be similar to those which affect other people. However, we cannot avoid the effect of chance variations on almost every phase of our life.

Chance variations affect the people that we meet, the people that we work for, the sex of a child, the house or apartment that we find to live in, the articles and books that catch our eye, the people that catch our eye. We recognize that there are influencing factors in many of these situations, but there is also a great deal of chance variation. Indeed, we expect a certain amount of variability in, say, the people we meet walking to work, or the students who are absent due to sickness. When this variability is not present as, for example, when one student is absent much more than others, we look for an underlying cause which we had not understood. This same type of reasoning is used when a manufacturer, say, of automobile pistons investigates a machine which has made several pistons that did not meet specifications. Statisticians have developed procedures for determining how large a sample of pistons (for example) need to be measured and also how much variation may be expected before looking to see what has gone wrong. These examples should suffice to indicate that statistical thinking underlies a great many aspects of our culture and indeed our personal lives. Also, it should be clear that statistical thinking is based upon a new type of mathematics. Rules are no longer sufficient to determine individual cases. Rather they imply that on the average certain events may be expected.

Statistical thinking is already gaining considerable significance in the control of the quality of manufactured products. Its importance in other aspects of our culture seems assured. Since it is a different type of mathematical thinking, recognition should be given to statistical thinking even in our general mathematics courses. Suggested topics from probability and statistics may be found in various experimental programs at the high-school and college freshman level. Detailed suggestions for the high-school curriculum reflecting experience gained in these programs may be expected in our professional literature in the near future.

The significance of mechanical computers should be easy to understand. Just as the adding machine removed the drudgery from many arithmetical calculations, the digital and analog computers can be expected to remove the drudgery from most advanced mathematical calculations. Furthermore, since these computers do not need to have an operator personally feed in each number as it i needed, the computers can be made to perform operations much faster than humans can react. The instructions (numbers and desired operations) for a digital machine are usually punched into a card or a long roll of narrow paper tape. These cards or tapes may be prepared without using the computer. Accordingly, it is not unusual to have fifty or even seventy trained technicians and mathematicians preparing problems for a single computing machine.

In view of the early prediction of a few uninformed observers that the machines would put mathematicians out of work, the actual impact of the machines should be emphasized. They have *not* reduced the number of mathematicians required. They have *not* solved all available problems. Techniques are being devised to extend the range of solvable problems almost indefinitely. The number of trained mathematicians is tragically low. In other words, as the president of one of our leading industrial research organizations said recently, we have not run out of problems, we have run out of mathematicians.

Tompkins has several suggestions in Chapter XII regarding topics that should be included in the high-school curriculum to prepare students for the type of thinking that is used in machine computation. Notwithstanding the complexity of the machines, there are in our elementary-school and junior high-school as well as in our senior high-school curriculums several related aspects of our mathematical program such as deductive thinking, estimation of results, computation with approximate data, significant digits, accuracy, and precision. Among new topics, binary arithmetic (the use of 2 as a base and only two digits, usually 0 and 1) and the algebra of sets provide excellent preparation for understanding and using machine computation.

Statistical thinking and machine computation represent new areas of mathematics that are now in their initial stage of exploration. We should expect a gradual effort to formalize the theories in these new areas and also, probably within the next twenty-five years, the acceptance of these areas on the basis of their underlying assumptions, which should be established between now and then. They are forerunners of new attitudes toward mathematics, new demands for mathematicians, and a broadening of the scope of mathematics. The progress of our country in many scientific areas depends upon our success in preparing our better students not only to be competent in traditional areas of mathematics but also to make use of statistical thinking and machine computation. They need to recognize individual measurements as elements of a larger population for which the normal expectancy and chance variations can provide information regarding the distribution of individual elements. They need to recognize the dependence of machine computation upon the fundamental mathematical concepts in their high-school mathematics curriculum.

TERMINOLOGY

For all students and at all levels a precise use of terminology is highly desirable to avoid confusion and to facilitate communication both at the student's present level and in the future.

New terms provide only mild difficulties compared with some of the terms that we have inherited from previous generations. For example, which one of the terms—variable, unknown, literal number symbol, pronumeral, unspecified number, representative of a set, general number—will be found in high-school texts twenty-five years from now? Which one has the most useful connotations? Even though no conjecture of an answer to these questions appears feasible, it should be obvious that in our search for an effective terminology we shall gain an increased understanding of the concept that is involved.

Terminology is a serious problem in mathematics. At the advanced levels we have reached a stage where terms such as "lattice" have entirely different meanings in different branches of mathematics. About fifteen years ago an attempt was made to improve this situation through the offices of the American Mathematical Society. However, the attempt was futile, probably because of the entrenched position of various terms in certain areas.

There is also another type of terminological difficulty that is very serious at the elementary- and secondary-school levels. I refer to terms which could be properly defined, but which are not usually properly defined, and which do not connote to the student the mathematical operations that are involved. For example, consider the terms "cancel," "transpose," and "cross multiply." I feel that such terms should be removed from our vocabulary for elementary mathematics as soon as possible. Perhaps you would like to suggest others to be added to this list. An operational approach to mathematics requires that the students recognize the operation involved and the conditions under which the operation is permissible. Terms such as "cancel," "transpose," and "cross multiply" seem to camouflage rather than to emphasize the mathematical operations.

Precision of terminology is not just an idle fancy of picayune individuals. It is necessary if we are to promote clear thinking and to develop the abilities of our students. Even though conceivably a fetish could be made of preciseness in this regard, most of us are in much greater danger of becoming sloppy in our terminology than in overemphasizing it. Modern mathematics with its logical foundations and its wide range of applications arising from different interpretations of mathematical systems must be based upon a precise terminology. Accordingly, teachers at all levels should emphasize the precise terminology of the mathematical systems under consideration in their classes.

GENERALIZATION, ABSTRACTION, AND ARITHMETIZATION

We now turn to three of the important processes underlying modern mathematics—generalization, abstraction, and arithmetization.

Arithmetization is the use of properties of numbers as a basis for other mathematical concepts. For example, it can be shown that Euclidean geometry is consistent if the set of real numbers is consistent. It can also be shown that the non-Euclidean geometries are consistent if the set of real numbers is consistent. In general, as in the case of coordinate systems, arithmetization provides a basis for our formal treatments of many aspects of modern mathematics.

Abstraction is the process used in identifying "2" as the common property of all sets that may be placed in one-to-one correspondence with the set $\{x, x\}$. Our concepts of cardinal numbers are based upon abstractions of common properties of sets of elements. Our concepts of colors, such as red, are developed in a similar manner; also our concepts of many words in our language. In advanced mathematics abstractions provide a basis for abstract points, abstract algebras, and other concepts. In most, if not all, elementary mathematical systems the axioms (postulates) originated as abstractions from observed properties of the physical universe.

Generalization is a process of extending our knowledge to include more extensive or intensive sets of elements. For example, we have already mentioned that algebra may be considered as a generalization of arithmetic through the use of symbols such as x which may be replaced by any element of a set of numbers. We may also generalize our concept of a point and thereby catch a glimpse of some of the generalized geometries (see Chapters IX and X) that are gaining increasing importance in modern mathematics and its applications. Indeed, with appropriate abstractions and generalizations we may use either algebraic or geometric techniques to study such sets as:

The set of all scheduled passenger airplane flights

The set of all telephone circuits in a city, in the United States, or in the world

The set of all rivers, lakes, and streams in North America

The set of all grains of sand on the beach at Miami

The set of all arteries and veins in a human body.

The list could be continued at great length, but the variety of the applications should be evident. Each of these sets has both algebraic and geometric properties. Both algebraic and geometric techniques may be used to increase our understanding of these sets.

How long will it be before the modern trend of teaching for generalizations will reach the stage of generalized points and generalized numbers? Teaching for generalizations seems to be an accepted goal among modern educators. So much lip service is paid to this goal that at times it appears to be a fad. We should teach for generalizations, but we must first understand them. More advanced techniques do not always involve generalizations. For example, the intercept form of the equation is not a generalization of the equation; plotting by intercepts is not a generalization of plotting from a table of values. Generalizations involve a reaching out to wider horizons. We generalize our concept of a "number" when we extend our concept to include not only counting numbers (positive integers) but also positive fractions, negative integers, the set of all rational numbers, the set of real numbers, the set of complex numbers, quaternions, etc. These successive stages of the development of our number concept represent successive generalizations. Most of these generalizations are an important part of our curriculum. Are they treated as generalizations? There is a definite trend in this direction.

Most students in general mathematics should understand the positive rational numbers as a generalization from the positive integers; students in algebra should understand a number x as a generalization of specific numbers. The practice of recognizing generalizations and encouraging students to seek generalizations must be developed from the multitudes of classroom incidents rather than from a special unit devoted to this purpose. In other words, teaching for generalizations is a way of teaching. It has not yet been satisfactorily expressed in a textbook although some texts are better suited than others for this purpose.

Generalizations usually arise as conjectures requiring further justification. Some are later accepted, many rejected. Students learn to generalize by developing the habit of making and testing conjectures that are based upon given situations but apply in a more general situation or another case of a more general situation. For example, a student who had considered areas of similar triangles might make a conjecture regarding the areas of similar quadrilaterals or even similar polygons. Another student might conjecture that if the sides of one quadrilateral are twice as long as the corresponding sides of another quadrilateral, then the two quadrilaterals are similar. Each student should be encouraged to test his conjectures. The second student might discover that his conjecture would hold for rectangles but not for some other quadrilaterals. Both students could increase their understanding of similar figures by their attempts to generalize the relationships that they had studied for triangles. From a very realistic point of view students who develop their ability to generalize situations and to abstract common properties of situations hold the key to our future in this scientific age.

GOALS AND CONTENT

A precise identification of the implications of modern mathematics depends upon the goals of our mathematics curriculum. In this regard we shall restrict our consideration to high-school mathematics. What should be the goal of our high-school mathematics curriculum? In this age of space ships and great-circle air navigation, we hear many statements that solid geometry has no place in our curriculum. We find many critics of the rigor in plane geometry, in algebra, and in other courses. In one sense mathematics deserves both of these criticisms. We have too often taught only the formalities of a geometric discipline of two thousand years ago while losing sight of both the social and mathematical implications of more recent developments—especially those of the last century. The major aim of this yearbook is to help mathematics teachers see beyond the formalities of mathematics to its underlying properties and implications. The goals for our high-school mathematics curriculum should reflect this insight.

First, we must recognize that we cannot have the same goals for all of our students. Students arrive in our hands with a wide range of ability, mathematical training, interest, and initiative. Most of us teach in schools where we are required to take the students as they arrive and do what we can for them. What should we do? Let us consider an answer to this question with reference to three groups of students—those needing remedial work in mathematics, those so-called average mathematics students who expect to use little mathematics in their life work or training, and finally those students who expect to enter college curriculums requiring at least college algebra or to enter technical training in which mathematics is used. For the first group, those needing remedial training, our path is clear—we must supply the remedial training. Except for the few students who are mentally incapable of completing the remedial work, we then find ourselves with essentially two groups—those whose life work will require extensive use of mathematics and those who need only enough mathematics to become good citizens in our scientific society.

What should our goals be for these two groups of students? Here again no firm answer is possible. However, for purposes of considering the implications of modern mathematics, let us take as our general goals the training of all these students:

(i) To understand operations with integers and rational numbers

(ii) To use properties of integers and rational numbers in practical situations

(iii) To formulate and work with elementary statements regarding numbers and variables

(iv) To visualize Euclidean geometry (plane and solid) as a mathematical system serving as an effective model of the physical universe

(v) To use elementary concepts of Euclidean geometry in practical situations

(vi) To develop moderate facility in reasoning from given assumptions and given sets of data in mathematics and other fields.

If these goals are accepted for the majority of our students as desirable for intelligent citizenship in our scientific society, then the problem of stating goals for the students who expect to make extensive use of mathematics is greatly simplified. Our goals for these students should certainly include all of the above with a more rigorous interpretation than for average students. In addition we should try to find a way in which to facilitate the development of our better students in whatever work or professions they may later find themselves. Such a goal becomes very confusing when we try to isolate the particular applications of mathematical principles that are used in the different fields of work which our students may enter. However, such a narrow analysis of the problem is neither necessary nor desirable. The center of our attention should be the mathematical systems, not their applications. Accordingly, we must develop our program in terms of mathematical systems and then use the applications as a means of broadening the student's understanding of mathematics and its applications in his particular type of activity.

Recognition of this emphasis upon mathematical systems makes our work much easier as well as making our classes useful to a greater number of students. The power of mathematics stems from the mathematical systems with their manifold applications rather than from a selection of applications. Thus we are substantially increasing the practical value of our courses when we center them around the mathematical concepts with a variety of applications to indicate to the students the extent of the applicability of the mathematical systems that we are considering.

What content should we consider for our better students after we agree that our major emphasis is upon the mathematical systems? Many schools, public schools, have found that all of their better students can cover most of the material associated with traditional courses through trigonometry and college algebra along with some additional material. (The restriction to "most of" the usual material rather than "all of" the usual material will be clarified in a moment.) From the point of view of modern mathematics all of our better students, including those who could do a semester of calculus in high school, should have an opportunity to broaden their understanding and concept of mathematics through an introduction to such topics as sets, the algebra of sets, use of sets in logic, deductive thinking, the need for precise terminology, limits, analysis of data, statistical thinking, expectation, application of statistical thinking to everyday problems such as the reports on the effectiveness of polio vaccine, methods for the representation and presentation of data, estimation, the use of coordinates and distance relationships in plane and solid geometry, and the existence of geometries without distance relationships (such as topology, Chapter X).

Many colleges are giving increased recognition to the importance of training students to reason mathematically, to make and to test conjectures, and to understand mathematical principles. Certainly drill is necessary, especially drill based upon processes that have been carefully established by reasoning. However, long lists of detailed but similar problems that may be worked simply by following the steps in an illustrative example seem to be losing their appeal to teachers as well as students. There is also another point of view regarding the long detailed problems. The very long and tedious manipulations are fast becoming the object of machine rather than human calculations. Our students need to understand mathematical operations even more than before, but they may not need to spend as much time pushing a pencil. We must teach our students to do the work that machines cannot do. We must teach our students to think in terms of mathematical operations.

We cannot teach our students to think in terms of mathematical operations without having them perform mathematical operations, understand mathematical operations, and recognize when specific mathematical operations are allowed. Under these conditions can we make any deletions from our traditional mathematics course content? Many mathematicians think that we can. They feel that Horner's method for solving equations should have been dropped years ago, that the formulas for solving general cubic and quartic equations have little if any place in our curriculum, that we over-emphasize trigonometric identities for multiple angles for functions different from sine and cosine, that the essential topics from solid geometry should be treated in connection with other courses in high school rather than as a separate course, and that we tend to overemphasize factoring, trigonometric identities, the trigonometric solution of triangles, and the use of logarithms. In the case of solid geometry, note that the above suggestion does not imply that anyone can teach all the geometric concepts (plane and solid) in a single year to an average class. Indeed there does not appear to be any possibility of combining plane and solid geometry into a single year of geometry. Rather it appears that some solid geometry should be done in conjunction with the study of plane geometry, while other topics from solid geometry, such as mensuration formulas, should be done inconjunction with algebra.

Briefly then, our goal for each of our students should be concerned primarily with mathematical operations and properties of mathematical systems with their applications serving to motivate and increase the understanding of the students. For many of our students the amount that we are able to accomplish will depend upon their recognition of their individual needs for mathematical concepts and the advice (at times far from enlightened) that they receive. For all students the above emphasis appears most promising. Through this approach we can broaden the student's understanding of traditional topics. We can attract students to areas of mathematics in which there are many opportunities but about which most students do not even hear until they are well advanced in other specializations. We can increase the possibility that students in other areas, including the social sciences, will use mathematical concepts in their work and thereby increase the effectiveness of both mathematics and other fields of study.

CONCLUSION

The preceding chapters have been concerned with concepts of modern mathematics, the present chapter with their implications for the curriculum. What service can such a yearbook render teachers of secondary school mathematics?

(i) Individuals and groups of teachers can use this yearbook for in-service review of mathematical concepts.

(ii) Groups of teachers considering curriculum revision can use the yearbook in their search for the point of view of contemporary mathematics.

It is hoped that the yearbook will facilitate a shortening of the lag between creation and application and will aid the development of a secondary-school mathematics program reflecting 20th century mathematics with its extensive use of the axiomatic method to find common generalizations of previous special theories and to discover new and important mathematical objects—infinite dimensional spaces, topological groups, non-associative algebras, etc.

One of the aims of this yearbook has been to obtain the points of view of several people at the frontiers of mathematical development regarding insights into modern mathematics that teachers should have as they introduce the youth of our country to mathematics. The points of view presented differ in many respects but show considerable agreement on the importance of developing mathematical systems and doing creative thinking in these systems.

An effort has been made in this concluding chapter to pull together the implications of the writers of the preceding chapters. The difficulties inherent in such a task are obvious, and it is expected that the chapter will serve most effectively in focusing the reader's attention upon certain aspects of the problem which should be kept in mind while considering the statements in previous chapters, the experiences of other teachers, and other reports. Many of the statements regarding methods of teaching, the organization of subject matter, and the modern point of view will be considered, and possibly implemented, by subsequent investigations, yearbooks, experiments, and teachers. Concept learning and the place of concepts are currently being investigated. The gradual development of mathematical concepts throughout our school program will be considered in another yearbook in the near future.

There are frequent articles in professional journals concerned with classroom procedures reflecting the point of view of 20th-Century mathematics (e.g., "The Concept of a Literal Number Symbol" by Max Beberman and B. E. Meserve in *The Mathematics Teacher* 48: 198–202). There are also frequent articles concerned with a contemporary approach to the mathematics curriculum (e.g., "The Goal is Mathematics for All" by H. F. Fehr in *School Science and Mathematics* 56: 109–120). Many more such articles can be expected in the future based upon the work of such groups as the Commission on Mathematics sponsored by the College Entrance Examination Board, the Curriculum Committee of the National Council of Teachers of Mathematics, the Committee on Mathematical Training of Social Scientists of the Social Science Research Council, the University of Illinois Committee on Secondary School Mathematics, the Mathematical Association of America, the Science Teaching Improvement Program of the American Association for the Advancement of Science, the Study on Admission to College with Advanced Standing, and the experiences gained by teachers and mathematicians in experimental programs. This activity is most encouraging for the future of our mathematics curriculum but discouraging relative to the selection of a bibliography for inclusion with this chapter. If articles from our professional journals were included, they would necessarily reflect ideas prevalent at least one or two years prior to the publication of this yearbook. They would place undue emphasis upon one era of our thinking rather than encourage readers to consider articles contemporary with their reading. Accordingly, with the expectation that even by the time this yearbook is published there will be several more articles presenting the point of view of contemporary mathematics and describing new procedures for implementing its implications, the following bibliography includes only a recognition of four very promising experimental programs—one at the high school level and three at the college freshman level with many implications for high school courses. These are included with the expectation that there will be future revisions of them and that they will continue to reflect new approaches to the mathematics curriculum and to be sources of tested procedures for considering both new and traditional topics. The restriction on the following bibliography is also in recognition of the fact that an alertness to contemporary articles in our professional journals is an important aspect of an awareness of the changing nature of mathematics and its implications for the curriculum. Such an alertness renders a detailed bibliography of articles from any one era a quickly outmoded superfluity.

BIBLIOGRAPHY

1. COMMITTEE ON THE UNDERGRADUATE PROGRAM OF THE MATHEMATICAL ASSOCIATION OF AMERICA. *Universal Mathematics*, Part I: Functions and Limits (1954); Part II: Structures in Sets (1955). Lawrence, Kans.: University of Kansas Book Store.
2. KEMENY, J. G.; SNELL, J. L.; and THOMPSON, G. L. *Introduction to Finite Mathematics*. New York: Prentice Hall, 1957.
3. UNIVERSITY OF CHICAGO, COLLEGE MATHEMATICS STAFF. *Concepts and Structures of Mathematics*. Chicago: University of Chicago Press, 1954.
4. UNIVERSITY OF ILLINOIS COMMITTEE ON SCHOOL MATHEMATICS. *First Course (1954, 1955) Second Course (1956), Third Course (1956)*. Urbana University High School. (Mimeo.)

Index

A

Abelian groups, 105–14
    definition, 106
Absolute value of complex number, 26
Abstract groups, 139–42
Abstract spaces, 315, 327–28, 330
    neighborhoods of, 328
Abstraction, 420
Accumulating register (*cf.* A-register)
Addend, 381
Adding register (*cf.* A-register)
Addition of vectors, 153
Addition rule for probability, 337
Addition table for digital computer, 381
Additive group, 111
Additive inverse, 103, 156, 188
Affine geometry, 196, 302
Affine group, 134
Affine transformations, 302
Aganis (*c.* 50 A.D.), 286, 289
Alberti (1435), 299
Aleph-null, $\aleph_0$, 63
Algebra:
    Boolean, 51, 66, 316, 387, 401
    fundamental theorem, 23
    geometric, 281
    laws, 100
    modulo n, 122–24
    number triples, 189
    points, 187, 190
    sets, 50, 66, 316
    transformations, 127
    vectors, 141
Algebraic numbers, 20, 24
Alternating sequence, 202
Alternation principle for directed segments, 150, 152
Amplitude of complex number, 26
Analysis:
    modern, 5
    numerical, 372, 402
    vector, 145
Analytic geometry, 167, 176
"And" circuit, 387
"And" function, 387
And/or, 77
Angle:
    of parallelism, 292
    of vector, direction, 186
    between vectors, 182
Apollonius (*c.* 225 B.C.), 279, 302
Approximations, rational, 21
Arabic numbers, 7–8

Archimedes (*c.* 225 B.C.), 279, 284
Area, 209–10
    between graphs, 234
    inner, 210
    outer, 209
A-register, 392–93, 396, 398–401
Aristotle (384–322 B.C.), 277
Arithmetic, 206
    computing machine, 397
    fundamental theorem, 14, 15
Arithmetization, 420
Associative law, 103, 155, 187
Augend, 381
Automatic decision methods, 377
Automatic digital computer, 392–99
Automation, 372
Average value, 346
Axiom of Pasch, 283
Axioms, 3, 69, 70, 306, 307
    addition and subtraction, 103
    equality, 100

B

Babylonian geometry, 273, 274, 279
Base:
    binary number system, 8
    decimal number system, 7
    logarithm, 251
Basis of a vector space, 193
Bernoulli, Jacob (Jacques) (1654–1705), 354
    law of large numbers, 360
    limit theorem, 354
Bessel, F. W. (1784–1846), 271
    functions, 271
Betweenness, 307
Bicompactness, 324
Binary analysis of decisions, 375–78
Binary digit (bit), 379
Binary number system, 8, 378
Binomial distribution, 350–61
Binomial expansion, 243
Bit (*cf.* Binary digit)
Bolyai, János (1802–1860), 3, 291, 292, 304
Boole, G. (1815–1864), 50
Boolean algebra, 51, 66, 316, 387, 401
Bound:
    lower, 325
    upper, 325
Boundary conditions, 267, 365, 366, 370
Brunelleschi (*c.* 1425), 299